ADVANCES IN CHEMICAL PHYSICS

VOLUME 136

EDITORIAL BOARD

ADVANCES IN CHEMICAL PHYSICS

VOLUME 136

Series Editor

STUART A. RICE

Department of Chemistry
and
The James Franck Institute
The University of Chicago
Chicago, Illinois

WILEY-INTERSCIENCE
A JOHN WILEY & SONS, INC. PUBLICATION

Published by John Wiley & Sons, Inc., Hoboken, New Jersey
Published simultaneously in Canada

Wiley Bicentennial Logo: Richard J. Pacifico

Library of Congress Catalog Number: 58-9935

ISBN 978-0-471-68232-5

Printed in the United States of America

10 9 8 7 6 5 4 3 2 1

CONTRIBUTORS TO VOLUME 136

SAVO BRATOS, Laboratoire de Physique Théorique des Liquides Université Pierre et Marie Curie, 75252 Paris Cedex, France

MARK S. CHILD, Physical and Theoretical Chemistry Laboratory, Oxford University, Oxford, 0X1 3QZ, United Kingdom

EVELYN M. GOLDFIELD, Department of Chemistry, Wayne State University of Michigan, 48202 USA

STEPHEN K. GRAY, Chemistry Division, Argonne National Laboratory, Illinois 60439 USA

VASSILIY LUBCHENKO, Department of Chemistry, University of Houston, Houston, Texas 77204-5003 USA

G. ALI MANSOORI, Departments of Biology and Chemical Engineering, University of Illinois at Chicago, Chicago, Illinois 60612 USA

PETER G. WOLYNES, Department of Chemistry and Biochemistry and Department of Physics, University of California at San Diego, La Jolla, California 92093-0371 USA

MICHAEL WULFF, European Synchrotron Radiation Facility, 38043 Grenoble Cedex, France

INTRODUCTION

Few of us can any longer keep up with the flood of scientific literature, even in specialized subfields. Any attempt to do more and be broadly educated with respect to a large domain of science has the appearance of tilting at windmills. Yet the synthesis of ideas drawn from different subjects into new, powerful, general concepts is as valuable as ever, and the desire to remain educated persists in all scientists. This series, *Advances in Chemical Physics*, is devoted to helping the reader obtain general information about a wide variety of topics in chemical physics, a field that we interpret very broadly. Our intent is to have experts present comprehensive analyses of subjects of interest and to encourage the expression of individual points of view. We hope that this approach to the presentation of an overview of a subject will both stimulate new research and serve as a personalized learning text for beginners in a field.

Stuart A. Rice

CONTENTS

QUANTUM DYNAMICS OF CHEMICAL REACTIONS

EVELYN M. GOLDFIELD

Department of Chemistry, Wayne State University, Detroit, Michigan 48202, USA

STEPHEN K. GRAY

Chemistry Division, Argonne National Laboratory, Argonne, Illinois 60439, USA

CONTENTS

Advances in Chemical Physics, Volume 136, edited by Stuart A. Rice

I. INTRODUCTION

Quantum mechanical effects—tunneling and interference, resonances, and electronic nonadiabaticity—play important roles in many chemical reactions. Rigorous quantum dynamics studies, that is, numerically accurate solutions of either the time-independent or time-dependent Schrödinger equations, provide the most correct and detailed description of a chemical reaction. While limited to relatively small numbers of atoms by the standards of ordinary chemistry, numerically accurate quantum dynamics provides not only detailed insight into the nature of specific reactions, but benchmark results on which to base more approximate approaches, such as transition state theory and quasiclassical trajectories, which can be applied to larger systems.

There are many facets to a successful quantum dynamics study. Of course, if comparison with experimental results is a goal, the underlying Born–Oppenheimer potential energy surface must be known at an appropriately high level of electronic structure theory. For nonadiabatic problems, two or more surfaces and their couplings must be determined. The present chapter, however, focuses on the quantum dynamics of the nuclei once an adequate description of the electronic structure has been achieved.

Section II discusses the real wave packet propagation method we have found useful for the description of several three- and four-atom problems. As with many other wave packet or time-dependent quantum mechanical methods, as well as iterative diagonalization procedures for time-independent problems, repeated actions of a Hamiltonian matrix on a vector represent the major computational bottleneck of the method. Section III discusses relevant issues concerning the efficient numerical representation of the wave packet and the action of the Hamiltonian matrix on a vector in four-atom dynamics problems. Similar considerations apply to problems with fewer or more atoms. Problems involving four or more atoms can be computationally very taxing. Modern (parallel) computer architectures can be exploited to reduce the physical time to solution and Section IV discusses some parallel algorithms we have developed. Section V presents our concluding remarks.

II. THE REAL WAVE PACKET METHOD

A. General Ideas

The real wave packet (RWP) method, developed by Gray and Balint-Kurti [1], is an approach for obtaining accurate quantum dynamics information. Unlike most wave packet methods [2] it utilizes only the real part of the generally complex-valued, time-evolving wave packet, and the effective Hamiltonian operator generating the dynamics is a certain function of the actual Hamiltonian operator of interest. Time steps in the RWP method are accomplished by a simple three-term Chebyshev iteration, involving per time step just one action of the Hamiltonian operator on the current (real) wave packet. One view of the RWP method is that it is a highly streamlined version of Tal-Ezer and Kosloff's well-known and highly successful Chebyshev expansion of the propagator [3]. The features of the RWP method noted suggest that it should be a good approach to use for four-atom reaction dynamics problems, which generally require a lot of computer memory and time.

The RWP method also has features in common with several other accurate, iterative approaches to quantum dynamics, most notably Mandelshtam and Taylor's damped Chebyshev expansion of the time-independent Green's operator [4], Kouri and co-workers' "time-independent wave packet" method [5], and Chen and Guo's Chebyshev propagator [6]. Kroes and Neuhauser also implemented damped Chebyshev iterations in the time-independent wave packet context for a challenging surface scattering calculation [7]. The main strength of the RWP method is that it is derived explicitly within the framework of time-dependent quantum mechanics and allows one to make connections or interpretations that might not be as evident with the other approaches. For example, as will be shown in Section IIB, it is possible to relate the basic iteration step to an actual physical time step.

Consider the time-dependent Schrödinger equation,

$$i\hbar \frac{\partial}{\partial t} \Psi(t) = \hat{H} \Psi(t) \tag{1}$$

with $\Psi(t)$ being a complex-valued wave packet and $\hat{H}$ being the Hamiltonian operator. The RWP method [1] arises from three surprising but simple facts concerning Eq. (1):

1. It is possible to propagate just the real part of a wave packet if the representation of $\hat{H}$ is real-valued, as it often is in chemical reaction dynamics.
2. Energy-resolved observables can be inferred with knowledge of just the real part of a wave packet.
3. A time-dependent Schrödinger equation with $\hat{H}$ replaced by $f(\hat{H})$ can be used to infer dynamics information about $\hat{H}$, with $f(\hat{H})$ being chosen for *computational* convenience.

The cosine iterative equation [8] is the ancestor of the RWP method. Consider propagating a wave packet at time t forward in time to $t + \tau$,

$$\Psi(t + \tau) = \exp(-i\hat{H}\tau/\hbar)\Psi(t) \tag{2}$$

and propagating it backwards in time to $t - \tau$,

$$\Psi(t - \tau) = \exp(+i\hat{H}\tau/\hbar)\Psi(t) \tag{3}$$

Adding Eqs. (2) and (3) leads to

$$\Psi(t+\tau) = -\Psi(t-\tau) + 2\cos(\hat{H}\tau/\hbar)\Psi(t) \tag{4}$$

Equation (4) is a three-term recursion for propagating a wave packet, and, assuming one starts out with some $\Psi(0)$ and $\Psi(\tau)$ consistent with Eq. (1), then the "iterations" of Eq. (4) will generate the correct wave packet. The difficulty, of course, is that the action of the cosine operator in Eq. (4) is of the same difficulty as evaluating the action of the exponential operator in Eq. (1), requiring many evaluations of $\hat{H}$ on the current wave packet. Gray [8], for example, employed a short iterative Lanczos method [9] to evaluate the cosine operator. However, there is a numerical simplification if the representation of $\hat{H}$ is real. In this case, if we decompose the wave packet into real and imaginary parts,

$$\Psi(t) = \mathrm{Q}(t) + i\mathrm{P}(t) \tag{5}$$

where $\mathrm{Q}(t) = \mathrm{Re}[\Psi(t)]$ and $\mathrm{P}(t) = \mathrm{Im}[\Psi(t)]$ are real valued, then upon insertion into Eq. (3) one has separate equations for Q and P. For example,

$$\mathrm{Q}(t+\tau) = -\mathrm{Q}(t-\tau) + 2\cos(\hat{H}\tau/\hbar)\mathrm{Q}(t) \tag{6}$$

Absorption of wave packet amplitude that approaches the edges of the computational grid can be accomplished by periodically damping out the wave packet in small regions at the grid edges [10]. Suppose $\hat{A}$ represents this (real) operation. Instead of Eq. (2), one has

$$\Psi(t+\tau) = \hat{A}\exp(-i\hat{H}\tau/\hbar)\Psi(t) \tag{7}$$

What is the equivalent of Eq. (7) in the cosine iterative scheme? Simply applying $\hat{A}$ to the result, Eq. (5), of a cosine iteration proved to be less stable than a different absorption algorithm, which can be "derived" on the basis of time-reversal ideas as follows [8]. Consistent with Eq. (7), the backward propagated result should be

$$\Psi(t-\tau) = \exp(i\hat{H}\tau/\hbar)\hat{A}^{-1}\Psi(t) \tag{8}$$

If $\hat{A}$ is "weak" enough such that we can commute it with $\hat{H}$, then

$$\hat{A}^{-1}\Psi(t+\tau) + \hat{A}\Psi(t-\tau) \approx 2\cos(\hat{H}\tau/\hbar)\Psi(t) \tag{9}$$

Multiplying both sides of Eq. (9) by $\hat{A}$ and rearranging gives

$$\Psi(t+\tau) = \hat{A}[-\hat{A}\Psi(t-\tau) + 2\cos(\hat{H}\tau/\hbar)\Psi(t)] \tag{10}$$

which was found empirically to be a more stable iterative scheme. The real part of the wave packet then satisfies

$$Q(t+\tau) = \hat{A}[-\hat{A}Q(t-\tau) + 2\cos(\hat{H}\tau/\hbar)Q(t)] \tag{11}$$

Equation (11) represents the first iteration of the RWP idea, but it is not the most efficient. It also represents the first appearance of the damping procedure as used in Mandelshtam and Taylor's Chebyshev iteration [4].

Consider a *modified* Schrödinger equation [1],

$$i\hbar\frac{\partial}{\partial u}\chi(u) = f(\hat{H})\chi(u) \tag{12}$$

where we assume that the function $f(E)$ is monotonic on the range of energies of interest. The dynamics of Eq. (12) can be related to the dynamics of the usual Schrödinger equation, Eq. (1). It is simplest to consider wave packets in a bound state problem, although the argument also applies to scattering states. A stationary state solution of Eq. (1) is also a stationary state solution of Eq. (12): if $\hat{H}\varphi_n = E_n\varphi_n$, then $f(\hat{H})\varphi_n = f(E_n)\varphi_n$. Therefore, if the initial condition for both Eq. (1) and Eq. (12) is the same, $\chi(0) = \Psi(0)$, then we have

$$\Psi(t) = \sum_n c_n\varphi_n \exp(-iE_nt/\hbar) \tag{13}$$

and

$$\chi(u) = \sum_n c_n\varphi_n \exp[-if(E_n)u/\hbar] \tag{14}$$

where the coefficients c_n are the *same* in both equations above.

With $\chi(t) = q(t) + ip(t)$, the corresponding cosine iterative equation (including absorption) for the real part of χ, following identical arguments that led to Eq. (11), is

$$q(u+\delta) = \hat{A}\{-\hat{A}q(u-\delta) + 2\cos[f(\hat{H})\delta/\hbar]q(u)\} \tag{15}$$

The choice [1]

$$f(\hat{H}) = -\frac{\hbar}{\delta}\cos^{-1}(\hat{H}_s) \tag{16}$$

with

$$\hat{H}_s = a_s\hat{H} + b_s \tag{17}$$

with a_s and b_s chosen such that the eigenvalues of $\hat{H}_s$ lie in $[-1, 1]$, leads to

$$q(u+\delta) = \hat{A}[-\hat{A}q(u-\delta) + 2\hat{H}_s q(u)] \tag{18}$$

which is also Mandelshtam and Taylor's damped Chebyshev recursion relation [4]. If the eigenvalues of $\hat{H}$ lie within $[E_{\text{min}}, E_{\text{max}}]$, then $a_s = 2/(E_{\text{max}} - E_{\text{min}})$ and $b_s = -1 - a_s E_{\text{min}}$. Note that, in numerical applications, $\hat{H}$ is approximated by a finite Hermitian matrix that has both lower and upper energetic bounds, allowing the Chebyshev iteration to be used as described here. The remarkable feature of Eq. (18) is that it *solves* the propagation problem—one need not construct any approximation to the exponential propagator (or the related cosine operator) to solve for the wave packet dynamics. It would appear to be the most efficient implementation of the RWP idea if accurate quantum dynamics information is desired. The reader may wonder why u has been used as the evolution or time variable in writing the equations of this section. (Reference [1] was a little loose in using t in both equations.) The reason for this will be made clear in the next subsection.

B. Inferring Observables

There are two issues that may be confusing in the development above. The first issue, which applies to any $f(\hat{H})$, including simply $f(\hat{H}) = \hat{H}$, is how to obtain correct scattering dynamics information if only the real part of the wave packet is available. The second issue is the relation of the wave packet dynamics generated by the $f(\hat{H})$ of choice in the RWP method, Eq. (16), to standard wave packet dynamics generated by $\hat{H}$. That is, can $\chi(u)$ be related to $\Psi(t)$ in a more explicit manner than in the discussion revolving around Eqs. (13) and (14)?

Consider $\Psi(t)$ generated by Eq. (1). Energy-resolved observables are obtained by Fourier transformation from time (t) into energy (E) space. An energy-resolved scattering state, from which such observables can be computed, is of the form

$$\begin{aligned} \Phi(E) &= c(E)\frac{1}{2\pi\hbar}\int_{-\infty}^{+\infty} dt \exp(iEt/\hbar)\Psi(t) \\ &= c(E)\delta(E-\hat{H})\Psi(0) \end{aligned} \tag{19}$$

where $c(E)$ is usually chosen to ensure delta function energy normalization [11]. Using the identity [12]

$$\delta(E-\hat{H}) = \left|\frac{df(E)}{dE}\right| \delta[f(E) - f(\hat{H})] \tag{20}$$

we can relate $\Phi(E)$ to wave packet dynamics under $f(H)$:

$$\begin{aligned}\Phi(E) &= c(E)\left|\frac{df(E)}{dE}\right|\delta[f(E) - f(\hat{H})]\Psi(0), \\ &= c(E)\left|\frac{df(E)}{dE}\right|\frac{1}{2\pi\hbar}\int_{-\infty}^{+\infty} du \exp[if(E)u/\hbar]\chi(u)\end{aligned} \tag{21}$$

The effective time integral over u in Eq. (21) involves the full complex wave packet, $\chi(u)$. However, it can be replaced by one involving just the real part, $q(u)$, of $\chi(u)$,

$$\int_{-\infty}^{+\infty} du \exp[if(E)u/\hbar]\chi(u) = 2\int_{-\infty}^{+\infty} du \exp[if(E)u/\hbar]q(u) \tag{22}$$

if the Fourier integral on the left hand side is "one-sided" (i.e., if it is nonzero only for $f(E) > 0$ or $f(E) < 0$ [1]). In practice, this condition is easy to satisfy. For example, with $f(E)$ consistent with Eq. (16), the condition will be satisfied if the wave packet has essentially no energy components in the upper half of the full energy range of the Hamiltonian. This upper range generally corresponds to the highly repulsive part of the potential and/or extreme kinetic energies, which are not of interest in chemical reaction dynamics.

Since all energy-resolved observables can be inferred from appropriate expectation values of an energy-resolved wavefunction, Eq. (21) shows that the RWP method can be used to infer observables. Specific formulas for S matrix elements or reaction probabilities are given in Refs. [1] and [13]. See also Section IIIC below.

Finally, we show how to relate the modified Schrödinger equation evolution $\chi(u)$ to the usual evolution $\Psi(t)$ [14]. Consider the modified Schrödinger equation, Eq. (12). We approximate $f(\hat{H})$ in this equation with a first-order Taylor series expansion,

$$f(\hat{H}) \approx f(\bar{E}) + \left.\frac{df(E)}{dE}\right|_{\bar{E}}(\hat{H} - \bar{E}) \tag{23}$$

where $\bar{E}$ is the average energy of the (usual) wave packet. With the particular choice given by Eq. (16), it is easy to reduce Eq. (12) to

$$i\hbar\frac{\partial\chi(u)}{\partial u} = [\alpha(\bar{E}) + \beta(\bar{E})\hat{H}]\chi(u) \tag{24}$$

with

$$\beta(\bar{E}) = \frac{a_s \hbar}{\delta \sqrt{1 - \bar{E}_s^2}} \tag{25}$$

where $\bar{E}_s = a_s \bar{E} + b_s$ is the scaled mean energy.

However, Eq. (24) above and Eq. (1) are now intimately related. A solution of Eq. (24) is indeed related to a solution of Eq. (1) via a simple phase factor and a time scaling:

$$\Psi(t) \approx \exp(i\alpha t/\beta)\chi(u = t/\beta) \tag{26}$$

with the approximate sign being used because of the linear approximation, Eq. (23). Equation (26) allows us to (approximately) equate physical times t and RWP (or Chebyshev) iterations. If u is measured in steps $k\delta$, then Eq. (26) implies [14]

$$t \approx \frac{a_s \hbar k}{\sqrt{1 - \bar{E}_s^2}} \tag{27}$$

Equation (27) also gives us some practical insight. Generally, one wishes to propagate to as long a physical time t as possible in order to resolve the effects of long-lived resonances and, in general, to obtain good energetic resolution. Carrying out, say, a fixed number of 1000 Chebyshev iterations ($k = 1, \ldots, 1000$) can correspond to very different maximum physical times and thus energetic resolutions, depending on the other variables in Eq. (27). In particular, a_s is related to the inverse of the spectral range of the matrix representation of $\hat{H}$ being used. Thus it is desirable to choose grid and basis set parameters (see Section III below) to keep the spectral range, $E_{\max} - E_{\min}$, as small as possible (i.e., largest possible a_s), but to still describe the relevant chemical physics correctly. In this manner, the longest possible physical time t can be found with the fewest number of iterations.

The magnitude of the discrete time step, δ, does not enter any final expressions for physical variables. For example, the physical time t in Eq. (27) involves k, the number of Chebyshev iterations taken, and not δ. Thus the core damped Chebyshev iteration, Eq. (18), may be taken to be

$$q(k+1) = \hat{A}[-\hat{A}q(k-1) + 2\hat{H}_s q(k)] \tag{28}$$

We now illustrate the utility of Eq. (27) in relating the RWP dynamics based on the arccosine mapping, Eq. (16), to the usual time-dependent Schrödinger equation dynamics, Eq. (1). We carried out three-dimensional (total angular momentum $J = 0$) wave packet calculations for the

$D + H_2 \rightarrow HD + H$ reaction. Three-atom Jacobi coordinates, with R being the distance from D to the center of mass of H_2, r being the H_2 internuclear distance, and an angle, γ, corresponding to the angle between vectors associated with R and r were used. Grids were used to represent R and r, and a rotational basis was used for γ, so that q is a vector indexed by the corresponding grid/basis set elements. (The evaluation of the action of the potential part of $\hat{H}_s$ on q, however, is accomplished with a transformation to angular grid points similar to that discussed for the four-atom case in Section III.) The potential surface and other details are similar to those used in Ref. [1], except that larger grid extents were used so that absorption was not a factor over the time scales discussed. With $\delta = \hbar = 1$, it is the case that $f(E) = -\cos^{-1}(E_s)$, which is shown in Fig. 1a. Consider an incoming Gaussian wave packet in the $D + H_2$ channel, with H_2 in its ground state, mean collision energy 0.7 eV, and approximate energetic spread 1 eV. The

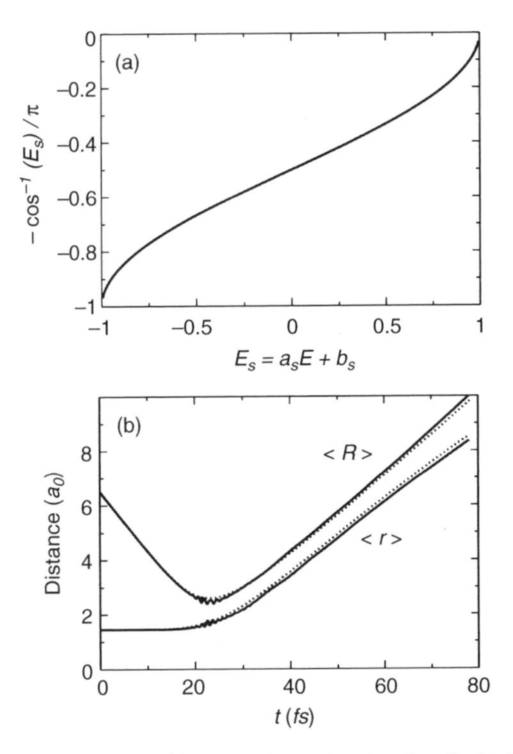

Figure 1. (a) The $f(E) = -\cos^{-1}(a_sEs + b_s)$ mapping ($\hbar = \delta = 1$). (b) Solid curves are an Ehrenfest trajectory for three-dimensional $D + H_2$ generated by the RWP method, that is, the modified Hamiltonian operator $f(\hat{H})$. Dotted curves in (b) correspond to the Ehrenfest trajectory determined by the usual Schrödinger equation. See text for further details.

mean scaled energy in this case is $\bar{E}_s = -0.92$, that is, close to the minimum of the allowed $[-1, 1]$ range. This is because the range of energies described by the Hamiltonian, $E_{\max} - E_{\min}$, is usually very large in accurate calculations in order, for example, to properly describe repulsive potential interactions. We calculated the average value, $\langle R \rangle$, of R, and the average value, $\langle r \rangle$ of r, in a straightforward manner at each iteration, k, of Eq. (28). We converted k to the corresponding physical time, t, from Eq. (27) and plotted the results in Fig. 1b as solid curves. We can view $\langle R \rangle(t)$ and $\langle r \rangle(t)$ as a quantum or Ehrenfest trajectory. $\langle R \rangle$ is initially large and decreases, reflecting the approach of D and H_2. Both $\langle R \rangle$ and $\langle r \rangle$ become large toward the final portion of the trajectory, consistent with a high reaction probability for forming HD + H, which is described by simultaneously large values of R and r. (Complete fragmentation to three separated atoms is extremely unlikely with this wave packet.) One can ask, however, how correct the detailed time-dependence of this trajectory is, since it is based on the approximation, Eq. (27).

To answer the question above, we carried out a similar calculation, but with ordinary complex wave packets based on Eq. (1). Technically, we used a symplectic propagator [15] although other propagator approximations [2, 16] could have been used. The resulting (physically correct) $\langle R \rangle(t)$ and $\langle r \rangle(t)$ are shown in Fig. 1b as dotted curves. It is remarkable how close these results are to the RWP results. For times up to 20 fs in this example, there are no visual differences on the scale of the figures. At later times some discrepancies are evident, but they are small. The RWP-based, Ehrenfest trajectory is so close to the correct one because the energy range that the grid and basis set representation of the problem must describe is much larger than the range of energies contained within the wave packet. Whenever accurate dynamical results are desired, necessitating large basis sets and small grid spacings, this will be the case. Thus reasonably accurate, time-dependent expectation values can be obtained from the RWP approach. Of course, we should emphasize that there is *no approximation* in the RWP energy-resolved quantities; that is, the arguments above concerning Eqs. (19)–(22) show that with appropriate (and trivial) weighting factors, $|df/dE|$, one can obtain physically correct, E-resolved observables.

III. FOUR-ATOM QUANTUM DYNAMICS

Six-dimensional, numerically accurate four-atom wave packet calculations were pioneered by Zhang and Zhang [17] and Neuhauser [18]. While numerous details of the present RWP implementation differ from these earlier approaches, it should be noted that many of the general ideas remain the same. In applications, finite-sized grids and basis sets are introduced to describe the wave packet, and the basic Chebyshev iteration becomes a matrix–vector equation,

$$\mathbf{q}(k+1) = \mathbf{A}[-\mathbf{A}\mathbf{q}(k-1) + 2\mathbf{H}_s\mathbf{q}(k)] \tag{29}$$

The evaluation of the action of the Hamiltonian matrix on a vector is the central computational bottleneck. (The action of the absorption matrix, $\mathbf{A}$, is generally a simple diagonal damping operation near the relevant grid edges.) Section IIIA discusses a useful representation for four-atom systems. Section IIIB outlines one aspect of how the action of the kinetic energy operator is evaluated that may prove of general interest and also is of relevance for problems that require parallelization. Section IIIC discusses initial conditions and final state analysis and Section IIID outlines some relevant equations for the construction of cross sections and rate constants for four-atom problems of the type $AB + CD \rightarrow ABC + D$.

A. Wave Packet Representation

We represent the four-atom problem in terms of diatom–diatom Jacobi coordinates: $\boldsymbol{R}$, the vector between the AB and CD centers of mass, and $\boldsymbol{r}_1$ and $\boldsymbol{r}_2$, the AB and CD bond vectors. In a body-fixed coordinate system [19, 20] with the z-axis chosen to $\boldsymbol{R}$, only six coordinate variables need be considered, which we choose to be R, r_1, and r_2, the magnitudes of the Jacobi vectors, and the angles θ_1, θ_2, and ϕ. Here θ_i denotes the usual polar angle of $\boldsymbol{r}_i$ relative to the z-axis, and ϕ is the difference between the azimuthal angles for $\boldsymbol{r}_1$ and $\boldsymbol{r}_2$ (i.e., a torsion angle).

The theory behind body-fixed representations and the associated angular momentum function expansions of the wavefunction (or wave packet) in terms of bases parameterized by the relevant constants of the motion and approximate constants of the motion is highly technical. Some pertinent results will simply be stated. The two good constants of the motion are total angular momentum, J, and parity, $p = +1$ or -1. An approximate constant of the motion is K, the body-fixed projection of total angular momentum on the body-fixed axis. For simplicity, we will restrict attention to the helicity-decoupled or centrifugal sudden (CS) approximation in which K can be assumed to be a constant of the motion. In terms of all its components, and the iteration number k, the real wave packet is taken to be [21]

$$q^{J,K,p}(R, r_1, r_2, \theta_1, \theta_1, \varphi, k) = \sum_{j_1,k_1,j_2} C^{J,K,p}_{j_1,k_1,j_2}(R, r_1, r_2, k) G^{J,K,p}_{j_1,k_1,j_2}(\theta_1, \theta_2, \phi) \tag{30}$$

where the angular momentum functions $G^{J,K,p}_{j_1,k_1,j_2}$ have definite values of J, p, and K. $q^{J,K,p}$ also depends on three Euler angles, which are suppressed in Eq. (30) for simplicity of notation [19]. The other rotational indices correspond to the "primitive" or uncoupled rotational quantum numbers associated with the two diatomics in the body-fixed frame (i.e., j_1 and j_2), the diatomic rotational quantum numbers, and k_1 and k_2, the corresponding projection quantum numbers on the body-fixed axis. (Since K is fixed, $k_2 = K - k_1$ and so it is not independent.) The use of such primitive, as opposed to coupled [17], quantum numbers or underlying basis functions has been suggested by several workers [19, 22, 23]. The parity adaptation in $G^{J,K,p}_{j_1,k_1,j_2}$ means that certain linear superpositions of primitive functions are used, which also can be chosen to lead to real basis functions [22].

The Hamiltonian operator can be written in the form

$$\hat{H} = -\frac{\hbar^2}{2\mu}\frac{\partial^2}{\partial R^2} - \frac{\hbar^2}{2m_1}\frac{\partial^2}{\partial r_1^2} - \frac{\hbar^2}{2m_2}\frac{\partial^2}{\partial r_2^2} + \frac{(\hat{J} - \hat{j}_1 - \hat{j}_2)^2}{2\mu R^2} + \frac{\hat{j}_1^2}{2m_1 r_1^2} + \frac{\hat{j}_2^2}{2m_2 r_2^2} + V(R, r_1, r_2, \theta_1, \theta_2, \phi) \tag{31}$$

with μ, m_1, and m_2 being the usual reduced masses. An iteration of Eq. (29) requires evaluating the matrix–vector product $\mathbf{H_s q} = (a_s \mathbf{H} + b_s)\mathbf{q}$; that is, $\mathbf{Hq}$ can be considered to be the fundamental operation. We decompose $\mathbf{Hq}$ into kinetic, $\mathbf{Tq}$, and potential, $\mathbf{Vq}$, contributions. Detailed expressions for the relevant matrix elements of Eq. (31) with respect to Eq. (30) can be found in Ref. [21] and in what follows we just outline the general ideas involved. The action of the three radial kinetic energy operators in Eq. (31) can be accomplished in a variety of ways. Radial coordinates that break or undergo large changes can be described with evenly spaced grids and treated with Fourier transform methods [24] or the finite difference technique [25] to be described in the next subsection. The actions of the three rotational kinetic energy operators in Eq. (31) are easily accomplished within the rotational basis of Eq. (30). With the basis chosen, composed of primitive angular momentum functions, there is some minor off-diagonal coupling coming from the first rotational term in Eq. (31) but it is computationally irrelevant [19, 21–23]. The two purely diatomic rotational kinetic energy terms in Eq. (31) are diagonal in the basis. Finally, the action of the potential energy on a wave packet is accomplished by first transforming to an angular grid (three successive transformations to go from j_1, j_2, k_1 to θ_1, θ_2, ϕ), multiplication by the diagonal potential on the grid, and back transformation to the rotational quantum number basis [21, 22]. We elaborate a bit more on the parity-adapted basis and the motivation for this type of evaluation of $\mathbf{Vq}$ in the two subsequent paragraphs.

The use of parity-adapted basis functions in Eq. (30) has several advantages: it permits us to use real sine and cosine basis functions for the torsional angle, ϕ ; it allows us to focus only on positive values of K and for the case of $K = 0$, it allows us to divide the calculation into two smaller calculations for each separate parity block. For $K = 0$, after integrating over Euler angles, the parity-adapted basis functions reduce to [21]

$$G^{J,K=0,p}_{j_1,k_1,j_2}(\theta_1, \theta_2, \phi) = \frac{1}{\sqrt{2(1+\delta_{k_1,0})}} P^{|k_1|}_{j_1}(\cos\theta_1) P^{|k_2|}_{j_2}(\cos\theta_2) \times \left(\frac{e^{ik_1\phi}}{\sqrt{2\pi}} + (-1)^{J+p} \frac{e^{-ik_1\phi}}{\sqrt{2\pi}} \right) \tag{32}$$

where in this case $k_2 = -k_1$. The basis functions separate into two parity blocks with sine and cosine functions, respectively. For $K > 0$, the basis functions have somewhat more complex but similar form [21]. The straightforward use of these basis functions, however, results in an extremely large and dense potential energy matrix. Such matrices are very inefficient, not only because they require huge amounts of memory for storage but also from a computational standpoint. Thus it is highly advantageous to transform to a basis in which the potential energy is diagonal (i.e., a grid basis in the angular coordinates).

We choose to use the primitive coupled basis, rather than the j_{12} coupled basis that is often employed, primarily because the primitive coupled basis has a direct product form. This form makes it easy to accomplish the transformation from j_1, j_2, k_1 to θ_1, θ_2, ϕ. The potential matrix, $\mathbf{V}$, may be written as a numerical quadrature over the angular grid points in the following matrix form:

$$\mathbf{V} = \mathbf{P}^T \mathbf{V_g} \mathbf{P} \tag{33}$$

where $\mathbf{P}$ is a transformation matrix and $\mathbf{V_g}$ is a diagonal matrix with the potential values evaluated on the angular grid ($\mathbf{V_g}$ is also parameterized by the three radial coordinates). Because of the direct product nature of the angular basis, $\mathbf{P}$ reduces to the product of very sparse block matrices, $\mathbf{P_1}$, $\mathbf{P_2}$, and $\mathbf{P_3}$, each of which is the function of only *one* angular coordinate.

To illustrate the computational efficiency of the method, we give a concrete example. Assume, for example, each of the radial coordinates has 100 grid points and that the maximum value of both j_1 and j_2 is 40. For $K = 0$, this gives rise to 23821 rotational basis functions. Thus the number of elements of $\mathbf{V}$ would be $(100)^3 \times (23821)^2 > 5 \times 10^{14}$, which in double precision requires 4×10^6 GB of storage, outstripping resources on even the largest supercomputer clusters. If we assume that there are 41 grid points for each of the angles, $\mathbf{V_g}$ has $(100)^3 \times (41)^3 < 7 \times 10^{10}$ elements, requiring 560 GB of storage, a dramatic reduction. Consider now the evaluation of $\mathbf{Vq}$. For each element of the radial grid, the straightforward matrix multiplication requires approximately 5.7×10^8 floating point operations (FLOPS) while the transformation given in Eq. (33) may be accomplished in only 1.7×10^7 FLOPS. In general, for low values of K, the transformation method will require between one and two orders of magnitude less computational time than the straightforward computation of $\mathbf{Vq}$ as a full matrix–vector product.

B. Dispersion-Fitted Finite Differences

Often the actions of the radial parts of the kinetic energy (see Section IIIA) on a wave packet are accomplished with fast Fourier transforms (FFTs) in the case of evenly spaced grid representations [24] or with other types of discrete variable representations (DVRs) [26, 27]. Since four-atom and larger reaction dynamics problems are computationally challenging and can sometimes benefit from implementation within parallel computing environments, it is also worthwhile to consider simpler finite difference (FD) approaches [25, 28, 29], which are more amenable to parallelization. The FD approach we describe here is a relatively simple one developed by us [25]. We were motivated by earlier work by Mazziotti [28] and we note that later work by the same author provides alternative FD methods and a different, more general perspective [29].

Following Colbert and Miller [27], let x be defined on the interval $[a, b]$ and discretize according to $x_k = a + k\Delta$, $k = 1, 2, \ldots, N$, with $\Delta = (b-a)/(N+1)$. We assume the function $y(x)$ is such that $y(a) = y(b) = 0$ and the N grid points do not include the end points a and b. These conditions are consistent with a sine or particle in a box basis. If $y = (y_1, y_2, \ldots, y_N)^{\mathrm{T}}$ denotes the corresponding array of function values, a $(2n+1)$-point FD approximation to the second derivative of $y(x)$ is

$$\mathbf{y}'' = \mathbf{D}\mathbf{y} \tag{34}$$

with $\mathbf{D}$ being the banded matrix

$$\mathbf{D} = \frac{1}{\Delta^2}\begin{pmatrix}
d_0 & d_1 & \ldots & d_n & & & & & & & & \\
d_1 & d_0 & d_1 & \ldots & d_n & & & & & & & \\
\ldots & d_1 & d_0 & d_1 & \ldots & d_n & & & & & & \\
d_n & \ldots & d_1 & d_0 & d_1 & \ldots & d_n & & & & & \\
 & d_n & \ldots & d_1 & d_0 & d_1 & \ldots & d_n & & & & \\
 & & d_n & \ldots & d_1 & d_0 & d_1 & \ldots & d_n & & & \\
 & & & d_n & \ldots & d_1 & d_0 & d_1 & \ldots & d_n & & \\
 & & & & d_n & \ldots & d_1 & d_0 & d_1 & \ldots & d_n & \\
 & & & & & d_n & \ldots & d_1 & d_0 & d_1 & \ldots & d_n \\
 & & & & & & d_n & \ldots & d_1 & d_0 & d_1 & \ldots \\
 & & & & & & & d_n & \ldots & d_1 & d_0 & d_1 \\
 & & & & & & & & d_n & \ldots & d_1 & d_0
\end{pmatrix} \tag{35}$$

For example, the well-known three-point Lagrangian FD has $d_0 = -2$ and $d_1 = 1$.

The idea behind dispersion-fitted finite differences (DFFDs) is that the eigenvalues of **D**, neglecting edge effects (i.e., the limit $N \gg n$), can be shown to be [30]

$$g(k_j) = \frac{1}{\Delta^2}\left[d_0 + 2\sum_{s=1}^{n} d_s \cos(sk_j\Delta)\right] \tag{36}$$

where the wavenumber $k_j = j\pi/[(N+1)\Delta]$, $j = 1, 2, \ldots, N$. Equation (36) is termed a dispersion relation. (Since $p_j = \hbar k_j$ corresponds to a momentum, the quantum view of Eq. (36) is that it yields the momentum representation of D.) From Fourier transformation theory or basic quantum mechanics, we know that the continuous limit wavenumber representation of d^2/dx^2 is simply $-k^2$. This suggests that when the second derivative is well approximated the dispersion relation really should be

$$g_{\text{ex}}(k_j) = -k_j^2 \tag{37}$$

Note that Eq. (37) also enters into a sine discrete Fourier transform approximation of the second derivative.

Interestingly, if one Taylor series expands Eq. (36) and equates the terms of the same order in k_j with Eq. (37) one can derive the standard Lagrangian FD approximations (i.e., require the coefficient of k_j^2 to be -1, and require the coefficient of all other orders in k_j up to the desired order of approximation to be 0.) A more global approach is to attempt to fit Eq. (36) to Eq. (37) over some range of $K_j = k_j\Delta$ values that leads to a maximum absolute error between Eq. (36) and Eq. (37) less than or equal to some prespecified value, ε. This is the essential idea of the dispersion-fitted finite difference method [25].

More specifically, the reduced variable $K_j = k_j\Delta$ is defined on $[0, \pi]$. Generally, the error in an FD approximation (or rather its dispersion relation, Eq. (36)) increases with K_j. The Taylor series approach outlined above, which leads to the standard Lagrangian FD approximations, is essentially perfect for very small K_j but quickly deviates from the correct, quadratic dependence, Eq. (37). The generic behavior is that the error increases monotonically with K_j. Instead of requiring that the fit be perfect in the limit of very small K_j, we require that the error be no greater than ε from $K_j = 0$ up to some $K_{\max}$ determined by ε. We still choose ε to be "small," for example, in the 10^{-3}–10^{-6} range. However, the result is that while the error in the lower K_j limit is not quite as good as that in the standard Lagrangian approximation, the error for higher K_j is much improved and, overall, since any given problem involves a range of K_j

values, superior results can be obtained for comparable grid spacings. In fact, in Ref. [25] we show that accuracy comparable to use of the Fourier method [24](“pseudospectral” accuracy) could be achieved.

C. Initial Conditions and Final State Analysis

A typical initial condition in ordinary wave packet dynamics is an incoming Gaussian wave packet consistent with particular diatomic vibrational and rotational quantum numbers. In the present case, of course, one has two diatomics and with the rotational basis representation of Eq. (30) one would have, for the full complex wave packet,

$$\chi(k=0) = G_{\varepsilon,\sigma}(R)\psi_{v_1}(r_1)\psi_{v_2}(r_2)G^{J,p}_{j_1,j_2,k_1}(\theta_1,\theta_2,\phi) \tag{38}$$

where $G_{\varepsilon,\sigma}(R)$ denotes the incoming complex Gaussian [1] consistent with collision energy ε and with position spread σ. (Of course, the smaller σ, is the larger the corresponding spread in energies present in the wave packet.) The diatomic vibrational states, which also may depend, parametrically, on j_1 and j_2, are denoted $\psi_{v_1}(r_1)\psi_{v_2}(r_2)$. In the RWP framework it is entirely possible to start with an initial real wave packet consistent with Eq. (38). One uses [1, 14]

$$\begin{aligned} q(k=0) &= \mathrm{Re}[\chi(u=0)] \\ q(k=1) &= \hat{H}_s q(k=0) - \sqrt{1-\hat{H}_s^2}\ \mathrm{Im}[\chi(u=0)] \end{aligned} \tag{39}$$

which is easily demonstrated by propagating $\chi(0)$ one iteration with $\exp[-if(\hat{H})\delta/\hbar] = \exp(i\cos^{-1}\hat{H}_s)$ and taking the real part. In practice, the action of the square root operator in Eq. (39) is accomplished with a series expansion. If a Chebyshev series is used, the number of terms required for reasonable accuracy is generally much less than the overall number of iterations required to achieve convergence of the relevant scattering properties.

Note that there are a number of quantum number indices in Eq. (38) and the computation of observables, depending on the degree of averaging associated with the observable, can require many separate wave packet propagations.

Section IIC showed how a scattering wave function could be computed via Fourier transformation of the iterates $q(k)$. Related arguments can be applied to detailed formulas for S matrix elements and reaction probabilities [1, 13]. For example, the total reaction probability out of some state consistent with some given set of initial quantum numbers, $I = (J, p, K, j_1, j_2, k_1)$, is [13, 17]

$$P_I(E) = \langle\Phi_I(E)|\hat{F}|\Phi_I(E)\rangle \tag{40}$$

where $\Phi_I(E)$ is an energy-resolved scattering function. $\hat{F}$ is a flux operator, usually involving some reaction coordinate s such that the condition $s = s_0$ separates the reactant and product regions:

$$\hat{F} = \frac{\hbar}{2m_s i}\left[\delta(s - s_0)\frac{\partial}{\partial s} - \frac{\partial}{\partial s}\delta(s - s_0)\right] \tag{41}$$

Often one of the diatomic bond distances r_1 or r_2 can be used as s. Insertion of Eq. (41) into Eq. (40), coupled with arguments such as those in Section IIC to connect $\Phi_I(E)$ to RWP iterates, then leads to an expression for Eq. (40) within the RWP framework [13]. The relevant reaction probability expression, Eq. (18) of Ref. [13], which need not be detailed here, involves Fourier transformation of $q|_{s=s_0}(k)$ and $\partial q/\partial s|_{s=s_0}(k)$ and so requires the real wave packet and its derivative with respect to s at $s = s_0$ to be stored as a function of k for subsequent analysis.

D. Cross Sections and Rate Constants

The reactive cross section for reactants $AB(v_1,j_1)$ and $CD(v_2,j_2)$ is [21]

$$\sigma_{v_1,j_1,v_2,j_2}(\varepsilon) = \frac{\pi}{2\mu\varepsilon}\sum_J (2J + 1)P^J_{v_1,j_1,v_2,j_2}(\varepsilon) \tag{42}$$

where μ is the reduced mass associated with R and ε is the collision energy, and

$$P^J_{v_1,j_1,v_2,j_2}(\varepsilon) = \frac{1}{(2j_1 + 1)(2j_2 + 1)}\sum_{K,p,k_1} P^{J,K,p}_{v_1,j_1,k_1,v_2,j_2}(\varepsilon) \tag{43}$$

is the average reaction probability for a given total angular momentum J.

The rate constant may be written

$$k(T) = \frac{g_{\rm el}(T)}{Q_r(T)}\sum_{v_1,j_1,v_2,j_2} g_{j_1,j_2}\exp(-\varepsilon_{v_1,j_1,v_2,j_2}/k_BT)k_{v_1,j_1,v_2,j_2}(T) \tag{44}$$

with state-resolved rate constants

$$k_{v_1,j_1,v_2,j_2}(T) = \left(\frac{8k_BT}{\pi\mu}\right)^{1/2}\frac{1}{(kT)^2}\int_0^\infty d\varepsilon\,\varepsilon\exp(-\varepsilon/k_BT)\sigma_{v_1,j_1,v_2,j_2}(\varepsilon) \tag{45}$$

and reactant partition function

$$Q_r(T) = \sum_{v_1,j_1,v_2,j_2} g_{j_1,j_2}\exp(-\varepsilon_{v_1,j_1,v_2,j_2}/k_BT) \tag{46}$$

In Eq. (44), $g_{el}(T)$ is the ratio of transition state and reactant electronic partition functions [31] and the rotational degeneracy factor $g_{j_1,j_2} = (2j_1 + 1)(2j_2 + 1)$ for heteronuclear diatomics, and will also include nuclear spin considerations in the case of homonuclear diatomics.

The rigorous evaluation of the above quantities involves calculations that include the Coriolis coupling between all of the K states that contribute to a given value of total angular momentum J. We present a method for doing these rigorous Coriolis-coupled calculations in Section IV [32]. For four-atom and larger systems, however, calculations of cross sections and rates, even within the helicity-decoupled approximation, can be a formidable task owing to the variety of total angular momenta J and other initial quantum numbers that need be summed over and requiring many wave packet propagations. J-shifting approximations, as championed by Bowman [33], wherein information from one or just a few total angular momenta J is used to interpolate and extrapolate to other J, are commonly employed and have been used in the RWP applications to $H_2 + OH$ [21] and $D_2 + OH$ [34]. Another approach is to use cumulative reaction probability and related ideas [12, 35], which can also reduce the computational burden. We outline the particular J-shifting approach that has proved to be useful in the $H_2 + OH$ and $D_2 + OH$ studies.

This J-shifting procedure allows us to estimate state-resolved cross sections or rate constants, and to later combine them to estimate $k(T)$. It is based on using not $J = 0$ information, but information for some larger J value, which may be more representative of the dynamics. The idea of using some nonzero J value is not new—see also Refs. [36–39]. For specific initial reactant quantum numbers, v_1, j_1, v_2, j_2, and some appropriately "typical" J, J_{ref}, we define

$$p_{v_1,j_1,v_2,j_2}(\varepsilon) \equiv P^{J_{ref}}_{v_1,j_1,v_2,j_2}(\varepsilon) \tag{47}$$

and construct either state-resolved cross sections, Eq. (42), or rate constants, Eq. (45), with the J-shifting approximation

$$P^J_{v_1,j_1,v_2,j_2}(\varepsilon) \cong p_{v_1,j_1,v_2,j_2}[\varepsilon' = \varepsilon + (E^{J_{ref}} - E^J)] \tag{48}$$

with E^J denoting the mean centrifugal barrier at the transition state,

$$E^J = \frac{1}{n_J} \sum_K [\bar{B}J(J+1) + (A - \bar{B})K^2] n_{JK} \tag{49}$$

and where n_{JK} is the number of possible initial conditions for given J and K. n_J is the total number of initial conditions for given J. (If $j_1 = j_2 = 2$, then the largest K can be is 4 and, for any $J \geq 4$ the sum of all the allowed k_1 and K combinations such that $k_1 + k_2 = K$ is $n_J = 25$ or the rotational degeneracy $(2j_1 + 1)(2j_2 + 1) = 25$.) Equation (49) assumes the transition state is a near prolate symmetric top; that is, if A, B, and C are the rotor constants at the transition state, $\bar{B} = (B + C)/2$. How would one choose J_{ref}? One approach is to choose J_{ref} based on considering which J contributes most to the quantity of interest. For example, one might choose J_{ref} as well as the initial rotational quantum numbers j_1 and j_2 based on those that are most relevant to a particular range of temperatures [21, 34]. One then does a series of helicity-decoupled calculations for each of the allowed K and k_1 combinations [21, 34]. (In principle, full Coriolis-coupled calculations could be used also.) It should be emphasized that $J = j_1 = j_2 = 0$ is often *not* the best set of reference angular momenta. Thus, in order to estimate $k(T)$, one need not sum over all possible reactant states but can restrict attention to those that are most relevant to the temperature range of interest [21, 34]; for example, one could replace the summations in Eqs. (42) and (44) with appropriately restricted summations.

IV. PARALLEL ALGORITHMS

A. General Overview

As stated earlier, quantum dynamics calculations for even relatively small systems can be computationally challenging, requiring large amounts of almost every computer resource including CPU time, memory, and disk space. In general, such calculations are quite suitable applications for parallel computing [22, 32, 40–44], as there is an enormous amount of inherent parallelism in wave packet propagations and other related iterative methods. The vast majority of computational effort is spent computing the action of the Hamiltonian matrix $\mathbf{H}$ upon a vector $\mathbf{v}$, $\mathbf{Hv}$. (Here $\mathbf{v}$ could be a full complex vector Ψ or the real-valued vector $\mathbf{q}$ of the RWP method.) By exploiting the structure of the Hamiltonian matrix, efficient and scalable parallel algorithms may be devised. For most wave packet methods, the patterns of communication between processors are known beforehand and are regular, static, and predictable. Nevertheless, the problems are far from embarrassingly parallel, since the inherent coupling in the mathematics of the problem will manifest itself in coupling between the various processors. In this section, we describe three parallel algorithms, ranging from the simple to the relatively complex. We have implemented each of these algorithms [22, 32, 40] and used them to solve challenging problems in wave packet dynamics [21, 45–54].

In this section we represent wave packets as vectors $\mathbf{v}$ and refer to the individual components of $\mathbf{v}$ in the computer-friendly form of multidimensional arrays with indices, for example, $v(i, j, k)$, where the indices i, j, and k refer to the various degrees of freedom. These vectors may be real, such as the real vector $\mathbf{v}$ of the previous sections, or they may be complex representations of the wave packet. The notation $\mathbf{v}(:, k)$ refers to the subvector of $\mathbf{v}$ containing all components of $\mathbf{v}$ with a particular value of the index k.

The parallel algorithms described here can be implemented on a variety of parallel architectures including distributed memory message passing computers (clusters), shared memory parallel computers, and clusters of shared memory computers. The algorithm described in Section IVD is designed to take particular advantage of clusters of shared memory processors (SMPs). We will refer to any computer that is linked to other computers via a network of some type as a *node*. These computers may consist of a single processor or two or more processors that share memory. Parallel jobs may distribute across the processors of a single node or they may be distributed across the nodes themselves.

Along with shared memory and distributed memory architectures, there are shared memory and distributed memory programming paradigms embodied in different parallel programming languages. The distributed programming language that we employ is MPI (Message Passing Interface)[55]. The first two algorithms we describe use MPI. MPI may be implemented on any type of architecture. The logical programming unit of MPI is called a *task*. Each task receives a copy of the same program and they run simultaneously processing different data. It makes no logical difference to MPI how these tasks are mapped onto the computer hardware. Tasks may be distributed to the processors of a single node or they may be distributed across a network. Performance, of course, may certainly depend on the particular mapping. Shared memory programming languages such as OpenMP [56], however, are designed to function only on the processors of a shared memory computer.

A measure of the efficiency of a parallel algorithm or its implementation in a computer code is its scaling properties. By scaling we mean the change in wall clock time as we increase the number of nodes. There are several ways to determine how a parallel algorithm scales. One is to increase the number of processors for a problem of a given size and measure the wall clock time. (One always assumes that the processors are dedicated, not shared with other jobs.) If the wall clock time decreases linearly with the number of processors, linear (and perfect) scaling has been achieved. Except for problems involving no or only trivial amounts of communication, there will be a maximum number of processors for which linear scaling holds. If one spreads the problem over too many processors such that the time involved in communication is comparable to the time needed for computation, linear scaling will break down. Another way to measure scaling is to increase both the size of the problem and the number of processors by the same amount. Thus, if one doubles the size of the problem and also the number of processors and the wall clock time does not change, then one also has perfect scaling. For very large problems it sometimes happens that a distributed job will experience a greater than linear speed-up. This occurs when the job on an individual node taxes the resource limits of the node, such as memory. Our discussion is written with distributed computing message-passing architectures in mind, but similar considerations hold for shared memory computers.

Assume that we have a program we will run on np processors and that this program has a serial portion and a parallel portion. For example, the serial portion of the code might read in input and calculate certain global parameters. It does not make any difference if this work is done on one processor and the results distributed, or if each processor performs *identical* tasks independently; this is essentially serial work. Then the time t_1 it takes the program to run in serial on one processor is the sum of the time spent in the serial portion of the code and the time spent in the parallel portion (i.e., the portion of the code that can be parallelized) is $t_1 = t_s + t_p$. Amdahl's law defines a parallel efficiency, P_E, of the code as the ratio of total wall clock time to run on one processor to the total wall clock time to run on np processors. We give a formulation of Amdahl's law due to Meijer [42]:

$$P_E = \frac{t_1}{\sum_{i=1}^{np} t_i} = \frac{t_s + t_p}{\sum_{i=1}^{np} [t_s + (t_p)_i + (t_c)_i + (t_w)_i]} \tag{50}$$

where t_i is the total wall clock time for processor i. $P_E = 1$ corresponds to perfect scaling. In the denominator in the third term of Eq. (50), $(t_p)_i$ is the time spent in the parallel part of the program on processor i, $(t_c)_i$is the time processor i spends on communication, and $(t_w)_i$ is the time it spends waiting (e.g., for other processors to finish their tasks). Note that the serial time is the same for each processor. Wait times are typically due to improper load balancing, which we discuss later. For a properly load-balanced program, $(t_w)_i = 0$, and each processor spends the same amount of time on communication. Thus we have

$$P_E = \frac{t_s + t_p}{np(t_s + t_c) + t_p} \tag{51}$$

Thus it is clear that for a load-balanced program, P_E can only approach unity if both the time spent in the serial portion of the code and the time spent in communication are small compared to the time spent in the parallel portion of the code. Additionally, good scaling with increasing the number of processors will occur only if, for each $(t_p)_i$, $(t_p)_i \gg (t_c)_i$. The time spent in computation must be greater than the time spent in communication. For a job of a given size, as we increase the number of processors, the amount of parallel work on each processor will decrease and will in general decrease faster than the time spent in communication. Thus it follows that we should not distribute a job of a given size "too finely."

In distributed computing there is often a trade-off between reducing wall clock time and using the processors or nodes most efficiently. Clearly, if a single processor has enough memory, it is most efficient to use only one processor since then there is no overhead associated with communication. By distributing the problem over several nodes, wall clock time may be considerably reduced. The reduction of wall clock time is, of course, one of the central reasons for using distributed computing. As we have noted, it is important that the ratio of computation to communication remain high. If a problem of a given size is spread over too many nodes, it is also possible not only to reduce efficiency but to actually increase the wall clock time. We have found that algorithms that involve communication between "nearest neighbor" processors are far more efficient than those that require communication between all of the processors.

Proper load balance is a major consideration for efficient parallel computation. Consider a job distributed over two processors (0 and 1) in such a way that wall clock time is reduced considerably. Nevertheless, it still may be that processor 0 has more work to perform so that processor 1 spends much time waiting for processor 0 to finish up a particular task. It is easy to see that, in this case, the scaling will, in general, not be linear because processor 1 is not performing an equal share of the work.

B. Coriolis-Coupled Processors

In order to accurately compute observables such as cross sections and rate constants, it is necessary to compute reaction probabilities as all relevant values of total angular momentum, J. The vast majority of quantum dynamics calculations, however, consider only $J = 0$ or use approximate methods to treat total angular momentum. The most sophisticated of these approximations, the helicity decoupling approximation, achieves its computational savings by ignoring the Coriolis terms that couple different projections of total angular momentum onto the body-fixed z-axis. The underlying assumption of this method, that the Coriolis terms will be small for all relevant configurations, breaks down for many important reactions, particularly those that go through floppy intermediates such as the combustion reaction $H + O_2$ [49–52]. For a more recent calculation on an updated potential energy surface of this system see Ref. [57].

The primary reason it is difficult to treat angular momentum rigorously is due to the "angular momentum catastrophe" [58]. As noted in Section III, cross sections and other experimental observables are sums over all relevant total angular momentum quantum numbers, J. Each J represents a quantum dynamics problem to be solved, and the size of the problem increases dramatically with J. For each J, there are N_K projections of K, where $N_K = K_{\max} - K_{\min} + 1$. For a three-atom system, the minimum value of $K, K_{\min}$, is a function of both J and p, such that $K_{\min} = 0$ when J and p are both even (or both odd); otherwise $K_{\min} = 1$ [32]. For a four-atom system, $K_{\min} = 0$ for both even and odd parity for all values of J [21, 22]. In most cases,$K_{\max} = J$. Thus the size of the problem grows roughly linearly with total angular momentum.

By exploiting the sparseness of the Coriolis coupling terms and the power of parallel computing, it is possible to compute rigorous scattering information for relatively high J values with computational times comparable to the $J = 0$ case [32]. For systems involving three or more atoms, the Coriolis terms arise from the rotational kinetic energy operator. For four-atom systems, for example, these terms arise from the expansion of the fourth term in Eq. (33). In a three-atom system, the analogous term is $(\hat{J} - \hat{j})^2/(2\mu_R R^2)$. Physically, these terms arise from the coupling of the total angular momentum with internal angular momentum of the system. They contain terms that are diagonal and off-diagonal in K. Thus, for each $|K| \leq J$, the Coriolis terms couple the vectors $\mathbf{v}(:, k)$ to $\mathbf{v}(:, k \pm 1)$, where k is the index that refers to quantum number K. The Coriolis terms are diagonal in all other variables and thus are extremely sparse. Since most of the Hamiltonian matrix is diagonal in K, we can decompose the Hamiltonian matrix into a block tridiagonal form with the dense matrices $\mathbf{H}^K$ representing the portion of the Hamiltonian corresponding to a particular value of K and the sparse off-diagonal elements corresponding to the Coriolis terms; for example,

$$\begin{bmatrix} \mathbf{H}^0 & \mathbf{c}^{0,1} & 0 & 0 & 0 \\ \mathbf{c}^{0,1} & \mathbf{H}^1 & \mathbf{c}^{1,2} & 0 & 0 \\ 0 & \mathbf{c}^{1,2} & \mathbf{H}^2 & \mathbf{c}^{2,3} & 0 \\ 0 & 0 & \mathbf{c}^{2,3} & \mathbf{H}^3 & \mathbf{c}^{3,4} \\ 0 & 0 & 0 & \mathbf{c}^{3,4} & \mathbf{H}^4 \end{bmatrix} \tag{52}$$

This decomposition suggests the strategy for parallel computing. In the simplest implementation, we distribute the job over N_K processors, where N_K is the number of K values in our problem. Each processor is labeled by an index, $k = 0, 1, \ldots, N_k - 1$, associated with a particular value of K. We distribute the wave packet vector, $\mathbf{v}$, over the processors such that processor k contains $\mathbf{v}(:, k)$ and is responsible for everything related to its propagation, in particular, for computing the action of $\mathbf{H}^K$ on $\mathbf{v}(:, k)$. Except for the work associated with the Coriolis terms, *all of the computation is done locally on each processor, proceeds independently, and involves no communication.*

With this distribution, each size of the job on each processor is less than or equal to the size of a $J = 0$ calculation. However, the computation of the Coriolis terms involves communication between processor k and processors $k \pm 1$. In fact, processor k will send $q(:, k)$ to each of its neighboring processors and will receive both $q(:, k - 1)$ and $q(:, k + 1)$ from its neighbors. (Processor 0 sends and receives data only from processor 1; processor $N_K - 1$ sends and receives data only from processor $N_K - 2$.) This communication is accomplished through the use of MPI calls, which are given in the appendix of Ref. [32]. One can see that *if the cost of communication can be kept to a minimum*, this simple model would require little more wall clock time than a $J = 0$ calculation. Therefore, if one has access to a parallel computer or cluster with the requisite number of processors, one can perform calculations for $J > 0$ in reasonable time.

Although this algorithm is quite straightforward, there are several important considerations regarding its efficiency. In general, the structure of the code itself will greatly affect its performance. To facilitate the distribution of the data over the various processors, it is important to structure the code so that the outer loop corresponds to varying k; thus in accordance with good FORTRAN programming, k should be the rightmost index in all relevant arrays. (The opposite is required in the C and C++ languages.) For small J, the simple method of one K per processor scales quite well and, until J gets quite large, achieves the maximum reduction in wall clock time. For this model, we use the second measure of scaling described earlier. The quantity t_J/t_0 is the ratio of the wall clock time required to perform a job for a given J on $J + 1$ processors to the time it takes to perform a comparable $J = 0$ job on one processor on the same computer. In the ideal case, $t_J/t_0 = 1$. For the cases we have looked at, which include studies of the predissociation of ArI_2 [45, 46], the complex-forming reactions $H + O_2$ [49–52] and $O(^1D) + H_2$ [53], good scaling is achieved for $J \leq 10$, but this will vary depending on the size of the basis and the particular computer. As J increases, t_J/t_0 increases slowly. In the case of ArI_2 we found that t_J/t_0 rises to ~ 2 in the $J = 39$–49 region [32]. This indicates that wall clock times are about twice the $J = 0$ case. Thus one can hope to complete a job for very large angular momentum using twice the wall clock time for a $J = 0$ job, albeit at the cost of using computer resources less efficiently than might be desirable.

The considerations of the last paragraph raise two questions. Why does efficiency degrade as we employ more processors? And is there an alternative strategy that might increase the efficiency without requiring large amounts of computer time? The answers to these two questions are of course related. The efficiency of the one J per processor model degrades due to both load imbalance and increased communication overhead. The communication overhead includes not only the time it takes to send and receive messages but also the synchronization time, that is, time spent waiting for other processors to finish their tasks. The communication overhead increases with the number of processors employed, largely in this case as a result of load imbalance as we shall see.

The load imbalance will grow as a function of J. The earlier discussion of scaling assumed that the amount of work that a processor must perform is independent of K. Let us consider a three-atom problem with j the quantum number corresponding to the internal rotational basis. In fact, this is not the case. The range of j depends on K since $K \leq j \leq j_{\max}$, where $j_{\max}$ is the largest value of j in the internal rotational basis. Suppose, for example, that the $J = 40$ and $j_{\max} = 79$ is a case where both even and odd js are included in the basis (no homonuclear symmetry). There will be 80 angular basis functions for $K = 0$ and only 40 for $K = 40$. The amount of work each processor does scales as the mth power of the number of angular basis function, where $1 \leq m \leq 2$. Thus processor 0 will have between 2 and 4 times more work to perform as processor 40. In fact, for this example, the entire amount of work that the computer must perform is between 24 and 30 times that of a $J = 0$ calculation. Thus it is wasteful to employ 40 processors on this task. A much better load balance can be achieved if more than one K block is placed on a given processor. When the number of K states, n, is even (odd parity case), it is easy to improve load balancing by placing two K states in each processor in *wrap-around* manner as shown in Fig. 2; for example, $K = 0,\ n \rightarrow$ processor 0, $K = 1,\ n - 1 \rightarrow$ processor $1, \ldots, K = (1/2)n - 1,\ (1/2)n \rightarrow$ processor $(1/2)n - 1$. Thus each processor will have exactly the *same* number of j states, and the amount of work

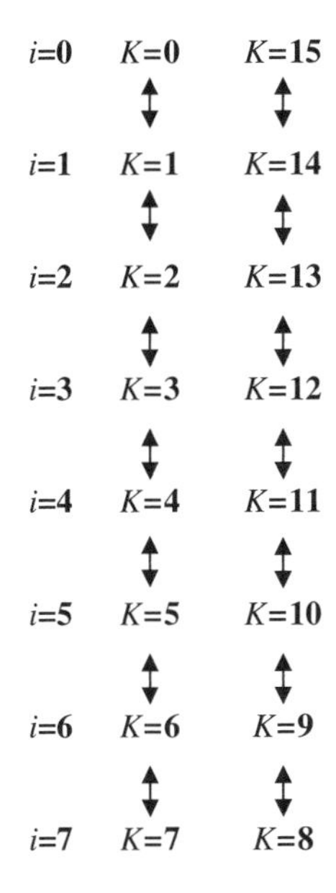

Figure 2. Diagram depicting interprocessor communication in the case of wrapped distribution of K states with two per processor. The figure illustrates $J = 15$, odd parity. The eight processors are labeled $i = 0, 1 \ldots, 7$ and the corresponding K states that they contain to the right of each i. Arrows indicate communication between processors for each of the K states.

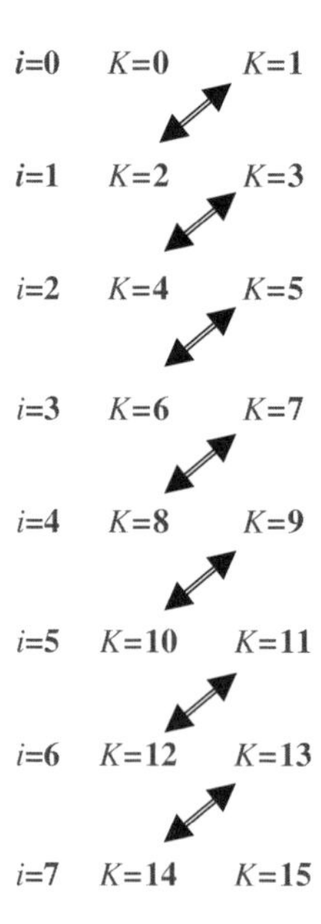

Figure 3. Diagram depicting interprocessor communication in the case of sequential distribution of K states with two per processor. The figure illustrates $J = 15$, odd parity. The eight processors are labeled $i = 0, 1 \ldots, 7$ and the corresponding K states that they contain to the right of each i. Arrows indicate communication between processors for each of the K states. Note that this model involves less communication than the wrapped model.

will be more nearly equal. For comparison, we show in Fig. 3 an alternative method of placing two K states on each processor in sequential fashion. Note that this method involves less communication than the wrapped method, but will certainly result in more overall load imbalance.

In the even parity case, with an odd number of K states, data distribution is not as straightforward. One approach is to "double up" the calculation. There are many possible variations on this theme and we just give a brief illustration. For example, both the $J = 8$ and $J = 9$ cases could be computed at the same time on 9 processors. In the case of even parity, for $J = 8$, the K states go from 0 to 8, while for $J = 9$, they range from 1 to 9. A logical way to distribute the work, in terms of (J, K) states, is as follows: (8, 0) and (9, 9) $\rightarrow$ processor 0, (8, 1) and (9, 8) $\rightarrow$ processor $1, \ldots, (9, 1) \rightarrow$ processor 8. Thus each processor will have the same total number of angular basis functions and perform roughly the same amount of work. Another approach successfully used in a study of the $H + O_2$ [49, 52] reaction is to wrap the problem as in the odd parity case, starting with processor 1. Processor 0 would then have less work to do than the other processors. Additional work is assigned to this processor, such as all the input/output and tasks related to an ongoing analysis of the wave packet.

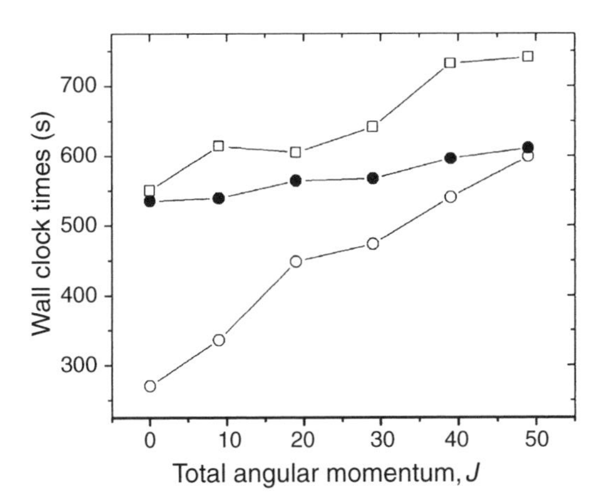

Figure 4. Comparing three Coriolis-coupled parallel model strategies. This particular run was for the dissociation of ArI_2 on an IBM SP2. Open circles denote the one K per processor model; filled circles denote the wrapped algorithm with two K states per processor; open squares denote the sequential assignment of two K states per processor. The basis set is modest in size with 64 grid points for R, 5 vibrational and 40 rotational states for I_2 (even symmetry: $j = 0, 2, 4, \ldots, 78$). Details will change depending on basis set size and type of computer, but, in general, the wrapped algorithm will outperform the other methods and will be the method of choice for large J.

In Fig. 4 we compare the timings for three different models, the simple one K per processor, the wrapped algorithm, and a model where two states are assigned per processor sequentially. Note that until $J = 50$ the one K per processor model job uses the smallest amount of wall clock time. It is clear, however, that this method does not make efficient use of computer resources. The wrapped model, however, scales very well and outperforms the sequential two K per processor model at every $J > 0$, a clear illustration of the degradation of performance due to load imbalance.

C. Distribution Over Radial Grids

The Coriolis-coupled model described earlier takes advantage of the sparse nature of the Coriolis terms. There are many problems, however, that benefit from parallelism involving tightly coupled degrees of freedom. For systems of four and more atoms, even a $J = 0$ computation may be too large for a one processor computer to handle. Thus we developed simple parallel models for four-atom codes [22]. The simplest such models involved distributing the wave packet according to one of the radial degrees of freedom. Let np be the number of processors, labeled as $i = 0, 1, \ldots, np - 1$. Then if nrt is the total size of basis in R, each processor will have $nr = nrt/np$ grid points in R. (We may restrict the calculation such that np divides evenly into nr; this makes the programming easier and is not generally a very severe restriction. Such a restriction, however,

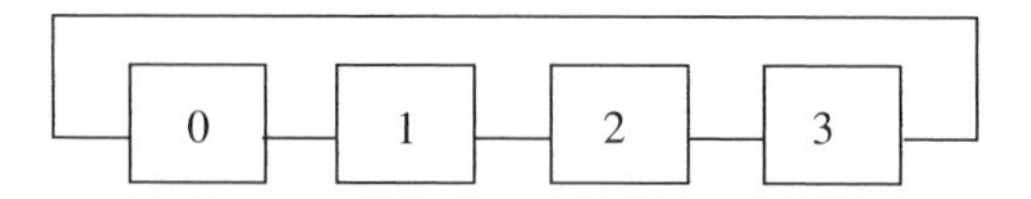

Figure 5. Schematic diagram of the topology of interprocessor communication on a four-processor ring. Data on each processor is sent to every other processor.

is not necessary.) The vectors corresponding to **v** and **Hv** to are partitioned into np equal sections, $\mathbf{v}_i$ and $(\mathbf{Hv})_i$, and distributed to the various processors along with the relevant portions of all other vectors and matrices. The portion of **v** and **Hv** indexed by $R(i \times nr + 1) \cdots R((i + 1)nr)$ is contained in processor i.

The calculation of **Hv** dominates the computation and may be computed simultaneously and for the most part independently on each processor. The computation of all but the first term in Eq. (31) is completely local to each processor. The action of the radial kinetic energy operator, $\hat{T}_R$, however, requires communication between the processors. Early implementations of this method employed sinc-DVRs [27] to compute the action of $\hat{T}_R$, where $\hat{T}_R$ is represented by a symmetric matrix of dimension $nrt \times nrt$. Matrix–vector communication requires communication between all of the processors. In our implementation, the vectors $\mathbf{v}_i$ are passed "round robin style" [22], for example, from right to left as shown in Fig. 5. Computation of **Hv** requires that each processor send and receive $(np - 1)/np$ of the wave packet. It is also possible, and often desirable, to distribute the wave packet according to more than one radial degree of freedom, for example, R and r_1 as shown in Fig. 6. Here each processor is labeled by two indices (i,j) referring to the particular chunk of $[R, r_1]$ for which it is responsible. The processors are logically connected as a torus and communication between processors occurs both vertically and horizontally along the torus. In both the 1D and 2D decompositions, we have lost the nearest-neighbor communication advantage of our Coriolis-coupled model [32]. The

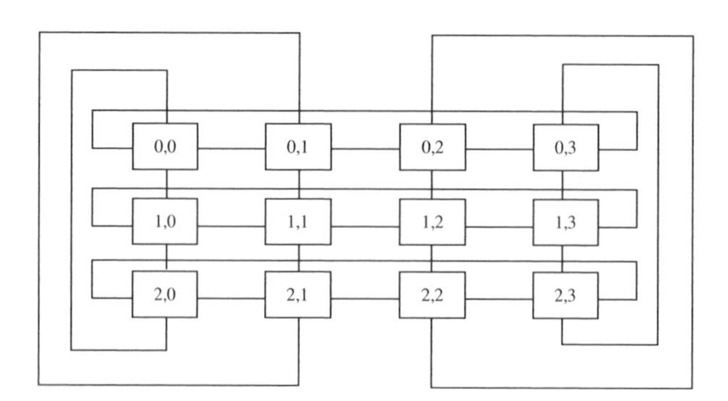

Figure 6. Schematic diagram of the topology of communication in a 3 by 4 torus. Each processor is labeled $[i, j]$, referring to the particular chunk of $[R, r_1]$ for which it is responsible. Communication between processors occurs both vertically and horizontally along the torus.

added communication time is reflected in the scaling properties of the code, which is considerably less optimal than that of the Coriolis-coupled model [22].

In general, the topology of interprocessor communication reflects both the structure of the mathematical algorithms being employed and the way that the wave packet is distributed. For example, our very first implementation of parallel algorithms in a study of planar OH + CO [47] used fast Fourier transforms (FFTs) to compute the action of $\hat{T}_R$, which also required all-to-all communication but in a topology that is very different from the simple ring-like structure shown in Fig. 5.

There have been several successful approaches [41, 42] to improving the scaling of parallel algorithms that employ DVRs or FFTs. These rely on more sophisticated methods of distributing the wave packet so as to reduce the amount of communication required. One should consult Refs. [41] and [42] for details.

Our approach has been, in fact, quite different. We have focused on developing an algorithm that would allow us to regain the advantages of the nearest-neighbor communication that we enjoyed in the Coriolis-coupled model. This was the original motivation for the development of the DFFD method [25] described in Section IIIB. The DFFD approach [25], like any finite difference approach, is a local method for obtaining second derivatives. (The distributed approximated functions developed by Kouri and co-workers is another local approach [59].) Thus, if we use a DFFD of order $(2n+1)$ to compute $\hat{T}_R$ at the kth grid point, only the points $k = k-n, \cdots, k-1, k, k+1, \cdots, k+n$ are required. As long as each processor holds at least n grid points, only information from neighboring processors will be required. The communication pattern for a two degree-of-freedom distribution of the wave packet using the DFFD method is shown in Fig. 7; notice that it is no longer a torus. The scaling properties of this algorithm are excellent as long as (1) we run

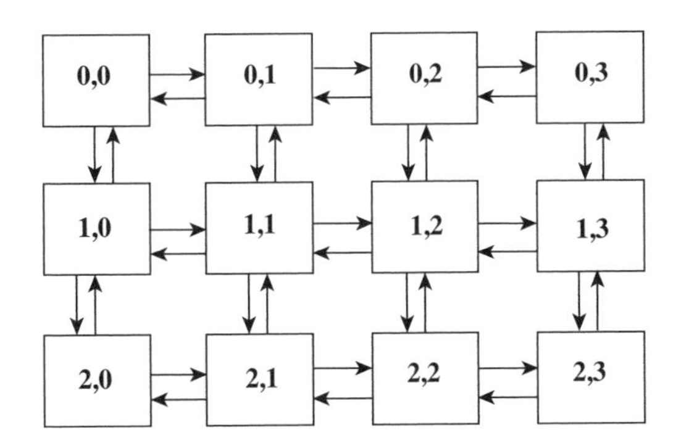

Figure 7. Schematic diagram of the topology of communication when the DFFD algorithm is employed. As in Fig. 5, each processor is labled $[i,j]$, referring to the particular chunk of $[R, r_1]$ for which it is responsible. However, communication is only between nearest neighbors; that is, processor (i,j) sends and receives data only from $(i \pm 1, j)$ and $(i, j \pm 1)$.

on a homogeneous system and (2) the parallelism is not too fine-grained. Under these conditions, it is relatively easy to distribute the wave packet such that load balance is nearly perfect. This method has been successfully applied for several systems including studies of the OH + H_2 [21] reaction and the decay of OH—CO entrance channel intermediates [54]. An OpenMP shared memory version of this method was used to study the OH + CO reaction [47].

D. Hybrid OpenMP and MPI Approach

The two conditions above for good scaling set limits on the number of processors that can be utilized effectively by the simple message-passing (MPI) algorithm described previously. The limitations are set by Eq. (50) because P_E decreases when the communication time or the wait time becomes comparable to the computation time. For a job of a given size run on a given system, there will be a maximum number of processors that can be made good use of. This is the requirement that the work not be distributed too finely.

Suppose, however, that we have a very large job that requires more processors than exist on a single multiprocessor node of our cluster. In a cluster, the nodes are connected to form a single system. Although the speeds of these connections vary widely, the communication time between processors on *different* nodes is generally much slower than the communication time between processors on a *single system* with shared memory. Therefore jobs that distribute across *processors* on more than one node will inevitably incur load imbalance as a result of waiting for the slower communication to finish, and this will be reflected in reduced efficiency. This is reflected in a degradation of performance of the simple MPI model developed in the previous section when distributed across the processors on different nodes.

To address the shortcoming of the simple MPI model, Medvedev and co-workers [40] developed a hybrid OpenMP/MPI method that takes advantage of both distributed and shared memory features of these clusters of multiprocessor nodes. The features of this model are:

1. To use MPI to distribute the wave packet across *nodes* rather than processors.
2. To use OpenMP on each node to implement the parallel work on each node.
3. *To overlap communication and computation.*

This method allows us to attain excellent scaling with a much larger number of processors than the simple MPI method described earlier. We give a brief outline of this method below; for details see Ref. [40].

The wave packet is represented by a vector with elements $v(j, i2, i1, ir)$, where j is a collective index for rotational states, and the indices, $i2, i1$, and ir

are associated with radial coordinates r_2, r_1, and R, respectively. We distribute this vector over nodes rather than processors according to values of the radial coordinates. As in the simple MPI method, our specific implementation allows for parallelization over 2D radial domains in *ir* and *i*1.

Hybrid OpenMP/MPI parallelization is used to compute **Hv**. The multi-threaded OpenMP library is used for parallelization within each node, while MPI is used for communication between nodes. The subroutine *hpsi* that computes **Hv** contains two arrays of the same size: the results of the portion of the calculation local to a node are stored in *hps_local*, while the results of the portion of the calculation that requires communication are stored in *hpsbuff*. On each node, a *dedicated* (master) OpenMP thread performs all the communication, while other threads perform local parts of the work. In general, one uses as many threads as there are processors on a given node. The following scheme presents an overview of the execution of subroutine, *hpsi*:

communication part :

  hpsbuff = 0

perform communication part, storing results of the calculation in hpsbuff

end communication part

local part :

  evaluate local parts of H^*v, *storing results in hps_local*

end local part

barrier

add hpsbuff to hps_local

Only the master thread performs communication, while all other threads perform the local part *at the same time*, leading to effective overlap of communication and computation. Once both local and communication parts are finished, array *hpsbuff* is added to the array *hps_local*. The *barrier* above denotes the place where all the threads are synchronized. It is necessary to ensure that addition of *hpsbuff* to *hps_local* will not start before both local and communication parts are finished.

The master thread performs all the internode communication using MPI. Only the action of the kinetic energy operators, $\hat{T}_R$ and $\hat{T}_r$ implemented with the DFFD approach, require such communication. While the master thread is executing the communication part and storing its results in *hpsbuff*, other threads perform local work storing results in the main array *hps_local*. The use of the two separate arrays is needed to avoid having to synchronize the threads. Moreover, if the communication part finishes before the local part, the master thread joins the other threads in the computation of the local part.

The shared memory OpenMP library is used for parallelization within each node. The evaluation of the action of potential energy, rotational kinetic energy, and r_2 kinetic energy are local to each node. These local calculations are performed with the help of a task farm. Each thread dynamically obtains a triple $(i2, i1, ir)$ of radial indices and performs evaluation of first the kinetic energy and then the potential energy contribution to *hps_local(:, i2, i1, ir)* for all rotational indices.

Once both local and communication parts of the Hamiltonian evaluation are finished, we add the buffer *hpsbuff* containing results of the communication part to the main array *hps_local*. However, it is necessary to ensure that this addition will not start before both local and communication parts are finished. A barrier directive forces all the threads to synchronize. A thread that has reached the barrier will not resume execution until all other threads have reached it too.

Good scaling as a function of the number of processors will occur when the ratio of the communication time to time spent in the local portion of the code is small. We give a good rough estimate of this ratio as [40]

$$nrot^{-1/3}\left(\frac{s_R}{nr}+\frac{s_1}{n1}\right) \tag{53}$$

where *nrot* is the size of the internal rotational basis, *nr* and $n1$ are the number of grid points in R and r_1, and s_R and s_1 are the numbers of nearest neighbors in the R and r_1 directions, respectively. Depending on the parallelization scheme, these numbers can be 0, 1, or 2. In particular, s_R is equal to 0 when there is no parallelization over R, is equal to 1 when R grid is parallelized over just two nodes, and is equal to 2 in all other cases. s_1 is defined in a similar way. Good scaling as a function of the total number of processors occurs when Eq. (53) is small. In this case, even though all the processors in the node perform the local part of the computation simultaneously, most of the time is still spent in the local part, and near perfect scaling is achieved. Thus larger basis sets favor good scaling. In the opposite case, that is, when communication dominates the running time, the total running time depends roughly on the number of nodes. As long as the local part of the work dominates the total running time, the exact distribution of the grid over different nodes is not important, provided that good load balance between the nodes is maintained. However, when the computational effort for the communication part becomes comparable to or exceeds that for the local part, this distribution becomes important, and one should distribute the grid over the nodes so as to minimize the communication time per node which is proportional to $s_R{*}n1 + s_1{*}nr$ [40].

In Fig. 8 we show a speed-up curve obtained from timing runs on the *Seaborg* cluster, at the National Energy Research Scientific Computing Center (NERSC) [60]. Each node is a 16 processor IBM RS6000. Nodes are connected by a high-speed switch. The particular calculation is for the energy levels of reactant channel complexes of OH—CO using the hybrid OpenMP/MPI

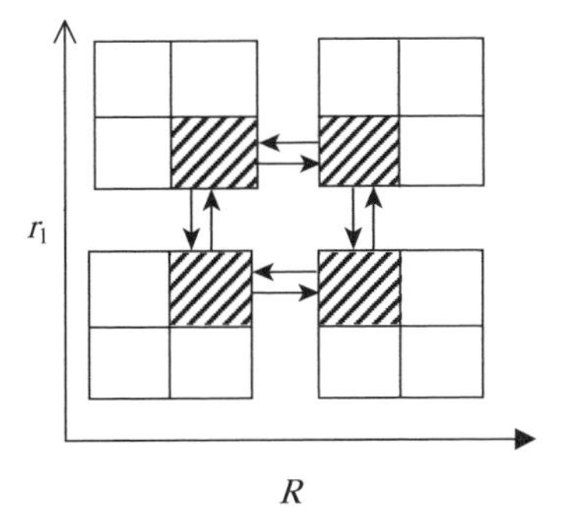

Figure 8. Schematic of distribution of the wave packet over the nodes and the communication pattern for the hybrid OpenMP/MPI method. The wave packet is distributed over four nodes, each of which contains four processors. All of the communication is handled by the shaded processors; the other processors are simultaneously performing computations. Note that there is only nearest-neighbor communication between nodes.

approach. The total size of the basis is ~1.9 GB. The solid line represents ideal speed-up: $t_1/t_n = n$, where n is the number of processors, t_1 is the wall clock time for a one processor run, and t_n is the wall clock time for the same run on n processors. The shape of this speed-up curve is typical for this method but the particulars will vary depending on many factors. Deviations from ideal speed-up occur when communication time becomes comparable to the computation time. Note that for $n \leq 16$, on Fig. 9, the processors are all on one node and there is

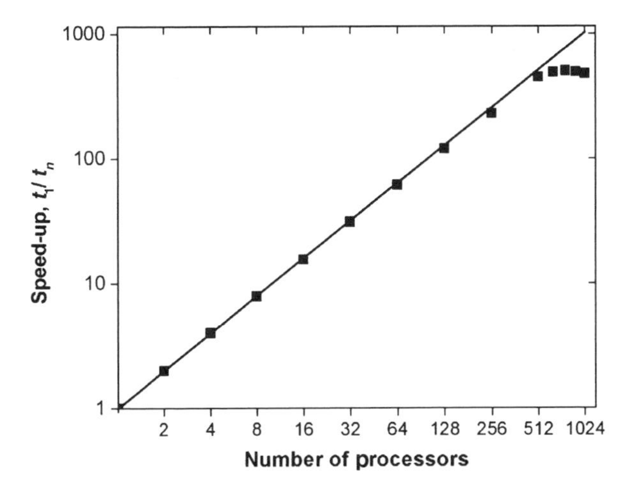

Figure 9. Typical speed-up curve for the hybrid OpenMP/MI approach. The symbols represent achieved speed-up; solid line shows ideal speed-up. The deviation from ideal speed-up occurs when computation time becomes comparable to computation time on the nodes.

no internode communication. For $n > 16$, two or more nodes are involved. The scaling is independent of the number of nodes employed, indicating successful overlap of communication and computation.

Recently, we [61] applied this method to model the $(HOCO)^-$ photodetachment experiments of Continetti and co-workers [62]. We are currently using the method to study the dissociation dynamics other four-atom systems, including hydrogen peroxide and formaldehyde.

V. CONCLUDING REMARKS

The methods presented in this chapter are designed to give accurate quantum dynamics information. Currently, such information is most readily obtainable for three- and four-atom systems. For such systems, the RWP method is a good choice because it leads to accuracy as good as Tal-Ezer and Kosloff's Chebyshev propagator [3], but involves somewhat less memory and computational effort. The RWP framework also allows one to use standard methods of wave packet analysis, and inspection of the evolving (real) wave packets can provide mechanistic information. Furthermore, the parallel computing algorithms we discussed enable the accurate numerical solution of computationally challenging problems such as complex-forming four-atom reactions involving several heavy atoms.

There is much current interest in the quantum dynamics of larger systems. The extension of RWP ideas to such systems is straightforward in principle. The difficulty is that the finite numerical representation of the wave packet increases dramatically with the addition of more degrees of freedom. Even with continued advances in computer technology, it is likely that approximate methods will have to be used for such systems. There are many approximate quantum and semiclassical methods for treating quantum dynamics and we can only touch on a few, in particular, those involving some form of wave packet propagation. The obvious approximation is to develop reduced-dimensionality models, such as those in Refs. [63] and [64]. An interesting application of reduced-dimensional wave packet methods to torsional motion in proteins was recently given [65]. A different and promising approach is the multiconfiguration time-dependent Hartree (MCTDH) method [66–69]. This approach may be used in an approximate manner but can also be converged to calculate rigorous rate constants for larger systems using cumulative reaction probability ideas [35]. For example, recently, a high quality *ab initio* potential energy surface was used in conjunction with this method to determine accurate thermal rate constants for the $H + CH_4$ reaction [69]. Other recent MCTDH work includes reduced-dimensionality studies on the photodissociation of a van der Waals complex [70], proton conduction along a chain of water molecules [71], and a nine-dimensional study of the vibrational energy redistribution in the highly excited fluoroform molecule [72]. For studying complex-forming reactions, recently developed wave packet based statistical approaches [73, 74] may be capable of being extended to larger systems. In addition to providing accurate dynamics information for few-atom systems, accurate methods such as the RWP method can be used to validate such approximate theories.

Acknowledgments

We have greatly benefited from collaborations and general scientific discussions with several individuals. The RWP method was developed in close collaboration with Professor Gabriel G. Balint-Kurti. Several challenging applications of the method were also carried out in collaboration with him. We were fortunate to have two collaborators with remarkable intellectual skills, both in terms of chemical physics and computer science. Anthony Meijer applied the Coriolis-coupled parallel method to the challenging $H + O_2$ reaction. Dmitry M. Medvedev championed much of the parallel four-atom work described here. This work was supported by the Office of Basic Energy Sciences, Division of Chemical Sciences, Geosciences, and Biosciences, U.S. Department of Energy, under Contract No. DE-AC02-06CH11357, and by the National Science Foundation, CHE-0315113.

References

1. S. K. Gray and G. G. Balint-Kurti, *J. Chem. Phys.* **108**, 950 (1998).
2. R. Kosloff, *Annu. Rev. Phys. Chem.* **45**, 145 (1994).
3. H. Tal-Ezer and R. Kosloff, *J. Chem. Phys.* **81**, 3967 (1984).
4. V. A. Mandelshtam and H. S. Taylor, *J. Chem. Phys.* **103**, 2903 (1995).
5. Y. H. Huang, S. S. Iyengar, D. J. Kouri, and D. K. Hoffman, *J. Chem. Phys.* **105**, 927 (1996).
6. R. Q. Chen and H. Guo, *J. Chem. Phys.* **105**, 3569 (1996).
7. G. J. Kroes and D. Neuhauser, *J. Chem. Phys.* **105**, 8690 (1996).
8. S. K. Gray, *J. Chem. Phys.* **96**, 6543 (1992).
9. T. J. Park and J. C. Light, *J. Chem. Phys.* **85**, 5870 (1986).
10. R. H. Bisseling, R. Kosloff, and J. Manz, *J. Chem. Phys.* **83**, 993 (1985).
11. D. H. Zhang and J. Z. H. Zhang, *J. Chem. Phys.* **101**, 3671 (1994).
12. K. M. Forsythe and S. K. Gray, *J. Chem. Phys.* **112**, 2623 (2000).
13. A. Meijer, E. M. Goldfield, S. K. Gray, and G. G. Balint-Kurti, *Chem. Phys. Lett.* **293**, 270 (1998).
14. G. G. Balint-Kurti, A. I. Gonzalez, E. M. Goldfield, and S. K. Gray, *Faraday Discuss.* **110**, 169 (1998).
15. S. K. Gray and D. E. Manolopoulos, *J. Chem. Phys.* **104**, 7099 (1996).
16. G. Nyman and H. G. Yu, *Rep. Prog. Phys.* **63**, 1001 (2000).
17. D. H. Zhang and J. Z. H. Zhang, *J. Chem. Phys.* **101**, 1146 (1994).
18. D. Neuhauser, *J. Chem. Phys.* **100**, 9272 (1994).
19. F. Gatti, C. Lung, M. Menou, Y. Justum, A. Nauts, and X. Chapuisat, *J. Chem. Phys.* **108**, 8804 (1998).
20. M. Mladenovic, *J. Chem. Phys.* **112**, 1070 (2000).
21. E. M. Goldfield and S. K. Gray, *J. Chem. Phys.* **117**, 1604 (2002).
22. E. M. Goldfield, *Comput. Phys. Commun.* **128**, 178 (2000).

23. R. Q. Chen, G. B. Ma, and H. Guo, *Chem. Phys. Lett.* **320**, 567 (2000).
24. D. Kosloff and R. Kosloff, *J. Comput. Phys.* **52**, 35 (1983).
25. S. K. Gray and E. M. Goldfield, *J. Chem. Phys.* **115**, 8331 (2001).
26. J. C. Light, I. P. Hamilton, and J. V. Lill, *J. Chem. Phys.* **82**, 1400 (1985).
27. D. T. Colbert and W. H. Miller, *J. Chem. Phys.* **96**, 1982 (1992).
28. D. A. Mazziotti, *Chem. Phys. Lett.* **299**, 473 (1999).
29. D. A. Mazziotti, *J. Chem. Phys.* **117**, 2455 (2002).
30. R. Guantes and S. C. Farantos, *J. Chem. Phys.* **111**, 10827 (1999).
31. D. Troya, M. J. Lakin, G. C. Schatz, and M. Gonzalez, *J. Chem. Phys.* **115**, 1828 (2001).
32. E. M. Goldfield and S. K. Gray, *Comput. Phys. Commun.* **98**, 1 (1996).
33. J. M. Bowman, *J. Phys. Chem.* **95**, 4960 (1991).
34. P. Defazio and S. K. Gray, *J. Phys. Chem. A* **107**, 7132 (2003).
35. W. H. Miller, *J. Phys. Chem. A* **102**, 793 (1998).
36. J. M. Bowman and H. M. Shnider, *J. Chem. Phys.* **110**, 4428 (1999).
37. H. B. Wang, W. H. Thompson, and W. H. Miller, *J. Phys. Chem. A* **102**, 9372 (1998).
38. S. L. Mielke, G. C. Lynch, D. G. Truhlar, and D. W. Schwenke, *Chem. Phys. Lett.* **216**, 441 (1993).
39. D. H. Zhang and J. Z. H. Zhang, *J. Chem. Phys.* **110**, 7622 (1999).
40. D. M. Medvedev, E. M. Goldfield, and S. K. Gray, *Comput. Phys. Commun.* **166**, 94 (2005).
41. S. Borowski and T. Kluner, *Chem. Phys.* **304**, 51 (2004).
42. A. Meijer, *Comput. Phys. Commun.* **141**, 330 (2001).
43. A. Lagana, S. Crocchianti, A. Bolloni, V. Piermarini, R. Baraglia, R. Ferrini, and D. Laforenza, *Comput. Phys. Commun.* **128**, 295 (2000).
44. H. Y. Mussa, J. Tennyson, C. J. Noble, and R. J. Allan, *Comput. Phys. Commun.* **108**, 29 (1998).
45. E. M. Goldfield and S. K. Gray, *J. Chem. Soc. Faraday Trans.* **93**, 909 (1997).
46. E. M. Goldfield and S. K. Gray, *Chem. Phys. Lett.* **276**, 1 (1997).
47. E. M. Goldfield, S. K. Gray, and G. C. Schatz, *J. Chem. Phys.* **102**, 8807 (1995).
48. D. M. Medvedev, S. K. Gray, E. M. Goldfield, M. J. Lakin, D. Troya, and G. C. Schatz, *J. Chem. Phys.* **120**, 1231 (2004).
49. E. M. Goldfield and A. Meijer, *J. Chem. Phys.* **113**, 11055 (2000).
50. A. Meijer and E. M. Goldfield, *J. Chem. Phys.* **108**, 5404 (1998).
51. A. Meijer and E. M. Goldfield, *J. Chem. Phys.* **110**, 870 (1999).
52. A. Meijer and E. M. Goldfield, *Phys. Chem. Chem. Phys.* **3**, 2811 (2001).
53. T. E. Carroll and E. M. Goldfield, *J. Phys. Chem. A* **105**, 2251 (2001).
54. Y. He, E. M. Goldfield, and S. K. Gray, *J. Chem. Phys.* **121**, 823 (2004).
55. W. Gropp, E. Lusk, and A. Skjellum, *Using MPI: Portable Programming with the Message-Passing Interface*, 2nd ed., MIT Press, Cambridge, MA, 1999.
56. OpenMP wesite: http://www.openmp.org.
57. C. X. Xu, D. Q. Xie, D. H. Zhang, S. Y. Lin, and H. Guo, *J. Chem. Phys.* **122** (2005).
58. D. G. Truhlar, *Comput. Phys. Commun.* **84**, 78 (1994).
59. Y. H. Huang, D. J. Kouri, M. Arnold, T. L. Marchioro, and D. K. Hoffman, *J. Chem. Phys.* **99**, 1028 (1993).
60. NERSC website: http://www.nersc.gov.

61. S. Zhang, D. M. Medvedev, E. M. Goldfield, and S. K. Gray, *J. Chem. Phys.* **125**, 164312 (2006).
62. T. G. Clements, R. E. Continetti, and J. S. Francisco, *J. Chem. Phys.* **117**, 6478 (2002).
63. Q. Cui, X. He, M. L. Wang, and J. Z. H. Zhang, *J. Chem. Phys.* **119**, 9455 (2003).
64. Q. Cui, M. L. Wang, and J. Z. H. Zhang, *Chem. Phys. Lett.* **410**, 115 (2005).
65. T. F. Miller, D. C. Clary, and A. Meijer, *J. Chem. Phys.* **122** (2005).
66. F. Huarte-Larranaga and U. Manthe, *J. Chem. Phys.* **113**, 5115 (2000).
67. R. van Harrevelt and U. Manthe, *J. Chem. Phys.* **123** (2005).
68. R. van Harrevelt, K. Honkala, J. K. Norskov, and U. Manthe, *J. Chem. Phys.* **122** (2005).
69. T. Wu, H. J. Werner, and U. Manthe, *Science* **306**, 2227 (2004).
70. B. Pouilly, M. Monnerville, F. Gatti, and H. D. Meyer, *J. Chem. Phys.* **122** (2005).
71. O. Vendrell and H. D. Meyer, *J. Chem. Phys.* **122** (2005).
72. C. Iung, F. Gatti, and H. D. Meyer, *J. Chem. Phys.* **120**, 6992 (2004).
73. S. Y. Lin and H. Guo, *J. Chem. Phys.* **122** (2005).
74. S. Y. Lin and H. Guo, *J. Chem. Phys.* **120**, 9907 (2004).

QUANTUM MONODROMY AND MOLECULAR SPECTROSCOPY

MARK S. CHILD

Physical and Theoretical Chemistry Laboratory, Oxford University, Oxford, OX1 3QZ, United Kingdom

CONTENTS

Advances in Chemical Physics, Volume 136, edited by Stuart A. Rice

I. INTRODUCTION

Quantum monodromy was introduced into the spectroscopic literature as a new way to interpret the reorganization of molecular energy levels around the barrier to linearity in quasilinear molecules [1]; but this is just one instance of a much wider phenomenon. Related considerations apply to a variety of angular momentum coupling models [2, 3], highly excited atoms in crossed fields [4–6], the bending rotating states of HCP [7], rotating dipolar molecules in electric fields [8], isomerizing states of LiCN [9] (but not HCN [10]), and Fermi resonant states of CO_2 [11–13] and other molecules [14]. The aim of this chapter is to bring out the threads that unify these diverse applications. The rich diversity of quantum monodromy in the electronic states of H_2^+ [15, 16] is also mentioned, although it is outside the scope of this chapter. To simplify the discussion, attention is restricted, unless otherwise stated, to systems with two degrees of freedom, including a conserved component of angular momentum, which means that the monodromy is displayed in the two-dimensional energy angular momentum plane. As an exception, the monodromy in Fermi resonant systems with conserved angular momentum (or 1:1:2 Fermi resonance [11–13]) is an intrinsically three-dimensional phenomenon.

To set the context from the viewpoint of molecular physics, it is convenient to recall the early work on quasilinear molecules by Johns [17] and Dixon [18], noting in particular that the "good vibrational quantum number" differs according to whether the origin of the vibrational band lies below or above the barrier to linearity. Low-lying bands, labeled by a "bent" quantum number, v_b, which counts the number of nodes in the vibrational wavefunction, have rotational levels that vary smoothly with k, which is the signed value of the a-axis rotational quantum number, K_a. However, bands with origins above the barrier show a smooth energy variation with k only for levels with common values of the "linear" vibrational quantum number $v_l = 2v_b + |k|$, where k is now identified with the bending angular momentum of a linear molecule, often denoted by l. It is remarkable to observe such behavior in the common eigenvalue lattice of two commuting operators. The resulting drastic reorganization of the level structure will be shown later to give rise to a characteristic distortion of the two-dimensional quantum eigenvalue lattice, which is the most readily observable consequence of quantum monodromy, and the most important for molecular spectroscopy.

Related work in the mathematics literature makes the point in a different way by regarding the barrier as a topological obstruction to the construction of a global single-valued system of angle–action variables [19–22]—the actions being the classical analogs of quantum numbers [23]. Those seeking mathematical detail are advised to consult the texts by Efstathiou [24] and Cushman and Bates [25]. The essence of the matter is that the simplest topological obstruction is an *isolated unstable* critical point in a four-dimensional classical phase space, the image of which appears as a point in the energy–angular momentum map. For example, the barrier to linearity is *isolated* in quasilinear systems because it is classically accessible only for zero angular momentum, $k = 0$; and the same is true for the top of the swing of a spherical pendulum (or the HPC conformation of HCP). However, the critical point need not be a physical barrier. In the case of spin–rotation coupling [2], it is one or the other of the special orientations with $S_z = \pm|\mathbf{S}|$ and $N_z = \pm|\mathbf{N}|$, while similar critical angular momentum configurations also dictate the quantum monodromy of H atoms in crossed fields [4, 5]. In addition, the isolated point generalizes to a line in the three-dimensional map associated with 1:1:2 Fermi resonant systems. In other words, the set of critical values of the relevant map for systems with N dimensions has codimension $N - 2$; a point for $N = 2$, a line for $N = 3$, and so on. Other more intricate situations can also arise. For example, HCP shows the quantum monodromy of a spherical pendulum [7], because there is no stable HPC isomer, but the general situation with regard to XCN isomerization is more complicated. As discussed later, the topological obstruction in the case of LiCN is a finite "fold" between the interpenetrating eigenvalues of LiCN and those of LiNC [9]. However, there is no such finite region in the case of HCN, and therefore no quantum monodromy [10].

The mathematical discussion centers on the connectivity of classical *invariant tori* [24–26], which are the analogs of quantum mechanical wavefunctions, as one moves over the energy and angular momentum plane. Each point in the plane has an associated two-dimensional torus in a four-dimensional phase space, which is characterized by two classical actions, I_v (or quantum numbers), determined by topologically distinct loop integrals, $\oint p\, dq$; by a time period, T, for one of the degrees of freedom; and by the angle change, $\Delta\theta$ (or *twist* [10] over the surface of the torus), in the second degree of freedom. The existence of a global system of angle–action variables then requires that one can follow smooth variations of the I_v, T, and $\Delta\theta$, as functions of energy, E, and angular momentum, L. In particular, transport of any torus around a closed loop in the energy and angular momentum plane must bring the above variables back to their initial values. Any such loop may be smoothly contracted to a point. On the other hand, situations can occur in which twist angles change discontinuously, by multiples of 2π, somewhere around the loop. The transported torus still comes back to its original form, but the twist discontinuity makes it impossible to contract the loop to a point. The simplest obstruction to such a contraction is found to be an isolated critical point, of the type discussed earlier. In the simple, quasilinear molecule case, discussed in Section II, the relevant twist angle discontinuity, $\Delta\theta_L$, in the angle conjugate to the angular momentum, is readily related to an abrupt change in the nature of the classical motions above and below the barrier to linearity. Moreover, $\Delta\theta_L$ also relates directly to changes in the directions of lines in the *EL* map plane because Hamilton's equations imply that $(\partial E/dL)_{vb} = \theta_L/T$, where T is the time period. The resulting sharp changes in $(\partial E/dL)_{vb}$ explain the characteristic distortion of the quantum lattice. Another way to interpret the distortion is to regard the quantum number v_b as a multivalued function of E and L, which is the mathematical signature of classical monodromy [19–22, 24, 25].

The aim of the chapter is to show how the above mathematical concepts dictate the organization of the quantum level structures for different systems, in a language appropriate to the chemical physics community. This means in particular that although the topology of the situation is most clearly displayed by a full Cartesian formulation with constraints [24, 25], the results will be obtained by methods more familiar to the molecular physics community. Section II introduces the main ideas in the context of the spectra of quasilinear molecules, including H_2O. Those simply seeking a description of what quantum monodromy means for molecular spectroscopy may wish to omit Sections IIC–E, which deal with the mathematical origin of multivalued quantum numbers and the validity of the classical–quantum correspondence for systems with quantum monodromy. Section III describes the quantum monodromy in highly excited states of HCP (as modeled by a spherical pendulum) and LiCN, and explains why the effect does not occur for HCN. It will be seen that the reorganization of the quantum level structure around the HPC conformation of HCP mimics that around the barrier to linearity in H_2O because both are dictated by the same type of isolated critical point of the classical mechanics. In LiCN, on the other hand, the role of the critical point is replaced by that of a finite *folded* part of the quantum spectrum, over which bending states of LiCN interleave with those of LiNC—finite in the sense that there are no bound states of LiNC for sufficiently high vibrational angular momentum. However, the form of the HCN potential energy surface is such that there appears to be no such angular momentum cut-off to the number of bending states of HNC—hence there is no cyclic path in the *EL* plane around the folded spectral region and no quantum monodromy.

The discussion of coupled angular momenta and hydrogen atoms in crossed fields, in Section IV, introduces a new line of reasoning. For example, the spin–rotation coupling model discussed in Section IVA is similar to the previous examples in having a good angular momentum quantum number, J_z, but the possible critical points are first related to the geometry of the angular momentum space, rather than the Hamiltonian. It is then shown how a change in the spin–rotation coupling strength can *isolate* a particular critical point over a limited region of its range [2]. The connection with the transition between Hund's cases (a) and (c) [27] is also discussed. A perturbation treatment of the H atom in crossed fields [4, 5] leads to a Hamiltonian that is qualitatively similar to that of the spin–rotation problem, with the special property that the magnitudes of the two coupled angular momenta are equal. As a consequence, there is a particular parameter range over which two overlapping critical points become isolated in the observable (E, J_z) projection of the angular momentum space, with special consequences for the quantum monodromy.

In order to emphasize the generality of the phenomenon, examples for different Fermi resonance coupling cases are discussed in Section V. One knows, from the papers of Xiao and Kellman [28, 29], that the standard spectroscopic Fermi resonance coupling Hamiltonian [30, 31] gives rise to four different cases, in a reduced two-parameter (fixed polyad) space—two of which give rise to quantum monodromy—one of the quasilinear molecule type, while the other is similar to that observed for LiCN/LiNC isomerization. In addition, the transition between the two cases involves geometrical considerations similar to those employed for the angular momentum coupling models. It should be noted, however, that detailed mathematical analysis [11–13] of the case of CO_2-like Fermi resonance, with exact zeroth order 1:1:2 resonance and relatively weak anharmonicity, demonstrates that the monodromy is an intrinsically three-dimensional phenomenon.

Section VI contains a summary of the conclusions and suggestions for future research.

II. QUASILINEAR MOLECULES

The central concepts are illustrated next for the case of the champagne bottle Hamiltonian [1, 22, 32],

$$\hat{H} = \frac{1}{2m}\left(\hat{p}_R^2 + \frac{\hat{p}_\theta^2}{R^2}\right) - AR^2 + BR^4 \tag{1}$$

in which the variable R may be interpreted as the height of the triangle of a quasilinear triatomic. The variable $p_\theta = k\hbar$ is the conserved angular momentum. It will also be convenient later to employ the substitutions $R = r/\gamma, p_R = \gamma\hbar p_r$ with $\gamma = (2mA/\hbar^2)^{1/4}$, in terms of which

$$\begin{aligned}\hat{h} = \hat{H}/[\hbar\omega^*] &= \frac{1}{2}\hat{p}_r^2 + \frac{k^2}{2r^2} - \frac{1}{2}r^2 + \beta r^4 \\ &= \frac{1}{2}(p_x^2 + p_y^2 - x^2 - y^2) + \beta(x^2 + y^2)^2\end{aligned} \tag{2}$$

where $\beta = B\hbar/\sqrt{8mA^3}$ and $\omega^* = \sqrt{2A/m}$. The Cartesian form is given for later reference. As a further point of notation, the symbols E and L will be used for the energy and angular momentum of the classical system, while the reduced energy, $\varepsilon = E/[\hbar\omega^*]$, and quantum number k will be employed for the scaled system. The classical motion is bounded below by the minima of the centrifugally corrected potential

$$v^{(k)}(r) = \frac{k^2}{2r^2} - \frac{1}{2}r^2 + \beta r^4 \tag{3}$$

or *relative equilibria*, which may be parameterized in the form

$$k = \pm z\sqrt{4\beta z - 1}, \quad \varepsilon = \frac{k^2}{2z} - \frac{1}{2}z + \beta z^2, \quad z \geq 1/4\beta \tag{4}$$

where $z = r^2$. Such relative equilibria play an important role in later sections.

A. Quantum Eigenvalue Lattice and Good Spectroscopic Quantum Numbers

The quantum mechanical eigenvalue spectrum is readily computed by diagonalizing $\hat{H}$ in a degenerate harmonic oscillator basis [32]. The typical spectrum, evaluated with parameters roughly appropriate to the bending states of H_2O, is shown in Fig. 1, in which the points are plotted at the eigenvalues. As discussed in Section I, the eigenvalues may be labeled by the angular momentum k and either by a bent quantum number, v_b, which is counted upward from zero in each k stack,

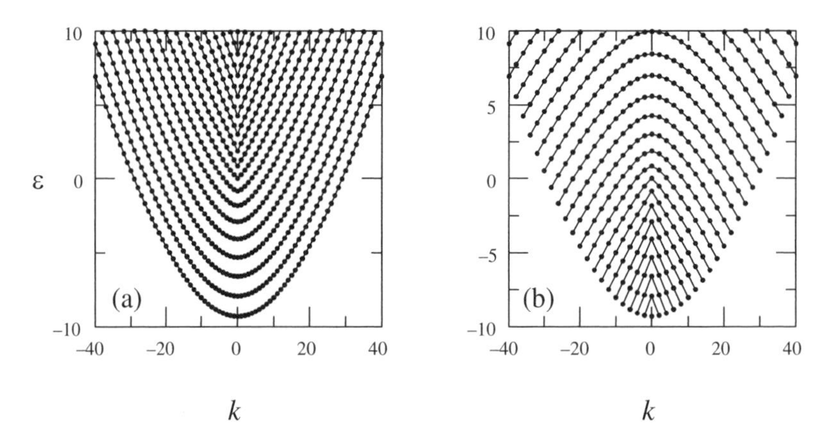

Figure 1. Eigenvalues of the scaled champagne bottle Hamiltonian (Eq. (2)) for $\beta = 0.00625$, in the energy, ε, and angular momentum, k map. The eigenvalues, represented by points, are joined (a) by lines of constant "bent" vibrational quantum number, v_b, and (b) by lines of constant "linear" quantum number, $v_l = 2v_b + |k|$.

or by a linear quantum number, $v_l = 2v_b + |k|$. The points are joined according to common values of these two labels in Figs. 1a and 1b, respectively, from which it is evident that the reduced energy, ε, in Fig. 1a varies smoothly with k at fixed v_b only for $v_b \leq 7$, after which the roughly quadratic variation with k gives way to a sharp "vee" shaped kink. Similarly in Fig. 1b, the points are joined by lines with common v_l values, which are seen to be smooth for the curves passing above the critical point at $(k, \varepsilon) = (0, 0)$ and kinked for those passing below it. It was this behavior that led Johns [17] to identify v_b as a good quantum number only for the lower bands, and to introduce v_l as a label for the higher ones, although it is evident, by comparison between the two panels in Fig. 1, that many points lie on both the smooth v_b and v_l curves. This underlines the important point that a smooth global system of quantum numbers (or classical action variables [23]) must not only provide quantum labels for the individual eigenvalues but must do so in a smooth unambiguous way over the entire spectrum.

To see the generality of the above pattern, Fig. 2 shows that a qualitatively similar reorganization of the quantum lattice occurs in the bending bands of H_2O computed by Partridge and Schwenke [33]. The relevant angular momentum

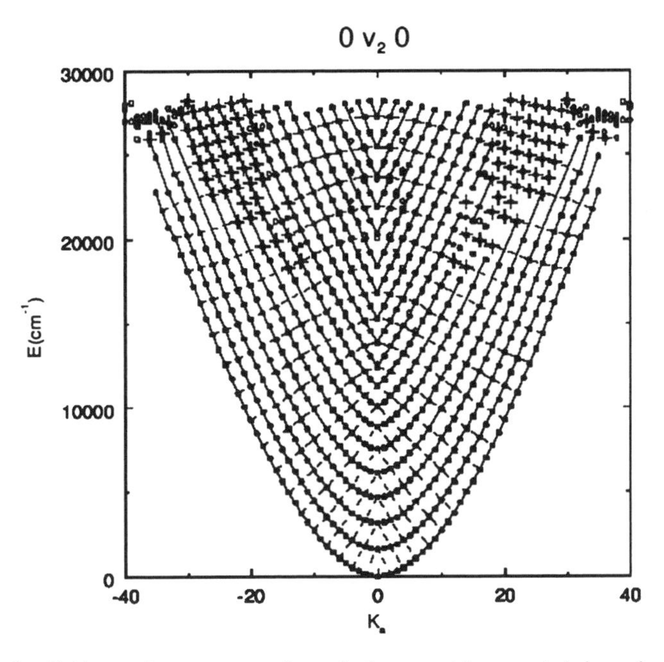

Figure 2. Evidence of quantum monodromy in the accurately computed eigenvalue spectrum of H_2O [33]. The energy points are plotted for the J_{J0} rotational levels of the $(0, v_2, 0)$ progression, with K_a taken equal to J. Crosses indicate levels that have been reassigned with respect to v_2. Open circles indicate reported, usually extra, states [20] that do not conform with the pattern. Taken from Ref. [1] with permission of Taylor and Francis, http://www.tandf.co.uk/journals.

in this case is the a-axis component, K_a, because the associated moment of inertia vanishes in the linear configuration. Hence the points are plotted for the J_{J0} rotational levels of the $(0, v_2, 0)$ progression with $K_a = J$ [34]. It is evident that the spectrum displays the same qualitative pattern as in Fig. 1a, with a smooth roughly quadratic energy variation with K_a, for $0 \leq v_2 \leq 7$, going over to a kinked variation for $v_2 \geq 8$, despite the fact that the Hamiltonian is very different from the simple form in Eq. (1). Moreover, similar patterns were found for five other bending progressions, with different levels of excitation in the stretching modes [1], despite the existence of increasingly strong stretch–bend coupling as the level of stretching excitation increases. Further confirmation comes from the firm assignment by Zobov et al. [35] of 134 experimental rotational energy levels, spanning $0 \leq J = K_a < 9$ for the bending states $5 \leq v_2 \leq 9$. The clear transition from smooth to kinked joining lines in the analog of Fig. 1a fixes the barrier to linearity at $11,114 \pm 5\,\mathrm{cm}^{-1}$. It is therefore evident that the monodromic pattern is remarkably robust to perturbations, at the level of resolution in Fig. 2.

In addition, Winnewisser et al. [36] have demonstrated quantum monodromy for the NCNCS molecule, not only in the (v_b, K_a) eigenvalues but also in the $B_{\mathrm{eff}} = (B + C)/2$ rotational constant, which varies smoothly with K_a, at fixed v_b for $v_b = 0$–2, but shows a kink for $v_b = 3$–5. This observation is a key to the assignment of the extremely congested rotational spectrum of this interesting quasilinear molecule.

B. Aspects of the Classical Dynamics

It is now shown how the abrupt changes in the eigenvalue distribution around the central critical point relate to changes in the classical mechanics, bearing in mind that the analog of quantization in classical mechanics is a transformation of the Hamiltonian from a representation in the variables $(p_R, p_\theta, R, \theta)$ to one in angle–action variables $(I_R, I_\theta, \theta_R, \theta)$ such that the transformed Hamiltonian depends only on the actions (I_R, I_θ) [37]. Hamilton's equations $(dI_R/dt = (\partial H/\partial \theta_R)$, etc.) then show that the actions are constants of the motion, which are related to the quantum numbers by the Bohr correspondence principle [23]. In the present case,

$$I_R = (v_b + 1/2)\hbar \quad \text{and} \quad I_\theta = L = k\hbar \tag{5}$$

Seen in this light, the curves in Fig. 1a may be interpreted as contours (or level lines) of the quantum number function $v_b(\varepsilon, k)$ or, in classical terms, of the radial action function $I_R(E, L)$. The aim is to use this classical insight to understand why the curves labeled by v_b have zero derivative, $(\partial H/\partial I_\theta)_{I_R} = 0$, at $I_\theta = k\hbar = 0$, for $H < 0$, but not for $H > 0$. To see this, note that

$$\left(\frac{\partial H}{\partial I_\theta}\right)_{I_R} = \frac{d\theta}{dt} = \frac{\Delta\theta}{\Delta t} \tag{6}$$

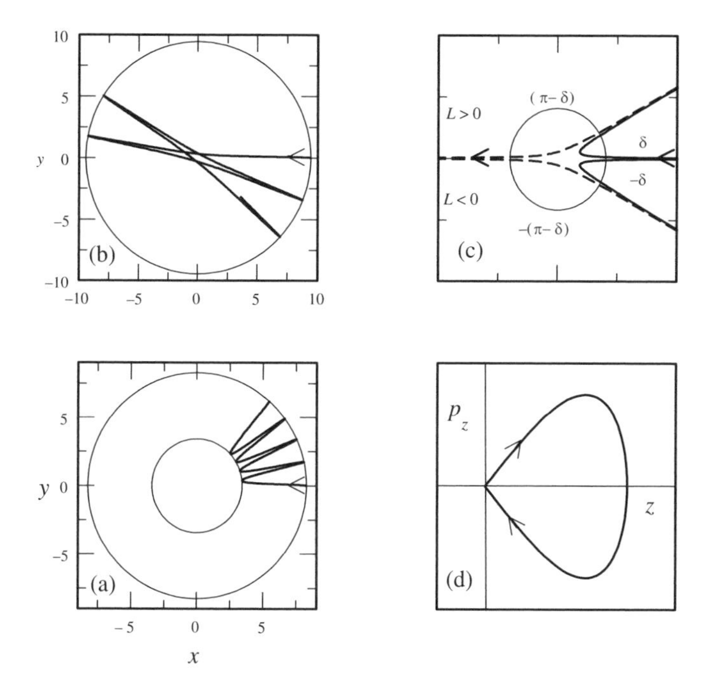

Figure 3. Classical trajectories of the champagne bottle Hamiltonian (Eq. (2)) at energies (a) below and (b) above the barrier maximum. (c) Trajectories of the local focus–focus Hamiltonian (Eq. (7)) at positive (dashed line) and negative (solid line) energies. (d) The singular cusped orbit of the champagne bottle Hamiltonian with $\varepsilon = k = 0$.

where $\Delta\theta$ is conveniently taken as the precessional change in θ during one radial cycle Δt. Its limiting form as $L \to 0$ may be deduced from the forms of the typical positive and negative energy trajectories, which are shown in Fig. 3, for small positive angular momenta, I_θ. The negative energy trajectory in Fig. 3a oscillates to and fro, while precessing around the groove in the champagne bottle (or Mexican hat) potential, in such a way that the precessional angle, $\Delta\theta$, falls to zero as $I_\theta \to 0$. However, the positive energy trajectory in Fig. 3b flies close to the central point of the potential, making an angular change, $\Delta\theta$, close to π as the radial variable passes from large to small to large values. Thus $\Delta\theta \to \pi$ in the limit $I_\theta = L \to 0^+$, while $\Delta\theta \to -\pi$ in the limit $I_\theta = L \to 0^-$. This argument shows that the character of the quantum eigenvalue lattice, around the central critical point in Fig. 1a, may be attributed to the sharp change in the nature of the classical trajectory in passing from energies below to energies above the potential maximum.

At this stage the argument only explains changes in the lines of constant *bent* quantum number v_b (corresponding to the action I_R) in Fig. 1a. The corresponding changes in Fig. 1b may be understood by noting that the change of representation to the linear quantum number $v_l = 2v_b + |k|$ corresponds to the definition of new tori, with new actions, $I_R' = 2I_R \pm L^{\pm}$ and $L' = L^{\pm}$, and new conjugate angles, $\theta_R' = \theta_R/2$ and $\theta_{\pm}' = \theta_{\pm} \mp \theta_R/2$.[1] Since $\theta_R = 2\pi$ after one cycle of the radial motion, it follows that $\Delta\theta_{\pm}' = \Delta\theta_{\pm} \mp \pi \rightarrow \mp\pi$ as $L^{\pm} \rightarrow 0^{\pm}$, when $\varepsilon < 0$. Similarly, $\Delta\theta^{\pm} \rightarrow 0$ as $L^{\pm} \rightarrow 0^{\pm}$ for $\varepsilon > 0$. This explains why the curves, labeled by common v_l values in Fig. 1b, kink downward at $k = 0$ for $\varepsilon < 0$ and cross the line $k = 0$ smoothly when $\varepsilon > 0$. Note, however, that these sharp changes are associated with particular definitions of I_R or I_R' such that $\Delta\theta$ is discontinuous at $L \rightarrow 0^{\pm}$ for either $\varepsilon > 0$ or $\varepsilon < 0$. An alternative approach, such that the radial action or local quantum number is locally smooth on a cycle around the critical point, leads more naturally, in Section IIC, to the connection with classical monodromy.

Two other features of the classical mechanics are illustrated in Fig. 3. The first is that the torque responsible for the angular deflection at low k comes from points close to $r = 0$ in the scaled form of Eq. (2), around which the quartic term can be ignored. Thus

$$h \simeq \tfrac{1}{2}[p_x^2 + p_y^2 - x^2 - y^2] = \varepsilon, \quad xp_y - yp_x = k \tag{7}$$

which defines what is known as a focus–focus singularity in the mathematics literature [40].[2] The trajectories lie on the hyperbolas in Figs. 3c, which closely approximate the inner segments of the trajectories in Figs. 3a and 3b. Since $\delta \rightarrow 0$ as $L \rightarrow 0^{\pm}$, it is evident that the deflection $\Delta\theta$ tends to zero for negative energies (solid line trajectories) and to $\pm\pi$ for positive energies (dashed line trajectories) as $L \rightarrow 0^{\pm}$. It should also be noted, for future reference, that the

[1]Readers familiar with canonical transformation theory [37] can confirm that these results follow from use of a type 4 generating function,

$$F_4 = -(2I_R \pm L^{\pm})\theta_R' - L^{\pm}\theta'$$

for which $\theta_R = -(\partial F_4/\partial I_R)$ and $\theta^{\pm} = -(\partial F_4/\partial L^{\pm})$.

[2]Strictly, Eq. (7) transforms under the canonical transformation

$$\begin{pmatrix} x_1 \\ p_1 \end{pmatrix} = \frac{1}{\sqrt{2}}\begin{pmatrix} 1 & 1 \\ -1 & 1 \end{pmatrix}\begin{pmatrix} x \\ p_x \end{pmatrix}$$

$$\begin{pmatrix} x_2 \\ p_2 \end{pmatrix} = \frac{1}{\sqrt{2}}\begin{pmatrix} 1 & 1 \\ -1 & 1 \end{pmatrix}\begin{pmatrix} y \\ p_y \end{pmatrix}$$

to

$$h = x_1p_1 + x_2p_2, \quad k = x_1p_2 - x_2p_1$$

which is the generic focus–focus form [40].

models in Sections IV and V give rise to Hamiltonians that may be approximated in the vicinity of critical points by the angle–action form,

$$h' \simeq AI + B\sqrt{I^2 - L^2}\cos 2\theta_I \tag{8}$$

where (I, θ_I) are conjugate variables and L is the conserved angular momentum. It is shown in the Appendix that h' may be transformed to the above Cartesian focus–focus form if $A^2 < B^2$.

In addition the Cartesian form of Eq. (2) implies that an infinitesimal displacement from the origin, $x = y = p_x = p_y = 0$, in the θ direction, will lead to motion along a *manifold* of the critical point, given by

$$p_z = \pm\sqrt{z^2 - 2\beta z^4} \tag{9}$$

where $z = x\cos\theta + y\sin\theta$ and p_z is the conjugate momentum. The shape of such a figure in Fig. 3d, with a conical point at $z = 0$, generalizes to a surface of revolution about the p-axis [25]. Another schematic representation is in the form of a *pinched torus* [10, 25], the horizontal section of which takes the form in Fig. 3d, while loops around the annulus span the angle θ. Examples of these two representations are given in Fig. 16 of Ref. [10]. Doubly pinched tori can also exist for more complicated systems [4, 5] in which the manifolds from one unstable point join with those of another. An example is discussed in Section VC.

C. Multivalued Quantum Numbers

1. *The Quantizing Integral*

Further classical insight may be obtained by reference to the action identity

$$I_R = (v + \tfrac{1}{2})\hbar = \tfrac{1}{2\pi}\oint p_R\, dR \tag{10}$$

which translates in the scaled variable system to

$$\begin{aligned}[v(\varepsilon, k) + 1/2] &= \frac{1}{\pi}\int_{r_1}^{r_2} \sqrt{2\varepsilon - k^2/r^2 + r^2 - 2\beta r^4}\, dr \\ &= \frac{1}{2\pi}\int_{z_1}^{z_2} z^{-1}\sqrt{2\varepsilon z - k^2 + z^2 - 2\beta z^3}\, dz\end{aligned} \tag{11}$$

where $z = r^2$ and the integration limits r_i and z_i, $i = 1, 2$, are the classical turning points. In addition, the twist angle on the torus is given by

$$\Delta\theta = -2\pi\left(\frac{\partial v}{\partial k}\right)_\varepsilon = k\int_{z_1}^{z_2} \frac{dz}{z\sqrt{2\varepsilon z - k^2 + z^2 - 2\beta z^3}} \tag{12}$$

The essential point is that the behavior of the final integrals in these equations is complicated by a coalescence between the inner turning point, z_1, and the singularity at the origin, as $(\varepsilon, k) \to (0, 0)$. The results may be expressed in terms of complete elliptic integrals [32], which change branches around the critical point, but the general implications for the function $v(\varepsilon, k)$ are most conveniently illustrated by adding and subtracting the integral

$$f(\varepsilon, k) = \frac{1}{2\pi}\int_{z_1}^{z_0} z^{-1}\sqrt{2\varepsilon z - k^2 + z^2}dz, \quad z_0 + \varepsilon \gg \sqrt{k^2 + \varepsilon^2},\ z_0 \gg |k^2/\varepsilon| \tag{13}$$

in order to eliminate any singular contribution to the difference function

$$u(\varepsilon, k, z_0) = v(\varepsilon, k) - f(\varepsilon, k) + (z_0 + \varepsilon)/4\pi \tag{14}$$

It follows by the use of standard tables [38] that

$$\begin{aligned} v(\varepsilon, k) &= u(\varepsilon, k, z_0) - \frac{1}{2\pi}\left[\frac{\varepsilon}{2}\ln\left(\frac{\varepsilon^2 + k^2}{4}\right) + k\arctan\left(\frac{\varepsilon}{k}\right) + \frac{k\pi}{2}\right] \\ &= u(\varepsilon, k, z_0) - \frac{1}{2\pi}\left[\mathrm{Im}(k + i\varepsilon)\ln\left(\frac{k + i\varepsilon}{2}\right) + \frac{k\pi}{2}\right] \end{aligned} \tag{15}$$

Similarly, the twist angle may be expressed as

$$\begin{aligned} \Delta\theta(\varepsilon, k) &= \frac{\pi}{2} + \arctan\left(\frac{\varepsilon}{k}\right) + kw(\varepsilon, k) \\ &= \frac{\pi}{2} + \mathrm{Im}\ \ln(k + i\varepsilon) + kw(\varepsilon, k) \end{aligned} \tag{16}$$

The important point is that $v(\varepsilon, k)$ and $\Delta\theta(\varepsilon, k)$ are *multivalued*, with a logarithmic branch point at $k = \varepsilon = 0$, while the residual functions $u(\varepsilon, k, z_0)$ and $w(\varepsilon, k)$ are single valued. The result, as discussed in more detail later, is that the value of the quantum number v depends on the chosen location of the branch cut and on which Riemann sheet is taken, bearing in mind that the branch of $\arctan(\varepsilon/k)$ must be taken according to the appropriate quadrant of the complex (k, ε) plane. Thus $\phi = \arctan(\varepsilon/k) + \pi/2$ increases smoothly from zero to 2π around a counterclockwise circle in the (k, ε) plane, starting at $k = 0$ and $\varepsilon < 0$.

The Bohr quantization condition, with quantum number v_b, corresponds to choosing the lowest Riemann sheet, which requires an additional phase correction of -2π on crossing the branch cut, which is taken along the positive ε-axis. Thus the phase term ϕ rises to π at $k = 0_+$ and $\varepsilon > 0$, drops abruptly to $-\pi$ on crossing the positive $k = 0$ axis, and returns to zero on the negative

k-axis. This sequence ensures that $v_b(\varepsilon, k)$ is independent of the sign of k, as required by the symmetry of the lattice in Fig. 1; and that the twist angle $\Delta\theta$ corresponds to the trajectory plots in Figs. 3a and 3b, which were used to explain the presence of the kinks in Fig. 1.

2. *The Monodromy Matrix*

It is, however, more revealing in the context of monodromy to allow $f(\varepsilon, k)$ to pass from one Riemann sheet to the next, at the branch cut, a procedure that leads to the construction in Fig. 4, due to Sadovskii and Zhilinskii [2], by which a unit cell of the quantum lattice, with sides defined here by unit changes in k and v, is transported from one cell to the next on a path around the critical point at the center of the lattice. Note, in particular, that the lattice is locally regular in any region of the (k, ε) plane that excludes the critical point; and that any vector in the unit cell such as the base vector, marked by arrows, rotates as the cell is transported around the cycle. Consequently, the transported dashed cell differs from that of the original quantized lattice.

This shape change may be related to the properties of the invariant tori by use of the tangent relation [21, 22]

$$\mathbf{x} = \mathbf{D}\mathbf{u} \tag{17}$$

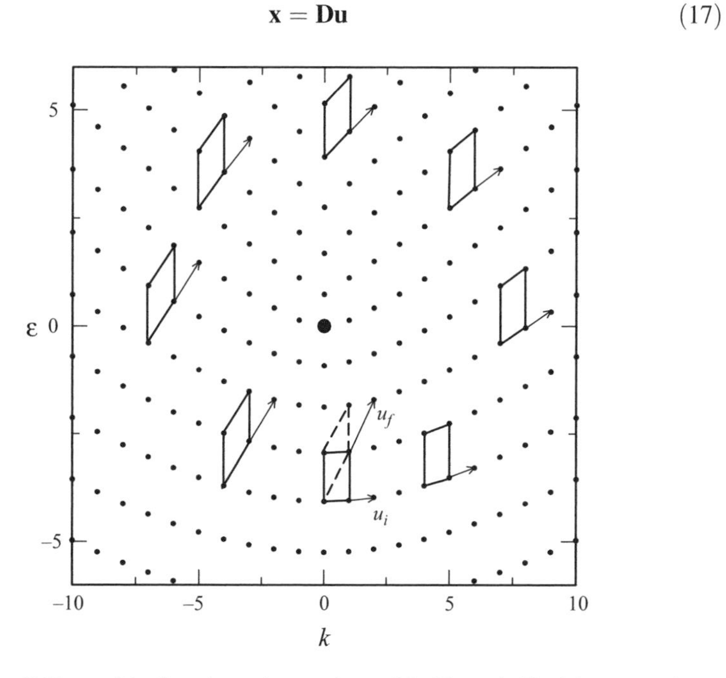

Figure 4. Evidence of the distortion to the central part of the *Ek* map in Fig. 1, by transporting an elementary unit cell around the isolated critical point, which is marked by a large dot at the origin. Note the presence of locally smooth lattice vectors, in the direction of the arrows, at every point on the cycle.

between vectors in the quantum number space, $\mathbf{x} = (\delta v, \delta k)^T$ and $\mathbf{u} = (\delta k, \delta \varepsilon)^T$ in the space of the conserved variables k and ε, where the matrix $\mathbf{D}$ is governed by the time period T and twist angle $\Delta\theta$ in the form

$$\mathbf{D} = \begin{pmatrix} (\partial v/\partial k) & (\partial v/\partial \varepsilon) \\ (\partial k/\partial k) & (\partial k/\partial \varepsilon) \end{pmatrix} = \frac{1}{2\pi} \begin{pmatrix} -\Delta\theta & T \\ 2\pi & 0 \end{pmatrix} \tag{18}$$

As an illustration, the base vector $\mathbf{x}$ in Fig. 4 is fixed as $\mathbf{x} = (1, 0)^T$ and $\mathbf{u} = \mathbf{D}^{-1}\mathbf{x}$ rotates in response to changes in $\Delta\theta$ and T around the cycle. Thus on completion of the cycle, $\mathbf{u}_f = \mathbf{D}_f^{-1}\mathbf{x}_i$, where the subscript i has been added to indicate the direction of $\mathbf{x}$ in the initial lattice. Similarly, the projection of the final vector $\mathbf{x}_f$ onto the original lattice is given by $\mathbf{x}_f = \mathbf{D}_i\mathbf{u}_f$, where $\mathbf{D}_i$ is the tangent matrix before translation. The overall change in Bohr quantized lattice vectors is therefore given by $\mathbf{x}_f = \mathbf{M}\mathbf{x}_i$, where the *monodromy matrix* is given by

$$\mathbf{M} = \mathbf{D}_i\mathbf{D}_f^{-1} = \begin{pmatrix} 1 & (\Delta\theta_f - \Delta\theta_i)/2\pi \\ 0 & 1 \end{pmatrix} = \begin{pmatrix} 1 & 1 \\ 0 & 1 \end{pmatrix} \tag{19}$$

because the discussion above shows that $\Delta\theta$ increases by 2π around the cycle. Expressed in another way, this means that

$$\begin{pmatrix} \delta v_f \\ \delta k_f \end{pmatrix} = \begin{pmatrix} 1 & 1 \\ 0 & 1 \end{pmatrix} \begin{pmatrix} \delta v_i \\ \delta k_i \end{pmatrix} \tag{20}$$

where the components of $\mathbf{x}_i$ and $\mathbf{x}_f$ are expressed as quantum number changes, to emphasize the connection with the quantum lattice. In particular, the horizontal lattice vector $\mathbf{x}_i = (0, 1)^T$ evolves to $\mathbf{x}_f = (1, 1)$, while the vertical, $\mathbf{x}_i = (0, 1)^T$, is invariant to transport around the cycle.

The above derivation was designed to demonstrate the influence of the torus twist angle on the form of the monodromy matrix. The following geometrical construction, due to Zhilinskii [39], relates it to the nature of the topological distortion of the lattice arising from the presence of the critical point. The idea, which is illustrated in Fig. 5a, is to generate the distortion of the final cell by applying a *wedge cut* to the regular lattice, which removes successively 1, 2, 3, and so on lattice points on moving away from the critical point. Points that are bisected by the cut are treated as half-points and glued together, as illustrated by the vertical arrows, to yield the distorted lattice in Fig. 5b. Two important points should be noted. First, the cuts must be made either to the right or left of the critical point, not the top or bottom, because the quantum number, k, must still take the same integer values after the distortion, whereas there is no such requirement on the energy. Second, the construction is designed to bring out the topological nature of the lattice in Fig. 4. The precise geometrical arrangement, with double degeneracy

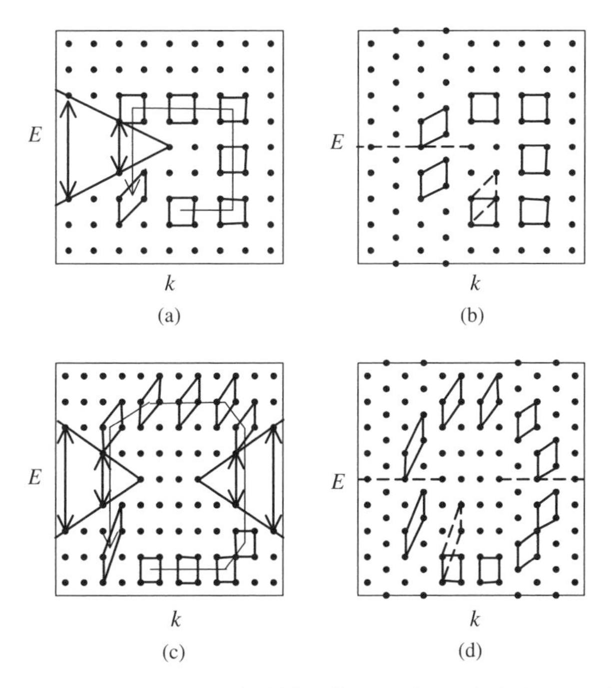

Figure 5. (a) A wedge cut construction [39] to illustrate the topological nature of the distortion arising from the presence of the focus–focus singularity in Figs. 1, 2, and 4. (b) The reconstructed lattice, in which the gluing line is marked as dashed. (c), (d) A similar construction with two wedge cuts.

with respect to the sign of k, would require subsequent energy adjustments that retained the same number of points above and below the cut.

Extensions of this approach are extremely valuable for determining the monodromy matrix in more complicated situations than those involving only a single isolated critical point in a two-dimensional energy–angular momentum map. The interesting three-dimensional example of 1:1:2 Fermi resonance [1, 2], with a second conserved variable, the polyad number, is discussed in Section VA. A simpler case involving two isolated critical points in two dimensions is illustrated in Figs. 5c and 5d. Each critical point has its own wedge cut and the original primitive unit cell suffers a shear distortion by two units, as appropriate to the monodromy matrix

$$M = \begin{pmatrix} 1 & 2 \\ 0 & 1 \end{pmatrix} \tag{21}$$

An example of this kind, in which the energy and angular momentum of the two critical points coincide, occurs for the hydrogen atom in crossed electric and magnetic fields (see Section IVC). The pinched torus then has two pinch points, and it is known, more generally, that the monodromy matrix for motion in two degrees of freedom has the general structure [40]

$$M = \begin{pmatrix} 1 & n \\ 0 & 1 \end{pmatrix} \tag{22}$$

where n is the number of pinch points (or focus–focus singularities).

3. Change of Basis

The tacit assumption above is that the monodromy matrix is defined with respect to the primitive unit cell, with sides $(\delta v, \delta k) = (0, 1)$ and $(1, 0)$, because the twist angle that determines the monodromy is given by $\Delta\theta = -(\partial v/\partial k)_\varepsilon$. However, situations can arise where other choices are more convenient. For example, the energy levels within a given Fermi resonance polyad are labeled by a counting number $v = 0, 1, \ldots$ and an angular momentum that takes only even or only odd values. Thus the convenient elementary cell has sides $(\delta v, \delta L) = (0, 2)$ and $(1, 0)$, and the natural basis, say, $\mathbf{y}$, is related to the primitive basis, $\mathbf{x}$, by

$$\mathbf{y} = A\mathbf{x}, \quad A = \begin{pmatrix} 1 & 0 \\ 0 & 2 \end{pmatrix} \tag{23}$$

Since the reduced Hamiltonian in this case has a single isolated critical point, the transformed monodromy matrix, in the $\mathbf{y}$ basis, becomes

$$A^{-1}MA = \begin{pmatrix} 1 & 0 \\ 0 & \frac{1}{2} \end{pmatrix} \begin{pmatrix} 1 & 1 \\ 0 & 1 \end{pmatrix} \begin{pmatrix} 1 & 0 \\ 0 & 2 \end{pmatrix} = \begin{pmatrix} 1 & 2 \\ 0 & 1 \end{pmatrix} \tag{24}$$

The origin of the off-diagonal 2 in Eq. (24) is quite different, however, from that in Eq. (21). The former arises simply from the change of basis, while the latter comes from the presence of two isolated critical points of the dynamics at the same energy and angular momentum.

This emphasizes that the monodromy matrix is defined only within a similarity transformation

$$M' = A^{-1}MA \tag{25}$$

and that any inference from the off-diagonal elements of M' that determine the shape change of the unit cell in Figs. 4 and 5 requires a careful specification of the relevant basis.

4. Relation to Geometric Phase

Questions are sometimes asked about the relationship between geometric phase [41, 42] and the construction in Fig. 4. Both involve transport on a closed loop around a critical point in the relevant space, but the characters of the critical points are quite different. In the case of geometric phase, the critical point is a pole of the nonadiabatic coupling, and the sign change in the real adiabatic eigenvectors around the loop may be canceled by taking a second loop. By contrast, in the case of monodromy, the critical point is a logarithmic branch point of the quantum number function, $v_b(\varepsilon, k)$. Hence successive circuits around the loop sample higher and higher Riemann sheets, and the monodromy matrix for p circuits becomes

$$M^p = \begin{pmatrix} 1 & np \\ 0 & 1 \end{pmatrix} \tag{26}$$

D. Local Corrections to the Bohr–Sommerfeld Rule

The previous sections have relied on the use of classical-angle theory to interpret the organization of the quantum mechanical eigenvalues in Figs. 1 and 2, on the basis of the Bohr–Sommerfeld correspondence $I_R = (v_b + 1/2)\hbar$ and $I_\theta = L = k\hbar$. The latter is exact in view of the cylindrical symmetry, but the former involves a significant approximation, because the Maslov term $\delta = \frac{1}{2}$ [43] only applies to situations where the classical turning points of the motion are well separated from any singularities, whereas it was shown in Section IIC that the monodromy itself arises from a confluence between the inner turning point and the centrifugal singularity. One way to take this into account [32] is to recognize that the equation

$$\left[-\frac{1}{2r}\frac{d}{dr}\left(r\frac{d}{dr}\right) + \frac{k^2}{2r^2} - \frac{1}{2}r^2\right]\Psi = \varepsilon\Psi, \quad \lim_{r\to 0} r^{-k}\Psi = \text{const} \tag{27}$$

describes the short-range form of the wavefunction, for small k and ε. Hence the known asymptotic solution to this Kummer equation [44],

$$\Psi \sim Cz^{-1/2}\cos\left[\frac{z}{2} + \varepsilon\ln z - (l+1)\frac{\pi}{4} - \arg\Gamma\left(\frac{|k| + 1 + i\varepsilon}{2}\right)\right] \tag{28}$$

where $z = r^2$, may be compared with the Langer corrected JWKB form [23],[3]

$$\Psi_{\text{JWKB}} \sim Cz^{-1/2}\cos\left(\frac{1}{2}\int_{z_1}^{z}\frac{\sqrt{z^2 + 2\varepsilon z - k^2}}{z}dz - \frac{\pi}{4}\right) \tag{29}$$

[3]A Langer correction is required by the presence of the r^{-2} singularity, arising from removal of the first derivative term in Eq. (27), even for $k = 0$.

to obtain a modified Maslov index. It follows by use of the expansion

$$\frac{1}{2}\int_{z_1}^{z}\frac{\sqrt{z^2+2\varepsilon z-k^2}}{z}dz \sim \frac{1}{2}z+\frac{\varepsilon}{2}\ln z+\frac{\varepsilon}{2}-\frac{k}{2}\arctan\left(\frac{\varepsilon}{k}\right)-\frac{\varepsilon}{4}\ln\left(\frac{\varepsilon^2+k^2}{4}\right)-\frac{|k|\pi}{4} \tag{30}$$

that the arguments of the cosines in Eqs. (28) and (29) differ by a term

$$\eta(\varepsilon,k)=\frac{\varepsilon}{4}\ln\left(\frac{\varepsilon^2+k^2}{4}\right)-\frac{\varepsilon}{2}+\frac{k}{2}\arctan\left(\frac{\varepsilon}{k}\right)-\arg\Gamma\left(\frac{|k|+1+i\varepsilon}{2}\right) \tag{31}$$

which corrects for neglect of the singular behavior at $z=0$. Consequently, an improved quantization condition may be expressed in the form

$$\int_{r_1}^{r_2}p(r)\,dr=\frac{1}{2}\int_{z_1}^{z}\frac{\sqrt{z^2+2\varepsilon z-k^2+2\beta z^3}}{z}dz=(v_b+\tfrac{1}{2})\pi-\eta(\varepsilon,k) \tag{32}$$

a correction that was independently derived by $V\bar{u}$ Ngoc [45].

The first-order effect of this correction, the form of which is shown in Fig. 6, translates into a discrepancy between the exact and JWKB eigenvalues of the form

$$\Delta\varepsilon=(\varepsilon_q-\varepsilon_{\mathrm{JWKB}})\simeq-[\eta(\varepsilon,k)/\pi](\partial\varepsilon/\partial v_b)_k \tag{33}$$

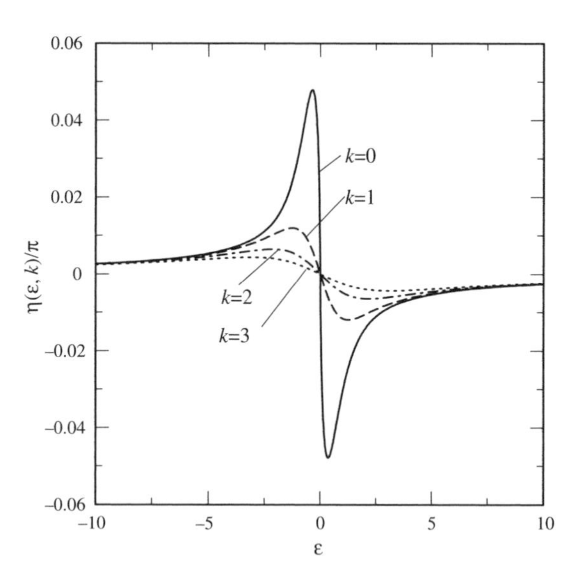

Figure 6. The Bohr–Sommerfeld phase corrections $\eta(\varepsilon,k)$ for $k=0$, 1, and 2. The ratio $\eta(\varepsilon,k)/\pi$ estimates of the error of primitive Bohr–Sommerfeld eigenvalues as a fraction of their local vibrational spacing.

where $(\partial\varepsilon/\partial v)_k$ may be approximated as the local energy separation, because v_b changes by one unit from one eigenvalue to the next. Numerical investigations [32] confirm the accuracy of this error estimate. Since the maximum value of $|\eta(\varepsilon, k)| = 0.15$ at $|\varepsilon| = 0.356$, this means that the maximum error in ε_{JWKB} corresponds to only 5% in of the local energy separation. Moreover, the error falls off rapidly as $|\varepsilon|$ and $|k|$ increase from zero. Consequently, it is justified to employ the classical arguments in the previous sections to interpret the organization of the quantum eigenvalues, provided that the cycle in Fig. 4 is taken well away from the critical point.

E. Summary

The main conclusions of this section are the following.

1. The presence of a cylindrically symmetric potential barrier (or isolated critical focus–fucus singularity) causes a characteristic distortion of the quantum mechanical eigenvalue lattice, which is sufficiently robust to perturbations to be observed in the computed bending progressions of H_2O, despite the influence of significant interactions between the bending and stretching modes.
2. The nature of the above distortion leads to an associated multivaluedness in the vibrational quantum number, which can be attributed to the fact that the twist angle of the classical invariant tori increases 2π on a cycle around the critical singularity.
3. It is important in defining the monodromy matrix, which quantifies changes in the unit cell in Figs. 4 and 5, to specify the lengths of the unit cell sides that define the basis. The monodromy theorem—that the monodromy index is equal to the number of pinch points on the pinched torus [40]—applies in a basis in which the cell sides represent unit changes in the relevant quantum number.
4. The validity of this classical interpretation is supported by an accurate estimate of the error arising from the Bohr–Sommerfeld correspondence principle.

III. SPHERICAL PENDULUM AND TRIATOMIC ISOMERIZATION

A. Spherical Pendulum

The spherical pendulum, which may be used to model the bending states of HCP [7] and the pendular states of dipolar molecules in strong electric fields [8], is of particular interest as the system in which classical monodromy was discovered [19]. The quantum Hamiltonian takes the form

$$\hat{H} = B\hat{J}^2/\hbar^2 + V\cos\theta \tag{34}$$

where $\hat{J}^2$ is the total angular momentum operator and B is an effective rotational constant for the motion of the H atom around CP, in the HCP molecule. The classical analog may be written

$$H = \frac{B}{\hbar^2}\left(p_\theta^2 + \frac{p_\phi^2}{\sin^2\theta}\right) + V\cos\theta \tag{35}$$

although the Cartesian form

$$H = B(p_x^2 + p_y^2 + p_z^2)/\hbar^2 + Vz \tag{36}$$

subject to the constraints

$$x^2 + y^2 + z^2 = 1 \quad \text{and} \quad \mathbf{r}\cdot p = xp_x + yp_y + zp_z = 0 \tag{37}$$

is preferred in the mathematics literature [10, 24, 25]. These references provide an excellent introduction to technical methods that go beyond what is appropriate for the present chapter, in which the discussion is centered around Eqs. (34) and (35).

In view of the cylindrical symmetry, the z component of angular momentum is again conserved, $p_\phi = M = m\hbar$; and there are again two fixed points—a stable equilibrium at $\theta = \pi$, with $m = 0$ and $E = -V$, and an isolated unstable critical point at $\theta = 0$, with $m = 0$ and $E = V$. The relative equilibria, given by $\partial H/\partial\theta = \partial H/\partial p_\theta = 0$ at fixed m, are parameterized by

$$E = \frac{V(3z^2 - 1)}{2z} \quad \text{and} \quad m = \pm\frac{\sqrt{V}(1 - z^2)}{\sqrt{-2Bz}} \tag{38}$$

where $z = \cos\theta$.

By the argument in Section IIB, the presence of a locally quadratic cylindrically symmetric barrier leads one to expect a characteristic distortion to the quantum lattice, similar to that in Fig. 1, which is confirmed in Fig. 7. The heavy lower lines show the relative equilibria and the point (0,1) is the critical point. The small points indicate the eigenvalues. The lower part of the diagram differs from that in Fig. 1, because the small amplitude oscillations of a spherical pendulum approximate those of a degenerate harmonic oscillator, rather than the a-axis rotations of a bent molecule. Hence the good quantum number is

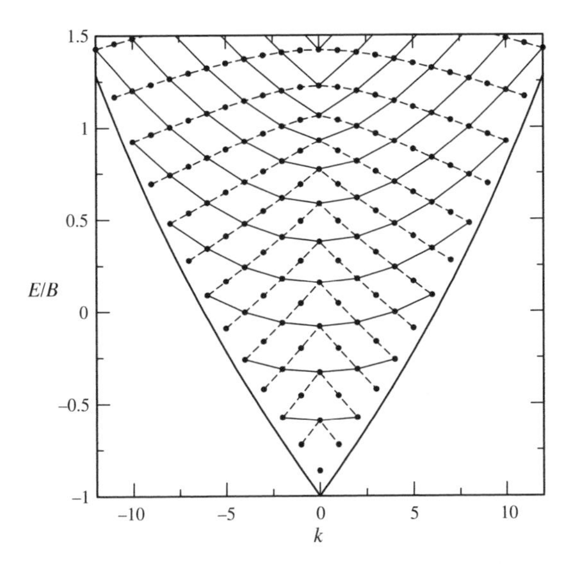

Figure 7. Eigenvalues of the spherical pendulum (points) joined by continuous lines of constant v_b and dashed lines of constant j.

$v_l = 2v_\theta + |m|$ [46]. However, it is seen that the smooth energy variation along the continuous lines, specified by common v_l values, again goes over to the characteristic "vee" shaped kink at energies above the critical point. Similarly, the good quantum number at high energies is the rotational quantum number $j = v_\theta + |m|$ (see Chapter 4 of Child [23] for an angle–action discussion) and the dashed lines joining points with common j values are seen to be smooth above the critical point and kinked below it. The reorganization of the spectrum around the critical point is therefore qualitatively similar to that for the champagne bottle Hamiltonian, despite the presence of a second, nonisolated singularity at $k = 0$ on the relative equilibrium line. A much more realistic model for the bending motions of HCP also shows the same qualitative behavior, which was used to explain and predict large changes in the spectroscopic vibration–rotation coupling parameters [7]. Since the essential features of the spectral pattern are again dictated by the presence of the isolated focus–focus critical point, the arguments of Sections IIB–E again apply with minor variants. The moving cell construction in Fig. 8 also mirrors that in Fig. 4.

B. LiCN and HCN Isomerization

The simplest way to model isomerization is to add a quadratic term to the spherical pendulum Hamiltonian [10, 24]. Thus

$$\hat{H} = B\hat{J}^2/\hbar^2 + V_0 + V_1 \cos\theta - \tfrac{1}{2} V_2 \cos^2\theta \tag{39}$$

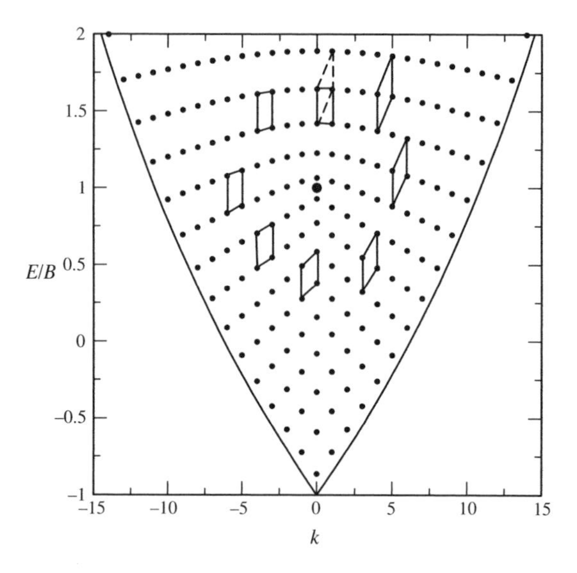

Figure 8. Evidence of quantum monodromy in the the spherical pendulum eigenvalue lattice. The heavy continuous lines are the relative equilibria, and the large dot indicates the critical point.

with $V_2 > V_1 > 0$. As shown by the heavy solid line in Fig. 9, there are then three stationary points—a minimum A at $\theta = \pi$ with $V_A = V_0 - V_1 - \frac{1}{2}V_2$, a secondary minimum B at $\theta = 0$ with $V_B = V_0 + V_1 - \frac{1}{2}V_2$, and a transition state maximum X at $\cos\theta = V_1/V_2$ with $V_X = V_0 + \frac{1}{2}V_1^2/V_2$. One should also note that the angular positions and energies of these stationary points change as m increases, leading to a coalescence between the secondary minimum and the transition state, at a critical value between $m = 10$ and $m = 15$ in Fig. 9.

These relative equilibria are parameterized by

$$E = V_0 + \frac{1}{2}\left(V_2 - \frac{V_1}{z}\right) + \frac{3}{2}V_1 z - V_2 z^2$$
$$m^2 = \frac{(1-z^2)^2}{2B}\left[V_2 - \frac{V_1}{z}\right] \tag{40}$$

where $z = \cos\theta$ and the resulting divisions of the Em map are shown in Fig. 10. The lower relative equilibria, which coalesce on V_A, are generated by the parameter range $-1 \leq z < 0$, while the upper triangular segment is given by $V_1/V_2 \leq z \leq 1$. The cusps at $\alpha_{\pm}$ occur at the $|m|$ values above which the centrifugally corrected curves in Fig. 9 fail to support a secondary minimum. The relevant z parameter is the value at which $dE/dz = dm/dz = 0$, so that the derivative dE/dm is undefined. The area within the triangle covers Em values for

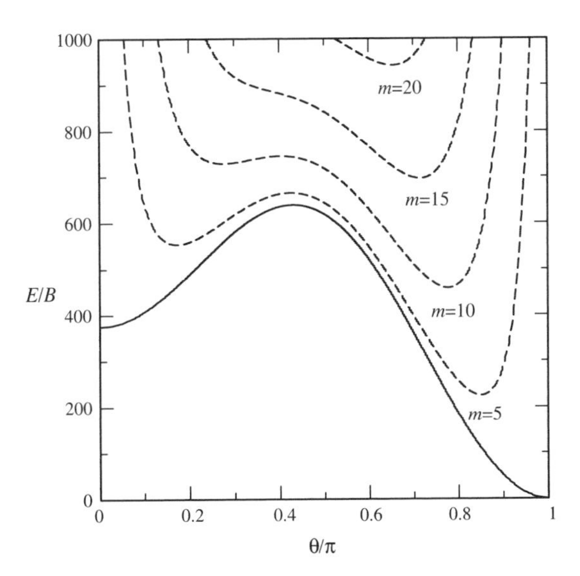

Figure 9. Effective potential energy curves $V^{(m)}(\theta) = V(\theta) + Bm^2/\sin^2\theta$ for the quadratic Hamiltonian in Eq. (39). The solid curve, for $m = 0$, indicates stable isomers at $\theta = 0$ and $\theta = \pi$ and a saddle point at $\theta \simeq 0.45\pi$. Note, however, that the secondary minimum disappears as the angular momentum increases.

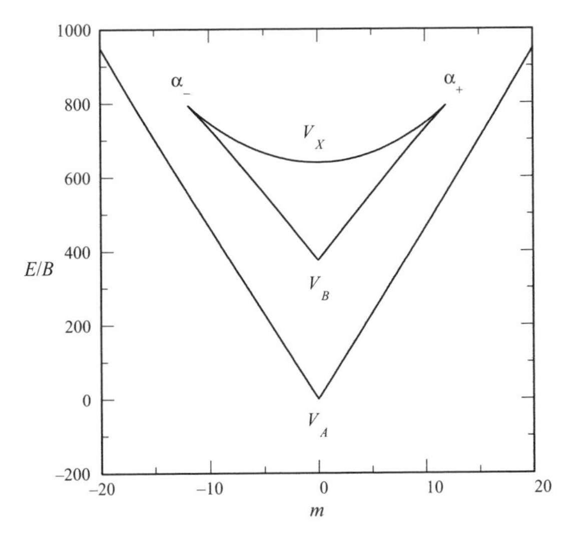

Figure 10. Divisions of the *Em* map for the quadratic Hamiltonian in Eq. (39). V_A, V_B, and V_X are the energies of the two isomers and that of the saddle point between them. The solid lines are relative equilibria, and the cusp points, $\alpha_{\pm}$, indicate points at which the secondary minimum and the saddle point in Fig. 9 merge at a point of inflection.

which a potential barrier separates states of the most stable isomer A from those of B. Thus the energy levels of the two isomers interleave in this folded region.

The form of the resulting quantum lattice for the coefficient choice, $V_0 = V_1 + \frac{1}{2}V_2, V_1 = 175B$, and $V_2 = 860B$, is shown in Fig. 11, in which states of the B isomer are distinguished by open rather than closed circles. The moving cell construction in Fig. 11, which excludes the B isomer states, shows a similar distortion of the unit cell to that in Fig. 8, demonstrating that the curved fold line, which joins the interleaving A and B type spectra, now acts as the topological obstruction responsible for distortion of the quantum lattice. It is also easily demonstrated that this fold shrinks to an isolated point as $V_2 \to V_1$ in Eq. (39), beyond which the secondary minimum lies in the nonphysical range $|\cos\theta| > 1$. Figure 12 shows a similar plot for bending states of LiCN computed on an ab initio potential surface [9, 47].

The situation with respect to HCN is rather different, because it is no longer valid to approximate the internal rotational constant, B, as independent of the angle θ. As discussed by Efstathiou et al. [10], the form of the most recent potential energy surface [48, 49] shows that the separation of the H atom from the CN center of mass decreases by about 30% between the HCN configuration and the T-shaped transition state, giving the optimized bending potential energy

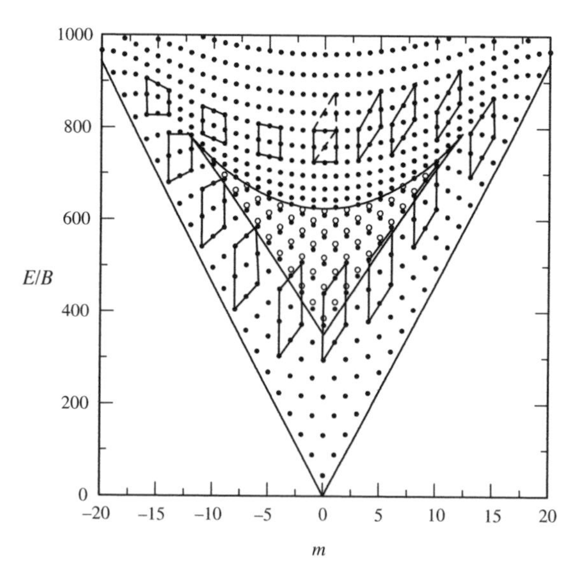

Figure 11. Quantum monodromy in the spectrum of the quadratic Hamiltonian of Eq. (38). The solid lines indicate relative equilibria. Filled circles mark the eigenvalues of the most stable isomer and those above the relevant effective potential barrier in Fig. 8. Open circles indicate interpenetrating eigenvalues of the secondary isomer. The transported unit cell moves over the filled circle lattice, around the curved fold line connecting the two spectra.

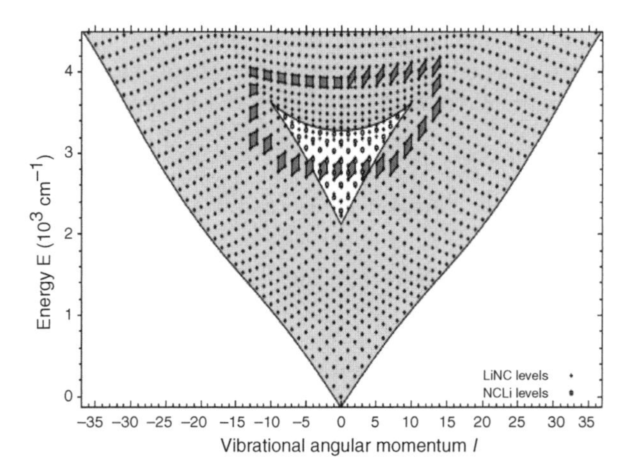

Figure 12. Quantum monodromy in the accurately computed spectrum of LiCN/LiNC. Taken from Ref. [9] with permission of Elsevier.

surface a peanut shaped, rather than a convex ellipsoidal, contour. One then finds that the cusp points in Figs. 10–12 disappear, because the energy of the barrier that separates the isomers increases with m at the same rate as the isomer minima. Consequently, the relative equilibrium line rising from the secondary HNC minimum never meets the corresponding line from the saddle point. The resulting image of the Em map therefore takes the form in Fig. 13, which means

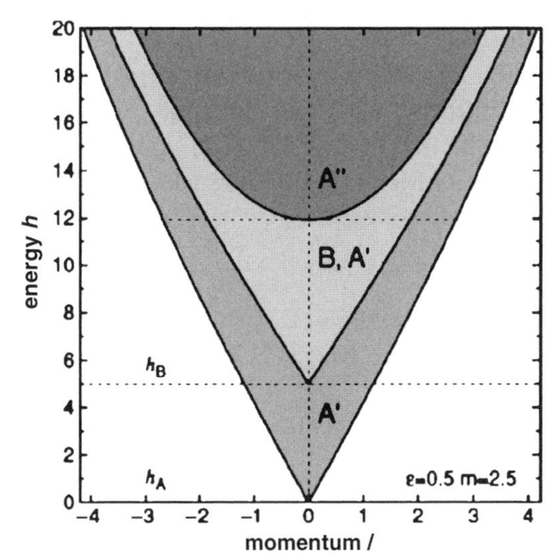

Figure 13. Divisions of the energy–angular momentum map for the isomerizing HCN/HNC system, for which the cusps, $\alpha_{\pm}$, in Figs. 10–12 lie at infinity. Taken from Ref. [10] with permission of the American Physical Society, Copyright 2004.

that the fold line has infinite length. Consequently, there is no finite region over which to make the tour, and no monodromy. As a simple model, Efstathiou et al. [10] suggest approximating the θ dependence of the optimized H to CN distance in the form

$$R(\theta) = R_0(1 - \varepsilon \sin^2 \theta) \tag{41}$$

in which case monodromy was found to occur for $\varepsilon < \frac{1}{3}$ but not for larger values.

C. Summary

The main points in this section are the following.

1. The spherical pendulum Hamiltonian has a simple focus–focus singularity at the top of its swing, leading to a similar distortion of the eigenvalue lattice to that observed for quasilinear species, except that the "good" quantum numbers are now the degenerate oscillator label, $v_l = 2v_\theta + |m|$, at energies below the barrier to rotation and the total angular momentum, $j = v_\theta + |m|$, at energies above [1, 10].
2. For some isomerizing species, such as LiCN/LiNC [9], quantum monodromy may be demonstrated by transporting a unit cell around a finite fold of the *Em* map, which joins the interleaving eigenvalues of the separate isomers. This fold therefore plays the role of the simple focus–focus singularity.
3. For other isomerizing species, such as HCN/HNC [10], there is no upper $|m|$ limit above which the centrifugally modified secondary well in Fig. 10 merges with the transition state. Hence the fold between the two interleaving spectra has no ends around which to transport the unit cell. Thus there is no monodromy.

IV. ANGULAR MOMENTUM COUPLING AND ATOMS IN CROSSED FIELDS

The previous sections concerned systems for which the isolated critical points, or folded spectral regions, were readily recognizable from the nature of the Hamiltonian. As discussed next, cases involving coupled angular momenta and atoms in crossed fields give rise to less obvious isolated critical points, which are best understood by exploring the geometry of the angular momentum space. The following section derives from a series of papers by Sadovskii, Zhilinskii, Cushman, and co-workers [2–6].

A. Spin–Rotation Coupling

The simplest example of angular momentum coupling is provided by the scaled spin–rotation coupling Hamiltonian

$$\hat{H} = \frac{1-\gamma}{S}\hat{S}_z + \frac{\gamma}{NS}\hat{\mathbf{N}}\cdot\hat{\mathbf{S}}, \quad 0 \leq \gamma \leq 1 \tag{42}$$

which was used by Sadovskii and Zhilinskii [2] as an early illustration of quantum monodromy. Here S and N are the magnitudes of the spin and rotational angular momenta and $\hat{J}_z = S_z + N_z$ is a constant of the motion [50].

It is interesting to follow changes in the qualitative eigenvalue pattern as the coupling strength γ increases from Fig. 14a to 14d. Here the chains of small dots are the eigenvalues, the large dots are images of limiting points of the classical

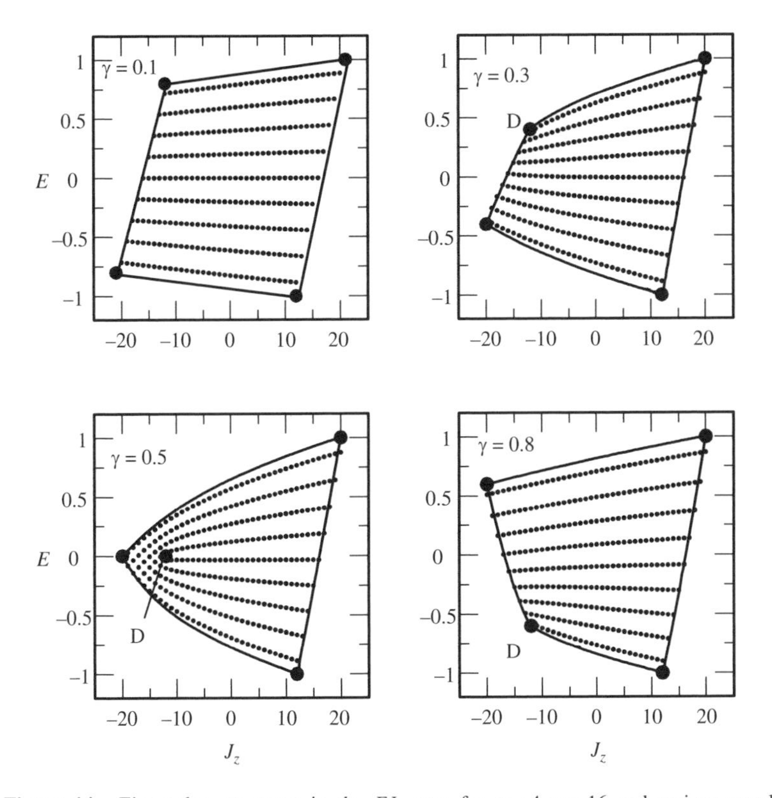

Figure 14. Eigenvalue structures in the EJ_z map for $s = 4, n = 16$ and various coupling strengths, $\gamma = 0.1$, 0.3, 0.5, and 0.8. The eigenvalues appear as strings of small dots. Heavy dots mark the images of the special points A, B, C, and D in the left-hand panel of Fig. 16. The connecting lines are the relative equilibria.

angular momentum space, at which $S_z = \pm S$ and $N_z = \pm N$, and the joining lines are images of the relative equilibria of the classical Hamiltonian. Results are shown for total angular momentum quantum numbers, $s = 4$ and $n = 16$. To understand this diagram, note first that the spin–rotation eigenvalues reduce in the Hund's case (a) limit $\gamma = 0$, to $2s + 1$ sets, each with degeneracy $2n + 1$. The distorted parallelograms for $\gamma = 0.1$ and 0.3 are organized in a similar way, although the degeneracy has been lifted. In addition, the total angular momentum, with values $j = n + s, n + s - 1, \ldots, |n - s|$, is a good quantum number in Hund's case (c) with $\gamma = 1$; and the associated $2j + 1$ degeneracies are reflected in the lengths of the chains shown for $\gamma = 0.8$. The migration of the special point D, which is seen to be *isolated* from the relative equilibria when $\gamma = 0.5$, is clearly responsible for the above eigenvalue reorganization.

The aim of this section is to show how these changes relate to the classical mechanics, along lines initiated by Sadovskii and Zhilinskii [2], except that we use angle–action methods, rather than vector algebra. The relevant angle–action forms for the angular momentum components are given by [23]

$$S_x = \sqrt{S^2 - S_z^2}\cos\theta_S, \quad S_y = \sqrt{S^2 - S_z^2}\sin\theta_S, \quad S_z \tag{43}$$

and similarly for N, where the total angular momenta take the semiclassical values $S = s + \frac{1}{2}$ and $N = n + \frac{1}{2}$ [23]. The canonical transformation to the variables

$$\begin{aligned} J_z &= S_z + N_z, \quad \theta_J = (\theta_S + \theta_N)/2 \\ K_z &= S_z - N_z, \quad \theta_J = (\theta_S - \theta_N)/2 \end{aligned} \tag{44}$$

is also employed later.

As a prelude to the arguments in subsequent sections, it should be recognized that the motion of the system is restricted to the space defined by the variables that commute with J_z, which may be taken as S_z, N_z (or J_z, K_z), and the two quantities

$$\begin{aligned} \xi &= \mathbf{N}\cdot\mathbf{S} = \sqrt{(S^2 - S_z^2)(N^2 - N_z^2)}\cos 2\theta_K + S_zN_z \\ \sigma &= N_xS_y - S_xN_y = \sqrt{(S^2 - S_z^2)(N^2 - N_z^2)}\sin 2\theta_K \end{aligned} \tag{45}$$

The resulting (ξ, σ, S_z, N_z) or (ξ, σ, J_z, K_z) space is constrained by the identity

$$\begin{aligned} \tilde{\sigma}^2 &= (1 - \tilde{S}_z^2)(1 - \tilde{N}_z^2) - (\tilde{\xi} - \tilde{S}_z\tilde{N}_z)^2 \\ &= 1 + 2\tilde{\xi}\tilde{S}_z\tilde{N}_z - \tilde{\xi}^2 - \tilde{S}_z^2 - \tilde{N}_z^2 \end{aligned} \tag{46}$$

TABLE I
Properties of the Limiting points A, B, C, and D in Fig. 16.

Point	N_z	S_z	ξ	H	J_z	K_z
A	N	S	NS	1	$N+S$	$-N+S$
B	$-N$	$-S$	NS	$2\gamma-1$	$-N-S$	$N-S$
C	N	$-S$	$-NS$	-1	$N-S$	$-N-S$
D	$-N$	S	$-NS$	$1-2\gamma$	$-N+S$	$N+S$

where $\tilde{\xi} = \xi/NS$, $\tilde{\sigma} = \sigma/NS$, $\tilde{S}_z = S_z/S$, and $\tilde{N}_z = N_z/N$. The surface $\tilde{\sigma} = 0$, which is seen in Fig. 15 to take the form of a tetrahedron with rounded edges, defines the limits of the accessible (ξ, S_z, N_z) space. The properties of the vertices A, B, C, and D, which appear as the large dots in Fig. 14, are collected in Table I.

This geometrical construction leads to the idea that the spectra in Fig. 14 may be viewed as a projection of the quantum states inside the tetrahedron onto the (E, J_z) plane—a procedure that is simplified by the fact that both the Hamiltonian

$$H = \frac{1-\gamma}{S} S_z + \frac{\gamma}{NS} \xi \tag{47}$$

and J_z are linear functions of $(\tilde{\xi}, \tilde{S}_z, \tilde{N}_z)$ in the present spin–rotation case. The projection plane is therefore normal to the vector $J_z \wedge H = (-(1-\gamma)N, \gamma N, -\gamma S)$ in the $(\tilde{\xi}, \tilde{S}_z, \tilde{N}_z)$ frame, which points to the octant containing the vertex D. The nature of the projection therefore depends on the

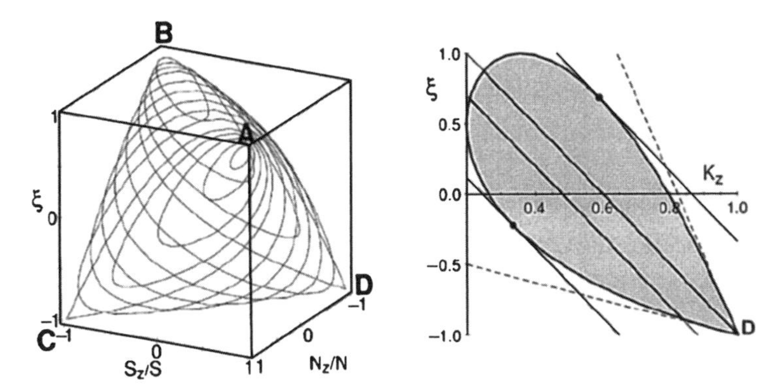

Figure 15. *Left*: Geometry of the surface $\tilde{\sigma} = 0$ in Eq. (46) with fixed total angular momenta S and N. Properties of the special points A, B, C, and D are listed in Table I. All other permissible classical phase points lie on or inside the surface of the rounded tetrahedron. *Right*: Critical section at J_z^D. Continuous lines are energy contours for $\gamma = 0.5$ and $N/S = 4$. Dashed lines are tangents to the section at D. Axes correspond to normalized coordinates ξ/NS and $K_z/(N+S)$. Taken from Ref. [2] with permission of Elsevier.

coupling strength γ and on the magnitudes N and S, which will mainly affect the image of D.

In order to establish conditions for the isolation of the image of point D under the *EM* map, the projection is performed by first taking a section through the surface $\sigma = 0$ at fixed J_z, an example of which is shown in the right-hand panel of Fig. 15, for the critical value $J_z^D = S - N$. The shaded area of the (K_z, ξ) plane defines the classically allowed range for the specified J_z value. The lines indicate energy contours for $\gamma = 0.5$. Those that touch the section correspond to relative equilibria of the Hamiltonian, whose (J_z, K_z, ξ) values determine points on the boundary of the projection onto the (E, J_z) plane. It is important to note that these tangent points lie far from the conical point D, which explains why this critical point is *isolated* from the relative equilibrium lines in Fig. 14, for $\gamma = 0.5$. However, the magnitude of the slope of the tangent lines, implied by Eq. (47), namely,

$$\left(\frac{\partial \xi}{\partial K_z}\right)_{J_z,H} = -\frac{(1-\gamma)N}{2\gamma} \tag{48}$$

will increase as γ decreases, causing the upper tangent point to move toward the point D, while an increase in γ will have a similar effect on the lower tangent point. Consequently, the condition for isolation of D is that the slope given by Eq. (48) should lie between those of the (dashed) tangents to the J_z^D section at D. The necessary condition on γ may be obtained with the help of the approximation

$$(S^2 - S_z^2)/S \simeq (N^2 - N_z^2)/N \simeq (N + S - K_z) \tag{49}$$

in the vicinity of D, so that the unscaled form of Eq. (46) simplifies, for $\sigma = 0$, to

$$\xi = S_z N_z \pm \sqrt{(S^2 - S_z^2)(N^2 - N_z^2)} \simeq \tfrac{1}{4}(J_z^2 - K_z^2) \pm \sqrt{NS}(N + S - K_z) \tag{50}$$

The required condition [2]

$$\frac{N}{2N + S + 2\sqrt{NS}} \leq \gamma \leq \frac{N}{2N + S - 2\sqrt{NS}} \tag{51}$$

follows, after combining Eq. (48) with the derivative of Eq. (50) and setting $K_z^D = N + S$.

The above geometrical argument may be supplemented by an explicit demonstration that the singularity at the point D is of the focus–focus type (Eq. (7)), when Eq. (51) applies. The procedure is to employ the local approximations

$$J_z \simeq (S - N + L), \quad K_z \simeq (S + N - I) \tag{52}$$

where the variables I and L are chosen to conform with the notation in the Appendix. Since the angle conjugate with I is the negative of θ_K, the leading terms in a local expansion of the Hamiltonian in Eq. (47) turn out, after some manipulation, to be

$$H \simeq AI + B\sqrt{I^2 - L^2}\cos 2\theta_I + \text{const}$$
$$A = \frac{1}{2NS}[-N + \gamma(2N + S)], \quad B = \frac{2\gamma\sqrt{NS}}{2NS} \tag{53}$$

where the constant term includes terms in the conserved variable L. The argument in the Appendix shows that H transforms to the Cartesian form

$$\tilde{H} = \tfrac{1}{2}[(A + B)(x^2 + y^2) + (A - B)(p_x^2 + p_y^2)] \tag{54}$$

which is of focus–focus type if $A^2 - B^2 < 0$—an inequality that is equivalent to Eq. (51).

The upshot of these considerations is that γ values in the range given by Eq. (51) lead to the presence of an isolated unstable fixed point of the system, analogous to those encountered above for quasilinear molecules and spherical pendulums. It is therefore natural to expect quantum monodromy, which is indeed demonstrated, for $\gamma = 0.5$, in Fig. 16. Panel (a) shows the familiar moving cell construction, while panels (b) and (c) indicate that j and s_z behave as good quantum numbers in the high and low energy regions of the spectrum, respectively. From another viewpoint, the upper smooth lines in Fig. 16b may be seen as incipient members of the eigenvalue lattice for $\gamma = 0.8$ in Fig. 14, and the kinked lines as remnants of the upper left-hand corner of the $\gamma = 0.3$ lattice.

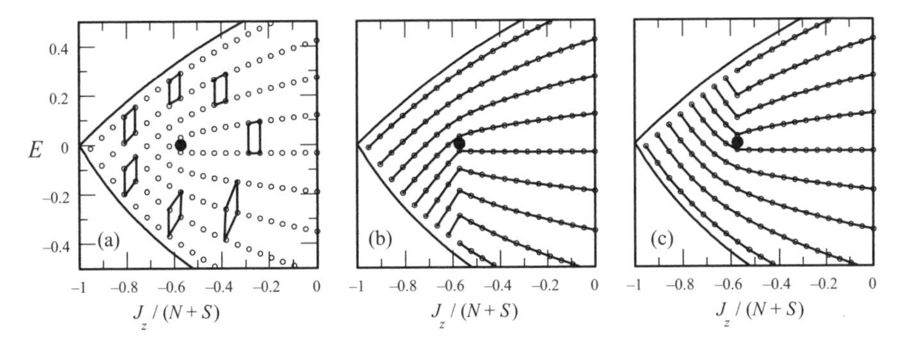

Figure 16. (a) Moving cell construction to illustrate quantum monodromy for $\gamma = 0.5$, $s = 4$, $n = 16$. The central dot is the image of the isolated critical point. Evidence of the characteristic distortion of the lattice is indicated in panels (b) and (c) by joining the points labeled by the total angular momentum j and the projection s_z, respectively.

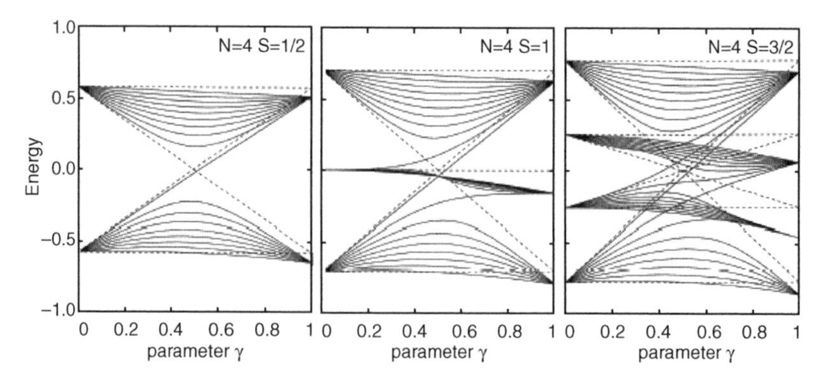

Figure 17. Quantum correlation diagrams (solid lines) for $S = \frac{1}{2}$, $S = 1$, and $S = \frac{3}{2}$, with $N = 4$ in each case. Taken from Ref. [2] with permission of Elsevier.

In other words, in passing from the upper to the lower boundary in Fig. 14, as γ increases, the point D pushes the corner of the low γ lattice ahead of itself and forms the new boundary of the high γ lattice in its wake. The eigenvalue reorganization between the limits $\gamma = 0$ and $\gamma = 1$ is therefore intrinsically tied up with quantum monodromy.

As emphasized by Sadovskii and Zhilinskii [2], this latter point is important for quantum systems for which the lattice is too small to allow the construction in Fig. 16a, because there is still a systematic reorganization of the spectra, involving transfer of individual levels or groups of levels from lower to upper bands, as γ increases from 0 to 1. Figure 17 shows examples for $n = 4$ and $s = \frac{1}{2}$, 1, and $\frac{3}{2}$, which illustrate the influence of quantum monodromy far from the classical limit.

B. Three Coupled Angular Momenta

The ideas in Section IVA have been extended to three coupled angular momenta by Grondin et al. [3], in the context of a Hamiltonian of the form

$$\hat{H} = \alpha(\hat{\mathbf{L}}_1 \cdot \hat{\mathbf{L}}_2) + (1 - \alpha)(\hat{\mathbf{L}}_1 \cdot \hat{\mathbf{L}}_3) \tag{55}$$

for which the magnitude of

$$\hat{\mathbf{J}} = \hat{\mathbf{L}}_1 + \hat{\mathbf{L}}_2 + \hat{\mathbf{L}}_3 \tag{56}$$

and J_z are constants of the motion. In this case the geometrical construction is based on the normalized scalar products

$$X_1 = \frac{\hat{\mathbf{L}}_2 \cdot \hat{\mathbf{L}}_3}{L_2 L_3}, \quad X_2 = \frac{\hat{\mathbf{L}}_3 \cdot \hat{\mathbf{L}}_1}{L_3 L_1}, \quad X_3 = \frac{\hat{\mathbf{L}}_1 \cdot \hat{\mathbf{L}}_2}{L_1 L_2} \tag{57}$$

which are subject to the auxiliary equation

$$\begin{aligned} Y^2 &= \det \begin{pmatrix} \mathbf{L}_1 \cdot \mathbf{L}_1 & \mathbf{L}_1 \cdot \mathbf{L}_2 & \mathbf{L}_1 \cdot \mathbf{L}_3 \\ \mathbf{L}_2 \cdot \mathbf{L}_1 & \mathbf{L}_2 \cdot \mathbf{L}_2 & \mathbf{L}_2 \cdot \mathbf{L}_3 \\ \mathbf{L}_3 \cdot \mathbf{L}_1 & \mathbf{L}_3 \cdot \mathbf{L}_2 & \mathbf{L}_3 \cdot \mathbf{L}_3 \end{pmatrix} (L_1 L_2 L_3)^{-2} \\ &= 1 + 2X_1X_2X_3 - X_1^2 - X_2^2 - X_2^2 \end{aligned} \tag{58}$$

where $L_i = |L_i|$, $i = 1$–3. It is readily seen, by comparison with Eq. (46), that the surface $Y = 0$ again takes the form of a rounded tetrahedron. Moreover, both the Hamiltonian and the conserved variable

$$J^2 = L_1^2 + L_2^2 + L_3^2 + 2(\mathbf{L}_2 \cdot \mathbf{L}_3 + \mathbf{L}_3 \cdot \mathbf{L}_1 + \mathbf{L}_1 \cdot \mathbf{L}_2) \tag{59}$$

are again linear functions of the polynomial invariants (X_1, X_2, X_3). Hence the boundaries of the quantum eigenvalue lattice and the positions of the critical points analogous to A, B, C, and D in Fig. 15 can always be generated by an appropriate projection onto the (J, E) plane. The complication is that two-dimensional SO(2) symmetry of the spin–rotation coupling model is now replaced by the three-dimensional SO(3) symmetry, with the result that the two previous possible forms of projection, with either zero or one isolated critical point, now proliferate to eleven possibilities, which are listed in Tables II and III of Grondin et al. [3].

C. Hydrogen Atom in Crossed Fields

The problem of a hydrogen atom in constant orthogonal magnetic and electric fields has generated a large literature (see Ref. [5] for a comprehensive reference list). Here we merely draw attention to the connection with Sections IVA and IVB by extracting some important symmetry aspects, relevant to quantum monodromy, from the papers of Cushman, Sadovskii, and Efstathiou [4–6]. The fields are assumed to be sufficiently weak to allow a perturbation expansion. The classical theory proceeds by first performing a Kustaanheimo–Stiefel (KS) transformation [51] to four-dimensional isotropic oscillator variables, in order to facilitate the perturbation analysis. A further transformation then expresses the Hamiltonian in terms of the components of angular momentum-like vectors, written here as $\mathbf{J}_1$ and $\mathbf{J}_2$, rather than $\mathbf{X}$ and $\mathbf{Y}$ in the parent text [4, 5]. Another minor difference is to use (x, y, z) rather than (1, 2, 3) to label the components, which are subject to the $SU(2) \times SU(2)$ Poisson bracket relations

$$\{J_{1i}, J_{1j}\} = \varepsilon_{ijk} J_{1k}, \quad \{J_{2i}, J_{2j}\} = \varepsilon_{ijk} J_{2k}, \quad \{J_{1i}, J_{2j}\} = 0 \tag{60}$$

and the normalization

$$J_{1x}^2 + J_{1y}^2 + J_{1z}^2 = J_{2x}^2 + J_{2y}^2 + J_{3z}^2 = N^2/4 \tag{61}$$

where N is the classical equivalent of the principal quantum number (see later discussion). For the unperturbed Coulomb problem, $\mathbf{J}_1$ and $\mathbf{J}_2$ are the sum and difference of the angular momentum and Runge–Lenz or modified eccentricity vectors (see Ref. [25], p. 400 for a historical note); but any orthogonal transformation that conserves the $SU(2) \times SU(2)$ symmetry is equally appropriate.

The second-order perturbation approach employed by Cushman and Sadovskii [4, 5] is similar to that given by Gourlay et al. [52] and Milczewski et al. [53, 54], except that the angular momenta are defined to ensure that the first-order term in the scaled Hamiltonian is a scalar multiple of the component, J_z, of the sum $\mathbf{J} = \mathbf{J}_1 + \mathbf{J}_2$, which is treated as a fast variable of the motion (the original authors employ the symbol c in place of J_z). The second-order term is then averaged over this fast motion to produce what is termed the second normal form Hamiltonian H_{snf}—the first normal form being the one obtained by the KS transformation. The effect of this averaging is that J_z becomes a constant of the motion under H_{snf}, with the implication that the space accessible to the residual motion under the second-order term must be expressible in terms of variables that commute with J_z. In order to highlight special features of the case $|\mathbf{J}_1| = |\mathbf{J}_2|$, the variables are taken as N, J_z and the three quantities

$$\begin{aligned} \pi_1 &= J_{1z} - J_{2z} = K_z \\ \pi_2 &= 4(J_{1x}J_{2x} + J_{1y}J_{2y}) \\ \pi_3 &= 4(J_{1x}J_{2y} - J_{2x}J_{1y}) \end{aligned} \tag{62}$$

where the notation K_z is adopted to conform with that in Section IVA. These five functions are related by the normalization conditions, Eq. (61), in the form

$$\pi_2^2 + \pi_3^2 = [(N - J_z)^2 - \pi_1^2][(N + J_z)^2 - \pi_1^2] \tag{63}$$

with the values of π_1, π_2, and π_3 subject to the restrictions

$$|\pi_1| = |K_z| \leq N - |J_z|, \quad |\pi_2| \leq N^2 - J_z^2, \quad |\pi_3| \leq N^2 - J_z^2 \tag{64}$$

The final step in the formulation of the model [4–6] is to recognize that the second-order term, $H^{(2)}$, say, must be a quadratic function of the angular momentum components, consistent with the symmetries of the crossed-field Hamiltonian, which allow terms only in the conserved variables N and J_z and combinations of π_1^2 and π_2. Thus the second-order term may be expressed as

$$\begin{aligned} H^{(2)} &= a\pi_2 - b\pi_1^2 \\ &= 4a(J_{1x}J_{2x} + J_{1y}J_{2y}) - bK_z^2 \end{aligned} \tag{65}$$

plus a possible additive term in N and J_z.

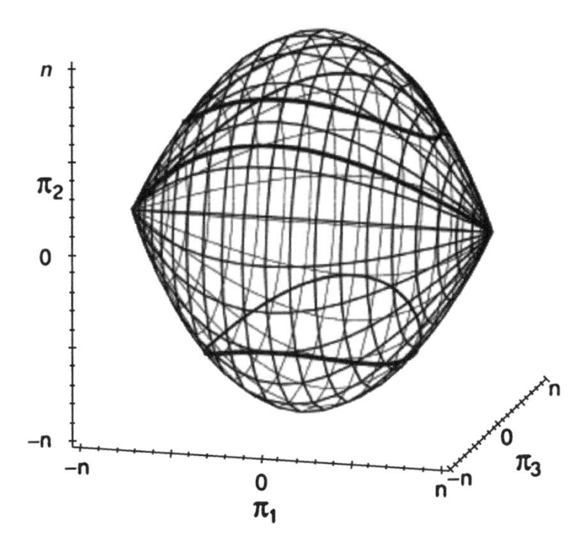

Figure 18. A graphical representation of Eq. (63). The surface corresponds to $J_z = 0$, with conical points that correspond to C and D in Table I. Taken from Ref. [4] with permission.

The special geometry of the case with $|\mathbf{J}_1| = |\mathbf{J}_2|$ is brought out by Fig. 18, which depicts the singular surface, $J_z = 0$, defined by Eq. (63) [4, 5], which encloses smooth ball-shaped surfaces for $J_z \neq 0$ (see Fig 3.1 of Ref. [24]). The two conical points at $(\pi_1, \pi_2) = (\pm N, 0)$ correspond to the points C and D in Table I, with $(J_{1z}, J_{2z}) = (\pm N/2, \mp N/2)$. Interested readers will find an extensive discussion of the geometry of the (π_1, π_2, π_3) space defined by Eq. (63) and its intersection with $H^{(2)}$ in the original papers [4–6]. Here we merely emphasize the connection with the earlier coupled angular momentum models, by replacing π_1 by K_z and expressing the Hamiltonian in terms of the scaled variables, $\tilde{\pi}_2 = 4\mathbf{J}_1 \cdot \mathbf{J}_2/N^2$, $\tilde{J}_z = J_z/N$, and $\tilde{K}_z = K_z/N$. Thus

$$\begin{aligned} H^{(2)} &= N^2[a\tilde{\pi}_2 - b\tilde{K}_z^2] \\ \tilde{\pi}_2^2 + \tilde{\pi}_3^2 &= [1 - (\tilde{J}_z - \tilde{K}_z)^2][1 - (\tilde{J}_z + \tilde{K}_z)^2] \end{aligned} \tag{66}$$

As mentioned earlier, the most interesting point, with regard to the quantum monodromy, is that the two critical points C and D in Table I now lie in the singular section of $\tilde{\pi}_2(J_z, K_z)$ at $\tilde{\pi}_3 = \tilde{J}_z = 0$, which means that the form with a single conical point in the right-hand panel of Fig. 15 goes over to one with two such points in Fig. 19, which is the case denoted as $s11s$ in Tables II and III of Grondin et al. [3].

It remains to establish conditions under which this doubly conical critical point is isolated from the relative equilibria in the EJ_z map. It is easy to see that the fixed points in Table I lie at $(\tilde{J}_z^X, E^X) = (1, 0), -(1, 0), (0 - bN^2)$, and $(0 - bN^2)$ for $X = A$, B, C, and D, respectively. In addition, since both $H^{(2)}$ and

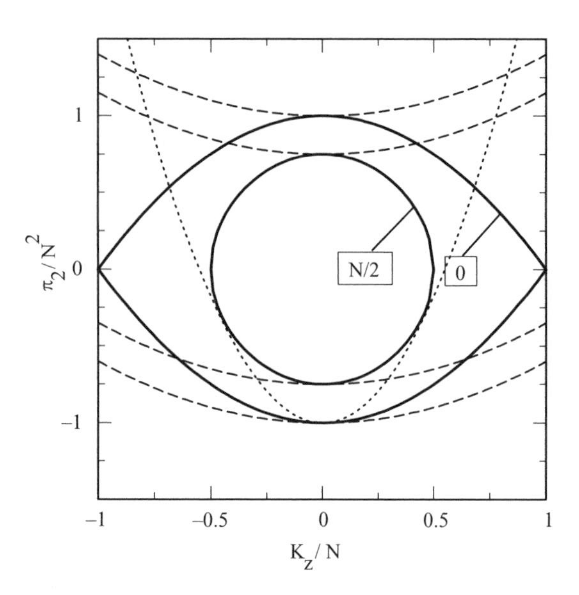

Figure 19. Sections defined by Eq. (66) for $\pi_3 = 0$ and $J_z = 0$ and $J_z = N/2$ (solid lines). The dashed lines show energy contours for the case $a^2 > b^2$, which touch constant J_z sections at $K_z = 0$. The dotted line is an energy contour for $b^2 > a^2$, which also touches the $J_z = N/2$ section at $K_z \neq 0$.

$\tilde{\pi}_2$ are even functions of $\tilde{K}_z$, contours of $H^{(2)}$ touch any constant $\tilde{J}_z$ section of $\tilde{\pi}_2$ at $\tilde{K}_z = 0$, to give relative equilibria of the form

$$E^{I}_{\text{rel}} = \pm N^2 a(1 - \tilde{J}_z^2) \tag{67}$$

The dotted contour in Fig. 19 shows that a second form of touching condition can also occur, for sufficiently large values of the parameter ratio b/a. Simple analytical manipulations show that touching conditions of this second type, with real K_z, give rise to relative equilibria described by

$$E^{II}_{\text{rel}} = N^2[2\sqrt{b^2 - a^2}|J_z| - b(1 + J_z^2)] \tag{68}$$

provided that $b^2 \geq a^2$ and $|J_z| \leq \sqrt{(b-a)/(b+a)}$.

It is evident that the projections of the fixed points A and B, in the EJ_z map, always lie on relative equilibria of type I, but that the position of the overlapping projections of C and D depends on the sign of $b^2 - a^2$. If $a^2 > b^2$ the double point is isolated between the two type I equilibria, and quantum monodromy is expected, for a sufficiently dense quantum lattice. If, on the other hand, $a^2 < b^2$ the critical point lies on the type II relative equilibrium line and no quantum monodromy is expected.

It may also be verified, by expressing the Hamiltonian in Eq. (65) in the angle–action form,

$$H = a\sqrt{[(N/2)^2 - J_{1z}^2][(N/2)^2 - J_{2z}^2]}\cos 2\theta_K - bK_z^2 \tag{69}$$

and generalizing the arguments of Eq. (52)–(54), that the singularities at both points C and D are of focus–focus type, when $b^2 \geq a^2$. Hence, by extension of the discussion in Section IIC, tori with $E \simeq E_{C,D}$ and $L \to 0^{\pm}$ are expected to pick up two twist angle differences of 2π, on passing close to C and D, to yield a monodromy matrix of the form in Eq. (22) with index $n = 2$.

Following Ref. [5], these conclusions may be verified by diagonalizing the quantum mechanical Hamiltonian

$$\hat{H} = 4a(\hat{\mathbf{J}}_1 \cdot \hat{\mathbf{J}}_2 - \hat{J}_{1z}\hat{J}_{2z}) - b(\hat{J}_{1z} - \hat{J}_{2z})^2 \tag{70}$$

The resulting spectra for $j_1 = j_2 = 5$ and $b/a = 0.2$ and 2 are shown in Figs. 20a and 20b, respectively, together with the positions of the critical points and relative equilibria. Following the normal semiclassical practice [23], the appropriate value of the equivalent total classical action is given by Eq. (61), with $J_i = j_i + 1/2$; thus $J_1 = J_2 = 5\frac{1}{2}$ and $N = 11$. It is evident from the moving cell construction in Fig. 20a that the unit cell side $(\Delta j_z, \Delta v) = (1, 0)$ transforms, after completing the circuit to $(\Delta j_z', \Delta v') = (1, 2)$. Thus

$$\begin{pmatrix} \Delta v' \\ \Delta j_z' \end{pmatrix} = \begin{pmatrix} 1 & 2 \\ 0 & 1 \end{pmatrix} \begin{pmatrix} \Delta v \\ \Delta j_z \end{pmatrix} \tag{71}$$

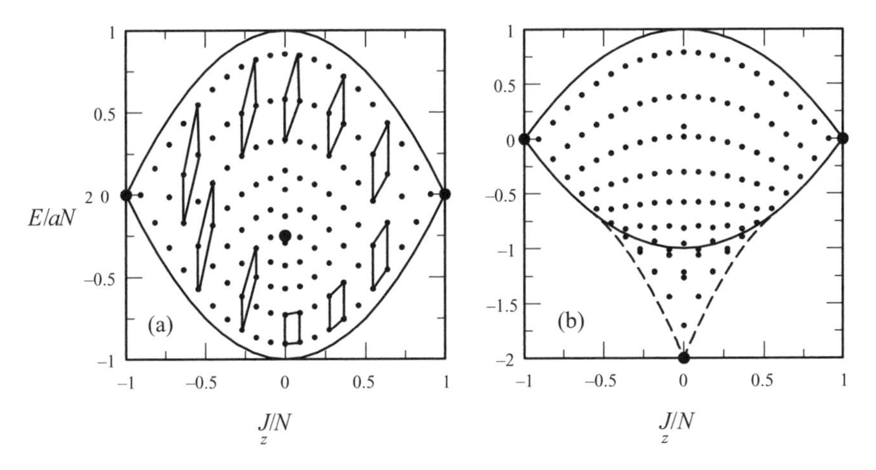

Figure 20. Quantum eigenvalue lattices for $j_1 = j_2 = 5$ and (a) $b/a = 0.2$, (b) $b/a = 2$. Solid and dashed lines are type *I* and type *II* relative equilibria, respectively. The large point at $J_z = 0$ is the overlapping projection of points C and D in Table I. Those at $J_z = \pm N$ are projections of points A and B.

as expected from Eq. (22). However, monodromy cannot occur in Fig. 20b because there is no classically allowed circuit when the critical points all lie on the boundary of the map.

D. Summary

The conclusions from this section are the following.

1. Critical points of the angular momentum space may or may not be isolated in the two-dimensional energy–angular momentum map (giving rise to quantum monodromy) depending on the coupling strength.
2. Changes in the degeneracy patterns between, for example, Hund's cases (a) and (c) can be attributed to monodromy.
3. The coupling of two angular momenta with equal magnitudes, as in the previous model of the perturbed H atom, is of special interest in leading to a monodromy matrix with index $n = 2$.

V. FERMI RESONANCE

The mathematical character of quantum monodromy in Fermi resonant systems, with conserved angular momentum, has been analyzed in detail for the special case of the so-called "resonant swing spring" [11–13], which roughly mimics CO_2 in that the equilibrium frequencies are taken to be in exact zeroth order 1:1:2 resonance, with weak diagonal anharmonicity. One of the major conclusions is that the monodromy of such systems is intrinsically three dimensional. There are three conserved actions, or quantum numbers, which may be taken as the angular momentum, L, the polyad number, N, and a vibrational quantum number, v; and the nature of the monodromy in any two-dimensional section through this three-dimensional space depends on how the section is chosen. This is not, however, the whole story because few molecules show the simple behavior of CO_2. As shown by Xiao and Kellman [28, 29] the standard effective spectroscopic Hamiltonian [30, 31] allows four main distinct Fermi resonant coupling regimes; and a given molecule may pass from one regime to another, as the polyad number increases. Consequently, Cooper and Child [14] find that the quantum monodromy within a given polyad takes different forms in the different regimes.

The next account starts by summarizing the three-dimensional monodromy of the resonant swing spring. It is then shown how the geometrical arguments of Section IV may be applied to examining the nature of the two-dimensional monodromy within a given polyad.

A. Three-Dimensional Quantum Monodromy of the Resonant Swing Spring

The resonant swing spring consists of a pendulum bob attached to a vertical spring, with the mass, length, and spring constant chosen so that the frequency of the vertical motion is exactly twice that of the degenerate small amplitude swing. The essence of the resulting classical [11] and quantum [12, 13] monodromy is reproduced by a simple model in which the stretching oscillator is coupled to the swing by a cubic coupling term that conserves the polyad number, N, and the vibrational angular momentum, L. The resulting classical motions are constrained to lie inside the horn-shaped surface in Fig. 21, in which the central thread is a singular line, in the plane $L = 0$, which replaces the isolated critical point of the previous two-dimensional models. The three-dimensional nature of the resulting monodromy may be illustrated by noting that the two-dimensional monodromy in different slices through Fig. 21 depends on how the slice is taken. Since the classical theory [11] is developed in terms of actions $N_1 = (N + L)/2$ and $N_2 = (N - L)/2$, it is convenient to define a sequence of two-dimensional sections such that

$$K_m = (m + 1)N + (m - 1)L = 2[mN_1 + N_2] = \text{const} \tag{72}$$

The examples in Fig. 22, for $m = 0$ and 2, which are taken from Ref. [13], show that the cyclic transport leads to vertical shears of the unit cell by one and three

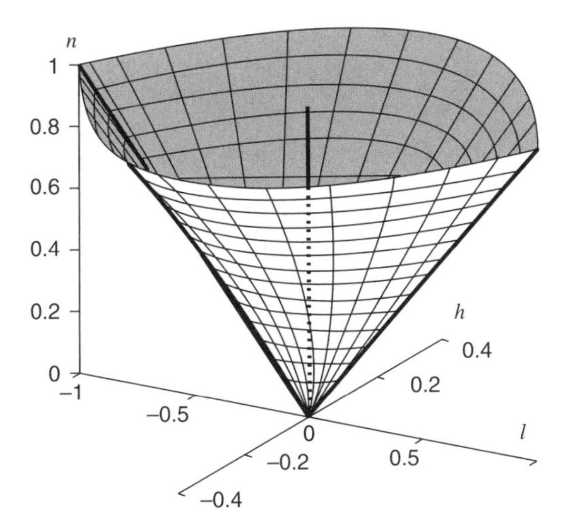

Figure 21. Boundaries of the energy–momentum map for the resonant 1:1:2 oscillator, with a central singular thread. Taken from Ref. [13] with permission of the American Institute of Physics, Copyright 2004.

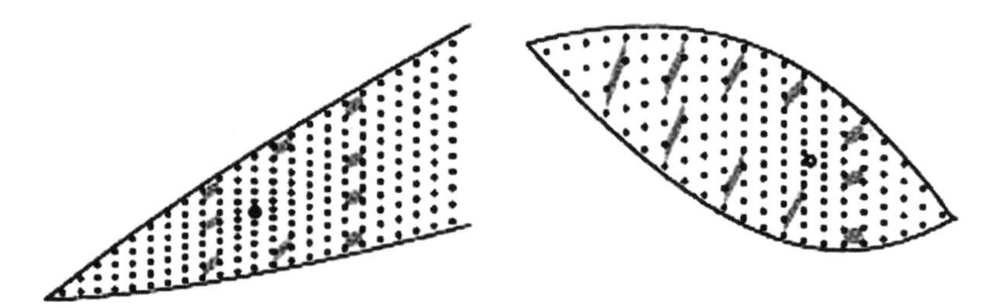

Figure 22. Quantum eigenvalues (dots) plotted against angular momentum, for two different slices through Fig. 21. *Left*: $n - l = 22$; *right*: $3n + l = 80$. The circle in each panel is the intersection with the singular thread, which lies at $L = 0$. Taken from Ref. [13] with permission of the American Institute of Physics, Copyright 2004.

units, respectively. In other words, one sees two-dimensional monodromy indices of one and three in the basis in which distances are measured in units equal to the sides of the initial unit cell. The general result is that two-dimensional monodromy index for the section $K_m = \text{const}$ is equal to $m + 1$.

Two derivations of the full three-dimensional monodromy matrix are available. Düllin et al. [11] generalize the approach outlined in Section IIC by following changes in the twist angles associated with actions N_1 and N_2 around three independent cycles in the *ELN* space, to obtain a classical monodromy matrix that is actually the inverse transform of the form that is used to interpret moving quantum cell construction (see Section VI of Ref. [13]). The alternative method [12, 13] is to employ a sequence of wedge cuts of the type shown in Fig. 5 (except that the cut space is now three dimensional), each such wedge being represented by a 3×3 matrix. A variety of similarity related forms is given, of which

$$M_\delta = \begin{pmatrix} 1 & 0 & 0 \\ 1 & 1 & -1 \\ 0 & 0 & 1 \end{pmatrix} \tag{73}$$

is defined in the basis in which the unit cell has sides $(\Delta n_1, \Delta v, \Delta n_2) = (1, 1, 1)$. The corresponding form for a representation with unit cell sides $(\Delta k_m, \Delta v, \Delta l) = (1, 1, 1)$, where

$$\begin{pmatrix} \Delta k_m/2 \\ \Delta l \end{pmatrix} = \begin{pmatrix} m+1 & -1 \\ 1 & -1 \end{pmatrix} \begin{pmatrix} \Delta n_1 \\ \Delta n_2 \end{pmatrix} \tag{74}$$

is found to be

$$M = \begin{pmatrix} 1 & 0 & 0 \\ 0 & 1 & 1 \\ 0 & 0 & 1 \end{pmatrix} \tag{75}$$

Equation (75) may be used to understand the computed results in Fig. 22, because a typical vector $\mathbf{x} = (x_v, x_l)^T$ in a two-dimensional section with $\Delta k_m = 0$ evolves around the cycle to

$$\begin{pmatrix} x_v' \\ x_l' \end{pmatrix} = \begin{pmatrix} 1 & 1 \\ 0 & 1 \end{pmatrix} \begin{pmatrix} x_v \\ x_l \end{pmatrix} \tag{76}$$

Moreover, a constant k_m section can only contain integer n and l values separated by $(\Delta l, \Delta n) = (m+1, -m+1)$, which means that the horizontal and vertical lattice vectors in Fig. 22 are given by $\mathbf{x} = (0, m+1)^T$ and $(1, 0)^T$, respectively. Hence an alternative basis may be defined with a unit cell having sides $(\Delta v, \Delta l) = (1, m+1)$, in which case the argument in Section IIC3 shows that the two-dimensional monodromy matrix for the fixed k_m section transforms to

$$M^{(2)'} = \begin{pmatrix} 1 & 0 \\ 0 & \frac{1}{m+1} \end{pmatrix} \begin{pmatrix} 1 & 1 \\ 0 & 1 \end{pmatrix} \begin{pmatrix} 1 & 0 \\ 0 & m+1 \end{pmatrix} = \begin{pmatrix} 1 & m+1 \\ 0 & 1 \end{pmatrix} \tag{77}$$

in agreement with the computed results in Fig. 22. Düllin et al. (see Section VI of Ref. [11]) reach the same conclusion by considering possible cycles over 2-tori of the reduced K_m space, rather than the quantum number (or action) differences that define the size and shape of the unit cell.

B. Two-Dimensional Quantum Monodromy in Fermi Resonance Polyads

1. *Effective Hamiltonian and Critical Points*

The effective spectroscopic classical Hamiltonian for Fermi resonance between a nondegenerate mode ν_1 and a doubly degenerate mode ν_2 takes the form [30, 31]

$$\begin{aligned} H = {} & I_1\omega_1 + I_2\omega_2 + x_{11}I_1^2 + x_{22}I_2^2 + x_{12}I_1I_2 \\ & + gL^2 + k_{122}[a_1^\dagger(a_d a_g) + a_1(a_d a_g)^\dagger] \end{aligned} \tag{78}$$

where the ω_i are the equilibrium frequencies (in the approximate ratio $\omega_1 \simeq 2\omega_2$), x_{ij} are anharmonicity parameters, g is the spectroscopic parameter that lifts the l degeneracy at given v_2, k_{122} is the intermode coupling strength and $(\hat{a}_d^\dagger, \hat{a}_d)$ and $(\hat{a}_g^\dagger, \hat{a}_g)$ are the right-hand and left-hand creation–annihilation operators associated with the degenerate mode [46]. The classical actions are related to the quantum numbers by $I_1 = v_1 + \frac{1}{2} = a_1 a_1^\dagger + \frac{1}{2}$ and $I_2 = v_2 + 1 = a_d a_d^\dagger + a_g a_g^\dagger + 1$, from which it follows that the constants of motion may be taken as the classical polyad number $N = 2v_1 + v_2 + 2$, and the angular momentum $L = a_d^\dagger a_d - a_g^\dagger a_g$.

To see the geometry of the situation, one may define

$$\begin{aligned}\zeta_1/\sqrt{2} &= a_1^\dagger(a_d a_g) + a_1(a_d a_g)^\dagger = \sqrt{I_1(I_2^2 - L^2)}\cos(\theta_1 - 2\theta_2) \\ \zeta_2/\sqrt{2} &= -i[a_1^\dagger(a_d a_g) + a_1(a_d a_g)^\dagger] = \sqrt{I_1(I_2^2 - L^2)}\sin(\theta_1 - 2\theta_2)\end{aligned} \tag{79}$$

both of which commute with N and L, where the angle–action expressions are obtained by combining Eqs. (A.2)–(A.5) with the single oscillator forms $a_1^\dagger = (a_1)^* = \sqrt{I_1}e^{i\theta_1}$ [23]. It is also convenient to employ the canonical transformation

$$\begin{aligned}J &= \tfrac{1}{2}(2I_1 + I_2), \quad \theta_J = \tfrac{1}{2}(\theta_1 + 2\theta_2) \\ K &= \tfrac{1}{2}(2I_1 - I_2), \quad \theta_K = \tfrac{1}{2}(\theta_1 - 2\theta_2)\end{aligned} \tag{80}$$

where the numerical factors have been chosen to make easy comparison with the discussion in Section IV. The actions (J, K) are defined to be twice the values of (I, I_z) employed by earlier authors [14, 28, 29], and the corresponding angles differ by a factor of $\frac{1}{2}$. It follows from Eqs. (79) and (80) that the variables (ζ_1, ζ_2, J, K) are related by

$$\zeta_1^2 + \zeta_2^2 = (J + K)[(J - K)^2 - L^2] \tag{81}$$

with $0 \leq |L| \leq 2J$ and $-J \leq K \leq J - L$.

Figure 23 shows sections of ζ_1, with J fixed, $\zeta_2 = 0$ and various values of L/N, where $N = 2J$ is the total polyad action. There is an obvious conical point, X, at $(K, L) = (J, 0)$. In addition, the concentric contours degenerate to points Y and Z at $(K, L) = (-J, \pm N)$. It is important for what follows to explore the character of the point X. As a preliminary, the Hamiltonian may be reduced from the above seven-parameter form, to one involving two essential parameters, both of which vary with the total action. The appropriate definitions [14, 28, 29], modified to conform with the present (J, K) notation, are

$$\begin{aligned}\beta &= \frac{\omega_1 - 2\omega_2 + (x_{11} - 4x_{22})J}{\gamma J} \\ \mu &= \frac{\sqrt{2}k_{122}}{\sqrt{J}\gamma} \\ \gamma &= x_{11} + 4x_{22} - 2x_{12}\end{aligned} \tag{82}$$

in terms of which

$$\begin{aligned}H &= H^{(0)} + H^{(1)} \\ H^{(0)} &= \tfrac{1}{2}(\omega_1 + 2\omega_2)J + \tfrac{1}{4}(x_{11} + 4x_{22} + 2x_{12})J^2 + gL^2\end{aligned} \tag{83}$$

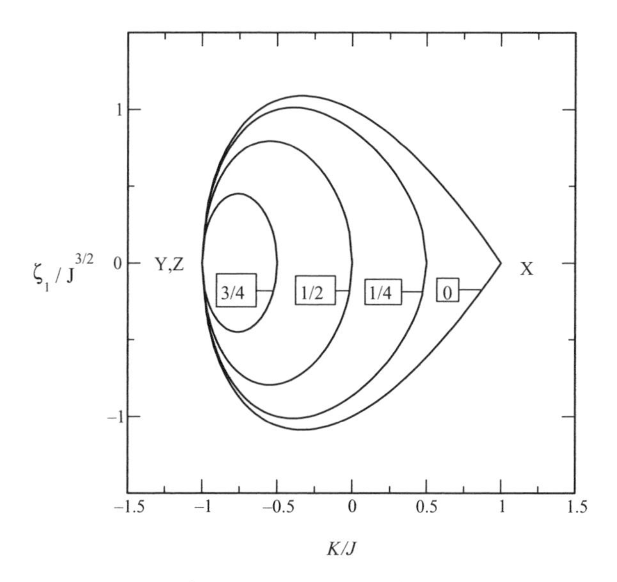

Figure 23. Sections of $\zeta_1/J^{3/2}$ at $L/N = 0, \frac{1}{4}, \frac{1}{2}$, and $\frac{3}{4}$, where $N = 2J$. X marks the conical point at $(K, L) = (J, 0)$, while Y and Z indicate critical points at $(K, L) = (-J, 0 \pm N)$.

and

$$H^{(1)} = \frac{\gamma}{2}\left\{\beta JK + \tfrac{1}{2}K^2 + 2\mu\sqrt{(J+K)[(J-K)^2 - L^2]}\cos 2\theta_K\right\} \tag{84}$$

Since the only angle dependence comes from θ_K, $H^{(0)}$ and the actions J, L are constant. From this point onwards we concentrate on motion under the reduced Hamiltonian $H^{(1)}$, which depends, apart from the scaling parameter γ, only on the values of scaled coupling parameter μ and the scaled detuning term β. In other words, we investigate the monodromy only in a fixed J (or polyad number $N = 2J$) section of the three-dimensional quantum number space.

Turning to the nature of the critical point, the present (J, K) notation was chosen to facilitate local reduction of $H^{(1)}$ to the form in Eq. (A.8), by the substitution $K \simeq J - I$. Thus, after neglect of a term in I^2,

$$\begin{aligned} H^{(1)} &\simeq \frac{\gamma}{2}\left\{\left(\beta + \frac{1}{2}\right)J^2 + AI + B\sqrt{I^2 - L^2}\cos 2\theta_I\right\} \\ A &= -(\beta+1)J, \quad B = 2\sqrt{2}\beta J \end{aligned} \tag{85}$$

It follows, by using the construction in the Appendix, that the critical point is an isolated focus–focus singularity in the two-dimensional fixed J section, if $A^2 < B^2$ or

$$-2\sqrt{2}|\mu| - 1 < \beta < 2\sqrt{2}|\mu| - 1 \tag{86}$$

In addition, one finds, by the methods outlined next, that the other critical points Y and Z always lie on relative equilibria.

2. *Catastrophe Map*

It was shown in the earlier sections that the existence or nonexistence of quantum monodromy in two-dimensional maps depends on the relative dispositions of the critical points and relative equilibria of the Hamiltonian, which involves a search for the stationary points of $H^{(1)}$ with respect to K and θ_K. For $L = 0$ there is a root at the critical point $K = J$, and other possible roots given by

$$(\beta + z)^2(1 + z) - \mu^2(3z + 1)^2 = 0 \tag{87}$$

where $z = K/J$. Such roots are physical if z lies in the real range $-1 \leq z \leq 1$, and the number and nature of these physical roots may be classified according to the divisions of the *catastrophe map* in Fig. 24 [28, 29]. The term comes from the fact that changes in the number of roots occur on lines in the (μ, β) plane where two or more roots coalesce—the algebra of such coalescence being the realm of catastrophe theory [55]. Xiao and Kellman show, in fact, that there is one real stable (elliptic) root of Eq. (87) for β values above and two real stable plus one real unstable (hyperbolic) root for those below the cusp-shaped line in Fig. 24. The associated transition is called a saddle node bifurcation,

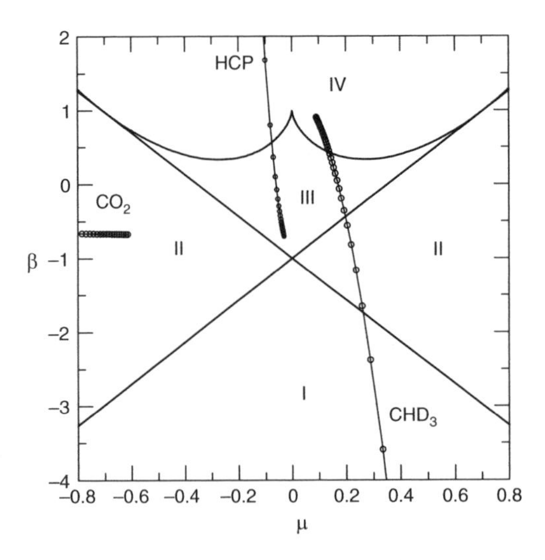

Figure 24. The Xiao–Kellman catastrophe map for $L = 0$. Regions I–IV differ according to the numbers and types of fixed point listed in Table II. The points on the curved trajectories mark the coordinates of successive polyads for the relevant molecule. Taken from Ref. [14] with permission of the PCCP Owner Societies.

examples of which are well known from classical and semiclassical studies of HCP on accurate potential energy surfaces [56, 57]. In addition, it is readily verified from Eq. (87) that one of the three real roots passes out of the physical range $|z| \leq 1$ as β decreases across the lines $\beta = -1 \pm 2\sqrt{2}\mu$. The resulting number and types of physical roots in the various regions are listed in Table II [28, 29].

TABLE II
Numbers and Types of Stationary Points in Different Regions of Fig. 24, for $L = 0$.

Region	Elliptic Roots	Hyperbolic Roots	Critical Point X
I	2	0	Elliptic–stable
II	2	0	Cusp–unstable
III	3	1	Elliptic–stable
IV	2	0	Elliptic–stable

The value of this map is that it allows one to predict the eigenvalue dispositions in different polyads for any given molecule [28, 29], because Eq. (82) shows that the (μ, β) values for successive polyads lie on a parabola. Illustrative examples, based on the spectroscopic parameters of CHD_3 [58], CO_2 [59], and HCP [60], are shown in Fig. 24, although one would typically require more than the seven parameters in Eq. (78) to have an accurate representation of the spectrum for the high polyads shown. Note, in particular, that CHD_3 moves up the map, with increasing polyad number, $N = 2J$, because the frequency of the $\omega_1 + 2I_1x_{11}$ tunes down toward $2(\omega_2 + 2I_2x_{22})$, while HCP moves down the map, because in this case the local bending frequency tunes down to half that of the CP stretch [67]. Finally, β is almost constant for CO_2, so that successive polyads always lie in the same region.

3. *Quantum Level Distributions and Quantum Monodromy in a Single Polyad*

Cooper and Child [14] have given an extensive description of the effects of nonzero angular momentum on the nature of the catastrophe map and the quantum eigenvalue distributions for polyads in its different regions. Here we note that the fixed points and relative equilibria, for nonzero $\lambda = L/2J$, are given by physical roots of the equation

$$f(z) = (\beta + z)^2(1 + z)[(1 - z)^2 - 4\lambda^2] - \mu^2[3z^2 - 2z - 1 - 4\lambda^2]^2 = 0 \quad (88)$$

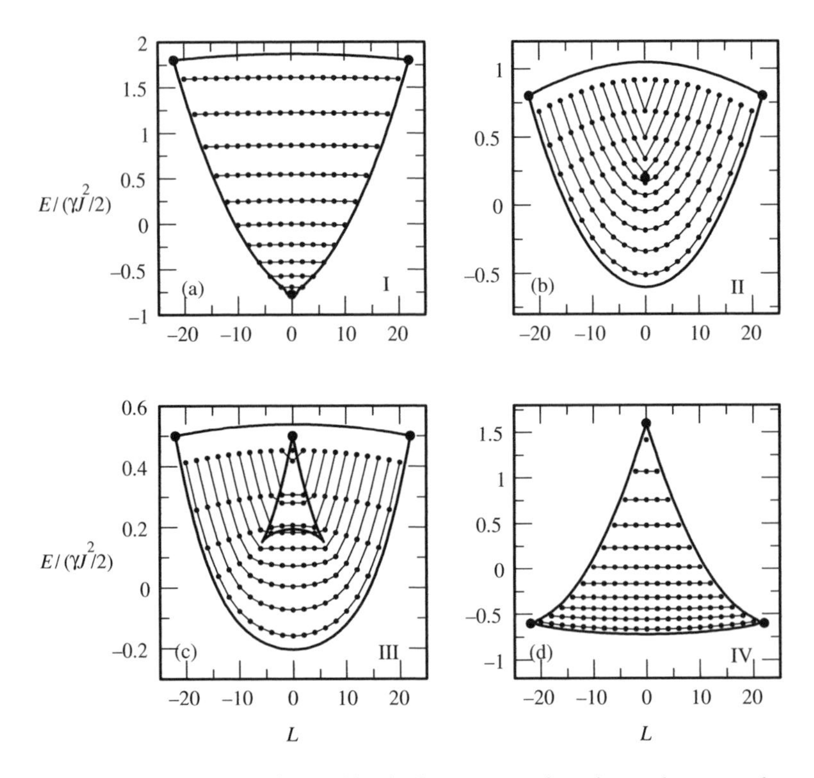

Figure 25. Scaled *EL* maps for $n = 20$ polyads at representative points on the catastrophe map. (a) ($\beta = 1.3, \mu = 0.1$), (b) (0.3, 0.3), (c) (0.0, 0.1), and (d) (1.1, 0.1). Small dots are the eigenvalues. Large dots mark the critical points, X at $L = 0$, and Y, Z at $L = \pm N = \pm(n + 2)$. Note the similarity between panel (b) and Fig. 1a and between panel (c) and Fig. 12.

which is readily verified to factor into $(1 - z)^2$ times Eq. (87) when $\lambda = 0$. This means that one can establish connections with the discussion in the previous sections by examining the patterns of relative equilibria, as well as the image of the critical point X in the (E, L) planes appropriate to different regions of the catastrophe map.

Figure 25 shows the scaled eigenvalues, $E/(\gamma J^2/2)$, of the quantum equivalent of $H^{(1)}$, in order to illustrate four characteristic types of behavior.[4] Panels (a) and (d) depict the simplest cases, for polyads in regions I and IV, in which the two modes are almost decoupled, due to a large local detuning, $|\beta|$, or small coupling, $|\mu|$. The two cases differ according to whether the point of the

[4]It should be noted that Fig. 25 is not exhaustive, because the catastrophe map for general L in Fig. 2 of Ref. [14] shows a narrow band that is bounded below by the junction between regions III and IV of the $L = 0$ catastrophe map in Fig. 24; and above by a line of cusps, parametric on L. The nature of the monodromy in this region remains to be investigated.

distribution, which corresponds to the $v_1 = n/2$, $v_2 = 0$ member of the polyad, lies lowest or highest in energy. The critical points X, Y, and Z all lie on relative equilibria, just as in panels (a), (b), and (d) of Fig. 14 for the spin–rotation coupling model of Section IVA. As discussed later, the behavior in region II (panel (b)) is qualitatively similar to that for a fixed polyad section of the three-dimensional resonant swing spring, but the form in region III cannot occur without introducing substantial anharmonicity.

The transition from Fig. 25a to Fig. 25b is analogous to that from Fig. 14b to Fig. 14c, in the sense that the critical point X has become isolated, and there is also a very close similarity with Fig. 1a, because the critical points in the two diagrams are both symmetrical about the zero angular momentum line. The only qualitative difference is that Fig. 25b is bounded by an upper relative equilibrium line, because any polyad contains only a finite number of eigenvalues.

There is also an obvious qualitative similarity between Fig. 25c and Figs. 11–13 appropriate to the LiCN/LiNC isomerization. Four distinct roots of $f(z) = 0$ at $L = 0$ give rise to the four roots in region III of Table II, the images of which appear in Fig. 25a as three relative equilibria at the top and bottom of the diagram and on the curved base of the triangle, plus the image of the critical point X, which is a repeated root. Two of the four relative equilibria come together as L^2 increases, to merge at the cusp points, which also correspond to repeated roots. The resulting relative equilibrium triangle encloses a "folded" region of the spectrum, containing two sets of quantum eigenvalues, an example of which is shown for $L = 0$ in Fig. 4c of Ref. [14]. The two inner verticals in this latter diagram come together as L^2 increases to coalesce at the L value at which the relevant diamond-shaped saddle node bifurcation line in Fig. 2 of Ref. [14] passes through the point (μ, β) of interest. Another way to look at the folded region is to note that the term in K^2 leads to a maximum or minimum in the first two terms of Eq. (84), within the physical range $-J < K < J$, for $-1 < \beta < 1$. Consequently, the zeroth order spectrum (without the term in μ) folds back on itself. The tridiagonal nature of the coupling matrix then causes the fold to be smoothed away at lower energies, but leaves a residual higher energy region, where two families of eigenvalues overlap. Finally, one should note that a change of parameters that moves the polyad from region III to region II causes the triangle to collapse to a point, just as a change in the ratio V_2/V_1 in Section IIIB was seen to have a similar effect on the triangle in Fig. 10.

To complete the picture, Figs. 26 and 27 give moving cell constructions for the polyads shown in Figs. 25b and 25c, respectively, which were incorrectly drawn in Figs. 7 and 8 of Cooper and Child [14]. It is evident by substituting $\mathbf{x}_i = (x_{li}, x_{vi}, 0)^T$ that the two unit shear in Fig. 26 is consistent with the three-dimensional monodromy matrix in Eq. (75), although this appears at first sight to violate the monodromy theorem [40] that the index of the two-dimensional

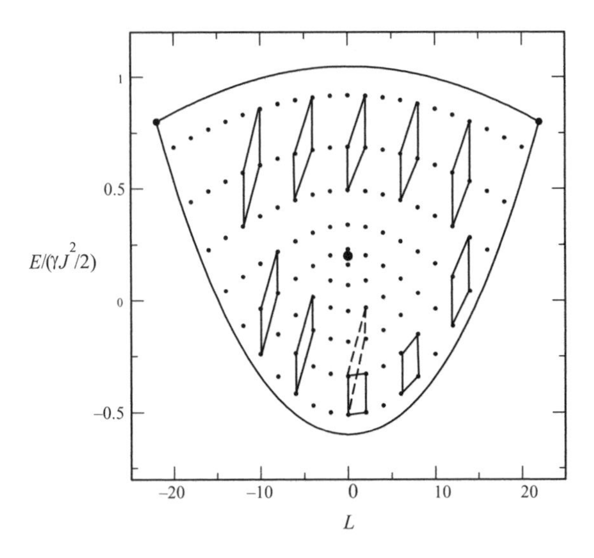

Figure 26. Quantum monodromy in region II of the catastrophe map in Fig. 24.

monodromy matrix is equal to the number of focus–focus singularities. As discussed in Section IIC4, the resolution of this paradox is that the monodromy theorem applies in the primitive basis, with unit cell sides $(\Delta v, \Delta l) = (1, 1)$, whereas the cells in Figs. 26 and 27 have sides $(\Delta v, \Delta l) = (1, 2)$ because l takes

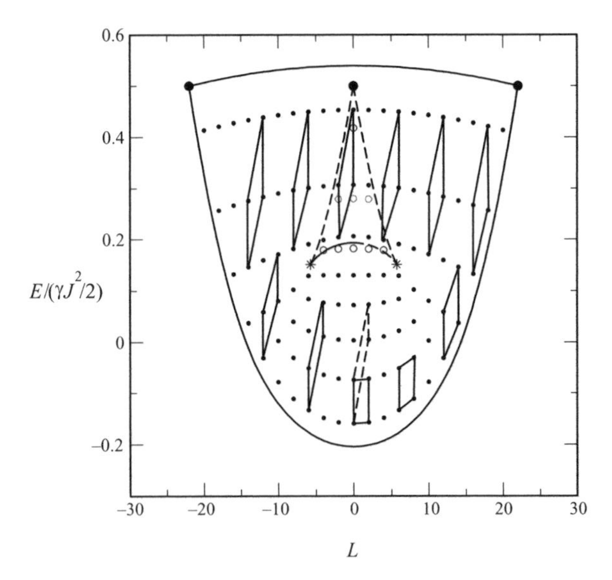

Figure 27. Quantum monodromy in region III of the catastrophe map in Fig. 24.

only even values. Thus the vector $\mathbf{x}_i$ transforms to $\mathbf{y}_i = (x_{vi}, 2x_{li})^T$ in the primitive two-dimensional basis, so that

$$\begin{pmatrix} y_{vf} \\ y_{lf} \end{pmatrix} = \begin{pmatrix} 1 & 1 \\ 0 & 1 \end{pmatrix} \begin{pmatrix} x_{vi} \\ 2x_{li} \end{pmatrix} = \begin{pmatrix} x_{vi} + 2x_{li} \\ 2x_{li} \end{pmatrix} \tag{89}$$

again with a two unit shear in the x representation.

C. Summary

1. Analysis of the prototypical resonant swing spring model [11–13] shows that Fermi resonance with conserved angular momentum is an intrinsically three-dimensional phenomenon. The form of the 3×3 monodromy matrix was given.
2. The standard effective spectroscopic Fermi resonant Hamiltonian allows more complicated types of behavior. The full three-dimensional aspects of the monodromy remain to be worked out, but it was shown, with the help of the Xiao–Kellman [28, 29] catastrophe map, that four main dynamical regimes apply, and that successive polyads of a given molecule may pass from one regime to another.
3. The nature of the monodromy within a given polyad was discussed. Two of the regimes allow no such monodromy; one conforms to the type displayed by the swing spring, which has qualitative similarities with that in quasilinear molecules. Finally, the monodromy in the fourth regime is of the folded type, with close qualitative similarities to that observed for LiCN isomerization in Section IIIB.

VI. CONCLUSIONS AND DISCUSSION

The first aim of this chapter was to describe the spectroscopically observable consequences of quantum monodromy, with particular relevance to the spectrum in the vicinity of a saddle point on the potential surface. Two broad patterns of behavior were observed. The simplest occurs in the presence of a cylindrically symmetric locally quadratic potential energy barrier (or more generally of an isolated focus–focus singularity). The quantum eigenvalue lattice, in energy and angular momentum space, then always takes on a characteristic distortion, such that smooth lines joining points of the second quantum number, on one side of the critical point, give rise to "vee" shaped kinks on the other. The best studied case occurs in the computed [1, 33] and experimental [35] bending progression of H_2O; and an interesting recent refinement is the observation of quantum monodromy in the effective *b*, *c*-axis rotational constant of the quasilinear molecule NCNCS [36]. Certain systems of a spherical pendulum type, such as the bending-rotational states of HCP [7] and the motions of dipolar molecules in strong electric fields [8], are also characterized by a locally quadratic cylindrically symmetric potential maximum, which gives rise to the same type of characteristic distortion in the quantum level structure around the barrier maximum. It was also demonstrated, in the case of HCP [7], that there is an associated large increase in the spectroscopic vibration–rotation coupling parameters, as the energy increases.

Angular momentum coupling models [2, 3] and the behavior of the hydrogen atom in crossed electric and magnetic fields [4–6] were also shown to give rise to quantum monodromy, associated with isolated critical points, for certain ranges of the Hamiltonian parameters. The latter case is particularly interesting, because equality between the magnitudes of the coupled angular momenta leads to a pair of isolated critical points, which overlap in the energy–angular momentum map (corresponding to two pinch points on the pinched torus). Hence the monodromy matrix has index 2.

The second type of quantum monodromy occurs in the computed bending-vibrational bands of LiCN/LiNC, in which the role of the isolated critical point is replaced by that of a finite "folded" region of the spectrum, where the vibrational states of the secondary isomer LiNC interpenetrate those of LiCN [9, 10]. The folded region is finite in this case, because the secondary minimum on the potential surface merges with the transition state as the angular momentum increases. However, the shape of the potential energy surface in HCN prevents any such angular momentum cut-off, so monodromy is forbidden [10].

The case of Fermi resonance with conserved angular momentum (or 1:1:2 resonance) is particularly interesting, because analysis of the *resonant swing spring* [11, 13], with close similarities to CO_2 [12], shows that the monodromy is only fully defined in the three dimensions specified by the angular momentum, polyad number, and a third vibrational quantum number that distinguishes the individual states. The full theory has been worked out for this special case. The general situation, however is more complicated because the standard Fermi resonance spectroscopic Hamiltonian [30, 31] allows four main types of coupling regime, only one of which conforms to the CO_2 case. Moreover, successive polyads for a given molecule may pass from one regime to another. Much work therefore remains to be done to establish the full three-dimensional picture. The discussion in Section VB indicates two types of two-dimensional monodromy in a given polyad, one with qualitative similarities to the monodromy in quasilinear molecules and simple spherical pendulums and the other with similarities to LiCN/LiNC isomerization. A promising case for the observation of this behavior is the molecule HCP, in which the occurrence of saddle node bifurcations [56, 57] is already known to give rise to an interpenetrated spectrum at zero angular momentum [61, 62].

Attention was also given to the influence of the classical mechanics on the quantum level structure. In quasilinear molecule and spherical pendulum-like models, the presence and properties of the required isolated critical point, P, are readily apparent. It was therefore relatively easy to demonstrate the connection between the quantum lattice distortion and an abrupt change in the limiting behavior of the torus twist angle as $L \to 0^{\pm}$, at energies above and below the critical point [1]. A more subtle argument, based on the connectivity of classical invariant tori on loops around the critical point in 2D then led to the derivation of a 2×2 monodromy matrix , which characterized the multivalued properties of the quantum numbers in the vicinity of P. The extension of these ideas to angular momentum coupling cases [2, 3], H atoms in crossed fields [4, 5], and Fermi resonance within a polyad [14] was greatly facilitated by first establishing the geometry of the relevant classical phase space—in particular, the nature and locations of its critical points, which might or might not be isolated according to values of certain parameters in the Hamiltonian. The idea that the spin–rotation eigenvalue patterns in Fig. 15 can be interpreted as geometrical projections of the quantum states supported by the tetrahedron in 16 is particularly appealing. It was also interesting to identify the special nature of the quantum monodromy in the perturbed H atom case, for which two coupled angular momenta have equal magnitudes.

Turning to the future, two lines of investigation are suggested. One concerns the water molecule, for which the observation of "damp spots on the sun" [63] and a general desire to understand the heating effect of water vapor in the atmosphere has already led to the observation and assignment of over 100 vibrational bands [64], including a recent experimental demonstration of monodromy [35]. There have also been substantial calculations of the vibrational–rotational energy levels, up to the dissociation limit [33, 65]. The challenge is to see whether the recent observation of monodromy in the *b*,*c*-axis rotational constant [36] can be used to aid the interpretation of this uniquely interesting complex spectrum.

The case of water is particularly convenient because the required high K_a states may be detected in the solar absorption spectrum. However, it is difficult to observe the necessary high vibrational angular momentum states in molecules, which can only be probed by dispersed fluorescence or stimulated emission techniques. On the other hand, it is now possible to perform converged variational calculations on accurate potential energy surfaces, from which one could hope to verify the quantum monodromy and assess the extent to which it is disturbed by perturbations with other modes. Examples of such computed monodromy are seen for H_2O in Fig. 2 and LiCN in Fig. 12.

Another interesting molecule is HCP, which is predicted to show two types of quantum monodromy—one arising from the spherical pendulum nature of its bending/rotational motions and the other from the predicted "folded" character of its higher Fermi resonance polyads. The recent review [61] of the available experimental and computational data indicates a wealth of experimental information up to $25,000\,\mathrm{cm}^{-1}$, which is close to the energy of the HPC conformation, but only for relatively low vibrational angular momenta. In addition, accurate variational calculations on a good ab initio potential surface [66] have been shown to reproduce the experimentally observed vibrational origins and rotational constants with reasonable accuracy. The effective spectroscopic Hamiltonian of Ishikawa et al. [60], which includes Fermi resonance terms, is even more accurate. The question as to whether the two types of monodromy might interfere has been addressed to some extent by model studies of Jacobson and Child [7, 67], though not in this explicit form. It was shown, for example, that the Fermi resonance polyad structure is expected to persist almost up to the energy of the potential maximum, which is close to the limit of the experimental data. It is also well established that the first saddle node bifurcation, indicative of a transition from region II to region III of the catastrophe map, occurs at $13,500\,\mathrm{cm}^{-1}$ [56, 57] and an interleaved energy level structure is observed for states with zero vibrational angular momentum for polyads above $n = 22$ [61], in the present notation. It therefore seems very likely that the pattern in Fig. 25c would be observed in these higher polyads, if the calculations were extended to higher vibrational angular momenta. Finally, it is known that the lowest members of each vibrational polyad have almost pure bending character [61], so that there is little doubt that the pattern in Fig. 9 will apply to such states. There is also indirect supporting evidence, in the fact that large observed changes in the spectroscopic vibration–rotation parameters [68, 69] are well reproduced by a modified spherical pendulum-like potential, deduced from these supposedly pure bending states.[7].

Extensive ab initio [70] and variational [71] calculations have also been performed for HOCl, which shows strong Fermi resonance between the OCl stretching and HOCl bending motions [72], such that the bending frequency tunes down to half that of the stretch as the polyad number increases. Thus the polyad trajectory starts in region IV of Fig. 24 and passes to region III, just like HCP. However, there are important differences because the equilibrium configuration of HOCl is bent. The bending motion is therefore nondegenerate, with no trivially conserved angular momentum. In other words, the Fermi resonance is of 1:2 rather than 1:1:2 type, and no monodromy of the type in Fig. 25 is expected. There is also the possibility of observing quantum monodromy around the barrier to linearity, and it would be interesting to examine the influence of the much stronger Fermi resonance, than that in H_2O.

APPENDIX: ANGLE–ACTION TRANSFORMATIONS IN TWO CARTESIAN DIMENSIONS

The following angle–action representations [14] are available for the classical equivalents of the right-hand and left-hand creation–annihilation operators [46]:

$$a_d = \tfrac{1}{2}(r + ip_r + p_\phi/r)e^{-i\phi} = \sqrt{(I+L)/2}e^{-i\theta_I - i\theta_L} \quad \text{(A.1)}$$

$$a_g = \tfrac{1}{2}(r + ip_r - p_\phi/r)e^{-i\phi} = \sqrt{(I-L)/2}e^{-i\theta_I + i\theta_L} \quad \text{(A.2)}$$

$$a_d^\dagger = \tfrac{1}{2}(r - ip_r + p_\phi/r)e^{i\phi} = \sqrt{(I+L)/2}e^{+i\theta_I + i\theta_L} \quad \text{(A.3)}$$

$$a_g^\dagger = \tfrac{1}{2}(r - ip_r - p_\phi/r)e^{i\phi} = \sqrt{(I-L)/2}e^{+i\theta_I - i\theta_L} \quad \text{(A.4)}$$

It follows that

$$I = a_d^\dagger a_d + a_g^\dagger a_g = \tfrac{1}{2}(r^2 + p_r^2 + p_{\phi^2}/r^2) = \tfrac{1}{2}(x^2 + y^2 + p_x^2 + p_y^2) \quad \text{(A.5)}$$

$$\sqrt{I^2 - L^2}\cos 2\theta_I = a_d^\dagger a_g + a_g^\dagger a_d = \tfrac{1}{2}(r^2 - p_r^2 - p_{\phi^2}/r^2)$$
$$= \tfrac{1}{2}(x^2 + y^2 - p_x^2 - p_y^2) \quad \text{(A.6)}$$

These identities may be used to obtain an angle–action equivalent of the quadratic form

$$h = \tfrac{1}{2}[(A+B)(x^2+y^2) + (A-B)(p_x^2+p_y^2)] \quad \text{(A.7)}$$

which is either of the focus–focus type or can be transformed into it by a canonical transformation, $(x, y, p_x, p_y) \to (p_x, p_y, -x, -y)$, that interchanges coordinates and momenta, provided that the coefficients $(A+B)$ and $(A-B)$ have opposite signs.[5] Thus the local angle–action form

$$h \simeq AI + B\sqrt{I^2 - L^2}\cos 2\theta_I \quad \text{(A.8)}$$

implies a singularity of focus–focus type if $A^2 < B^2$—an inequality that is employed in Sections IV and V to identify the conditions for isolation of various critical points.

Acknowledgments

It is a pleasure to acknowledge stimulating discussions with Richard Cushman and Boris Zhilinskii, who introduced the author to this interesting field. He is particularly grateful to Zhilinskii for constructive comments on an initial draft, which led to substantial improvements to the chapter.

Readers may obtain a complementary view of quantum monodromy from Ref. [73], which is addressed to the chemical physics community.

[5]Note that the generic form employed by Zung [40] (see footnote to Eq. (7)) avoids the awkward ambiguity between coordinates and momenta, but at the cost of losing a clear separation of the Hamiltonian into kinetic and potential terms.

References

1. M. S. Child, T. Weston, and J. Tennyson, *Mol. Phys.* **96**, 371 (1999).
2. D. A. Sadovskii and B. Zhilinskii, *Phys. Lett. A* **256**, 235 (1999).
3. L. Grondin, D. A. Sadovskii, and B. Zhilinskii, *Phys. Rev. A* **65**, 012105 (2001).
4. R. H. Cushman and D. A. Sadovskii, *Europhys. Lett.* **47**, 1 (1999).
5. R. H. Cushman and D. A. Sadovskii, *Physica D* **142**, 166 (2000).
6. K. Efstathiou, R. H. Cushman, and D. A. Sadovskii, *Physica D* **194**, 250 (2004).
7. M. P. Jacobson and M. S. Child, *J. Chem. Phys.* **114**, 262 (2001).
8. I. N. Kozin and R.M. Roberts, *J. Chem. Phys.* **118**, 10523 (2003).
9. M. Joyeux, D. A. Sadovskii, and J. Tennyson, *Chem. Phys. Lett.* **382**, 439 (2003).
10. K. Efstathiou, M. Joyeux, and D. A. Sadovskii, *Phys. Rev. A* **69**, 032504 (2004).
11. H. R. Dullin, A. Giocobbe, and R. H. Cushman, *Physica D* **190**, 15 (2004).
12. R. H. Cushman, H. R. Dullin, A. Giacobbe, D. D. Holm, M. Joyeux, P. Lynch, D. A Sadovskii, and B. Zhilinskii, *Phys. Rev. Lett.* **93**, 024302 (2004).
13. A. Giacobbe, R. H. Cushman, D. A. Sadowskii, and B. I. Zhilinskii, *J. Math. Phys.* **45**, 5076 (2004).
14. C. D. Cooper and M. S. Child, *Phys. Chem. Chem. Phys.* **7**, 2731 (2005).
15. H. Waalkens, A. Junge, and H. R. Dullin, *J. Phys. A* **36**, L307 (2003).
16. H. Waalkens, H. R. Dullin, and P. H. Richter, *Physica D* **196**, 265 (2004).
17. J. W. C. Johns, *Can. J. Phys.* **45**, 2639 (1967).
18. R. N. Dixon, *Trans. Faraday Soc.* **60**, 1363 (1964).
19. J. J. Duistermaat, *Commun. Pure Appl. Math.* **33**, 687 (1980).
20. R. H. Cushman, *Cent. Wisk. Inform. Newsletter* **1**, 4 (1983).
21. R. H. Cushman and J. J. Duistermaat, *Bull. Am. Math. Soc. New Ser.* **19**, 475 (1988).
22. L. R. Bates, *J Appl. Math. Phys.(ZAMP)* **42**, 837 (1991).
23. M. S. Child, *Semiclassical Mechanics with Molecular Applications*, Oxford University Press, Oxford, 1991.
24. K. Efstathiou, *Metamorphoses of Hamiltonian Systems with Symmetry, Lecture Notes in Mathematics* **1864**, Springer-Verlag, New York, 2004.
25. R. H. Cushman and L. Bates, *Global Aspects of Classical Integrable Systems*, Birhauser, Basel, 1997.
26. A. J. Lichtenberg and M. A. Lieberman, *Regular and Stochastic Motion*, Springer-Verlag, New York, 1983.
27. G. Herzberg, *Spectra of Diatomic Molecules*, 2nd ed., Van Nostrand, New York, 1950.
28. L. Xiao and M. E. Kellman, *J. Chem. Phys.* **93**, 5805 (1990).
29. L. Xiao and M. E. Kellman, *J. Chem. Phys.* **93**, 5821 (1990).
30. M. E. Kellman and E. D. Lynch, *J. Chem. Phys.* **85**, 7612 (1986).
31. M. E. Kellman and E. D. Lynch, *J. Chem. Phys.* **88**, 2205 (1986).
32. M. S. Child, *J. Phys. A.* **31**, 657 (1998).
33. H. Partridge and D. W. Schwenke, *J. Chem. Phys.* **106**, 4618 (1996).
34. G. Herzberg, *Infra-red and Raman Spectra*, Van Nostrand, New York, 1945.

35. N. F. Zobov, S. G. Shirin, O. L. Polansky, J Tennyson, P.-F. Coheur, P. F. Bernath, M. Carter, and R. Colin, *Chem. Phys. Lett.* (in press).
36. B. P. Winnewisser, M. Winnewisser, I. R. Medvedev, M. Benke, F. C. DeLucia, S. C. Ross, and J. Koput (in preparation).
37. H. Goldstein, *Classical Mechanics*, 2nd ed., Addison-Wesley, Boston 1980.
38. I. S. Gradshteyn and I. M. Ryzhik, *Tables of Integrals, Series, and Products*, Academic Press, New York, 1980.
39. B. I. Zhilinskii, *Acta Appl. Math.* **87**, 281 (2005).
40. N. T. Zung, *Differ. Geom. Appl.* **7**, 123 (1997).
41. G. Herzberg and L. H. Longuet-Higgins, *Discuss. Faraday Soc.* **35**, 77 (1963).
42. M. V. Berry, *Proc. R. Soc. London Ser. A* **392**, 45 (1979).
43. V. P. Maslov, *Theorie des perturbations et methodes asymptotiques*, Dunod, Gautiers Villars, Paris, 1972.
44. M. Abramowitz and I. A. Stegun, *Handbook of Mathematical Functions*, Dover, London, 1965.
45. San Vũ Ngoc, *Comm. Math. Phys.* **203**, 465 (1999).
46. C. Cohen-Tannoudji, B. Diu, and F. Laloe, *Quantum Mechanics*, Vol. I, Wiley-Interscience, Hoboken, NJ, 1977.
47. R. Essers, J. Tennyson, and P. E. S. Wormers, *Chem. Phys. Lett.* **89**, 223 (1982).
48. T. van Mourik, G. J. Harris, O. L. Polansky, J. Tennyson, A. G. Császár, and P. J. Knowles, *J. Chem. Phys.* **115**, 3706 (2001).
49. G. J. Harris, O. L. Polansky, and J. Tennyson, *Spectrochimica Acta Part A* **58**, 673 (2002).
50. R. N. Zare, *Angular Momentum*, Wiley, Hoboken, NJ, 1988.
51. P. Kustaanheimo and E. Steifel, *J. Reine Angew. Math.* **218**, 204 (1965).
52. M. J. Gourlay, T. Uzer, and D. Farrelly, *Phys. Rev. A* **47**, 3113 (1992).
53. J. von Milczewski, G. H. F. Dierksen, and T. Uzer, *Phys. Rev. Lett.* **73**, 2428 (1994).
54. J. von Milczewski, G. H. F. Dierksen, and T. Uzer, *Int. J. Bifurc. Chaos* **4**, 905 (1994).
55. T. Poston and I. Stewart, *Catastrophe Theory and Its Applications*, Pitman, 1978.
56. M. Joyeux, D. Sugny, V. Tyng, M. Kellman, H. Ishikawa, R. W. Field, C. Beck, and R. Schinke, *J. Chem. Phys.* **112**, 4162 (2000).
57. M. Joyeux, S. C. Farantos, and R. Schinke, *J. Phys. Chem. A* **106**, 5407 (2002).
58. M. Lewerenz and M. Quack, *J. Chem. Phys.* **88**, 5408 (1988).
59. Z. Chila and A. Chedin, *J. Mol. Spectrosc.* **40**, 337 (1971).
60. H. Ishikawa, Y.-T. Chen, Y. Oshima, et al., *J. Chem. Phys.* **105**, 7383 (1996).
61. H. Ishikawa, R. W. Field, S. C. Farantos, M. Joyeux, J. Koput, C. Beck, and R. Schinke, *Annl. Rev. Phys. Chem.* **50**, 443 (1999).
62. M. Joyeux, D. Sugny, V. Tang, and M. E. Kellman, *J. Chem. Phys.* **112**, 4162 (2000).
63. J. Tennyson and O. L. Polyansky, *Contemp. Phys.* **39**, 283–294 (1998).
64. J. Tennyson, N. F. Zobov, R. Williamson, O. L. Polyansky, and P. F. Bernath, *J. Phys. Chem. Ref. Data* **30**, 735–831 (2001).
65. S. V. Shirin, N. F. Zobov, O. L. Polyansky, A. G. Csaszar, P. Barletta, and J. Tennyson (in preparation).
66. C. Beck, R. Schinke, and J. Koput, *J. Chem. Phys.* **112**, 8446 (2000).
67. M. P. Jacobson and M. S. Child, *J. Chem. Phys.* **9114**, 250 (2001).

68. A. Cabana, Y. Doucet, J.-M. Garneau et al., *J. Mol. Spectrosc.* **96**, 7383 (1996).

69. H. Ishikawa, C. Nagao, N. Mikami, and R. W. Field, *J. Chem. Phys.* **106**, 2980 (1998).

70. S. Skokov, K. A. Peterson, and J. M. Bowman, *J. Chem. Phys.* **109**, 2662 (1998).

71. S. Skokov, Q. Jianxin, J. M. Bowman, C.-Y. Yang, S. K. Gray, and V. Man-delstam, *J. Chem. Phys.* **109**, 10273 (1998).

72. R. Jost, M. Joyeux, S. Skokov, and J. M. Bowman, *J. Chem. Phys.* **111** 6807 (1999).

73. D. A. Sadovskii and B. Zhilinskii, *Mol. Phys.*, **104**, 2595 (2006).

THE MICROSCOPIC QUANTUM THEORY OF LOW TEMPERATURE AMORPHOUS SOLIDS

VASSILIY LUBCHENKO

Department of Chemistry, University of Houston, Houston, Texas 77204-5003, USA

PETER G. WOLYNES

Department of Chemistry and Biochemistry and Department of Physics, University of California at San Diego, La Jolla, California 92093-0371, USA

CONTENTS

Advances in Chemical Physics, Volume 136, edited by Stuart A. Rice

I. INTRODUCTION

During the past several decades, it has gradually been recognized in the condensed matter and materials science community that amorphous materials, while sharing many characteristics with the more common crystalline solids, represent a distinct solid state of matter. On the one hand, glasses exhibit rigidity and elastic response on humanly relevant time scales, thus qualifying them as solids for many practical purposes. In fact, until the relatively recent advent of systematic studies of the materials' response to mechanical and electromagnetic perturbation, as well as of their detailed microscopic structure, the only commonly known distinct attributes of amorphous substances had been their optical properties and the low magnitude and isotropic character of their thermal expansion. Those properties still undergird the main technological importance of amorphous materials. On the other hand, there are many ways in which glasses are fundamentally different from crystals. This is most noticeable in their properties at cryogenic temperatures.

We presently know very well that an amorphous solid is in reality a liquid caught *locally* in a small set of metastable free energy minima [1], each of which is separated from the much lower free energy crystalline arrangement by high barriers. Therefore the glass transition, as manifest in the laboratory, is not strictly speaking a phase transition in the regular thermodynamic sense and is not accompanied by a symmetry change or appearance of a free energy singularity. In contrast, a liquid that was cooled below its melting point fast enough so as to avoid crystallization (i.e., has become supercooled) experiences a crossover to (highly viscous) activated transport. As the temperature is lowered further, the relaxation barriers grow in a very dramatic fashion, thus confining the molecules in their metastable arrangements long enough to give the appearance of shear elasticity in the sample on the technologically relevant frequency scales. A quantitative understanding of the physics behind the glass transition has recently been achieved with the random first-order transition (RFOT) theory of glasses [1, 2]. This theory has provided a microscopic picture of molecular motions in supercooled liquids, such as first principle predictions of the length scales of these motions and the cooperativity lengths and the barrier heights of the activated transport. At any given time, a supercooled liquid is a mosaic of cooperatively rearranging regions, whose size becomes larger as the temperature is lowered. This chapter describes how the RFOT theory also provides the necessary microscopic input to understand the cryogenic anomalies observed in glasses.

In spite of the absence of periodicity, glasses exhibit, among other things, a specific volume, interatomic distances, coordination number, and local elastic modulus comparable to those of crystals. Therefore it has been considered natural to consider amorphous lattices as nearly periodic with the disorder treated as a perturbation, oftentimes in the form of defects, so such a study is not futile. This is indeed a sensible approach, as even the crystals themselves are rarely perfect, and many of their useful mechanical and other properties are determined by the existence and mobility of some sort of defects as well as by interaction between those defects. Nevertheless, a number of low-temperature phenomena in glasses have persistently evaded a microscopic model-free description along those lines. A more radical revision of the concept of an elementary excitation on top of a unique ground state is necessary [3–5].

Let us give a brief historical overview of some of the most outstanding issues in low-temperature amorphous state physics. It was already noted in the 1960s that the thermal conductivity of amorphous solids is significantly lower than that of crystals. A low-temperature experimentalist using epoxy in his apparatus knew that its thermal conductivity at liquid helium temperatures went roughly as constant $\times T^2$, where the constant was practically the same for other amorphous substances as well [6]. Surprisingly, this had not particularly alarmed anyone, even though one would not priori expect low-temperature properties of disordered solids to be different from crystals, as the appropriate thermal phonon length is much larger than the molecular scale, which was presumed to characterize the relevant heterogeneity scale. It was not until Zeller and Pohl published their classic paper [7] that it became generally known that both the heat capacity and thermal conductivity of glasses were significantly different from those of crystals, and that these anomalies were correlated. The heat capacity turned out to be approximately linear in temperature and larger than the T^3 phononic contribution up to temperatures ~ 10 K. The challenge to the theorists was soon met by the so-called standard tunneling model (STM) [8, 9] in which one assumes that, due to a disordered pattern of molecular bonds in glasses, there are a number of defects in the lattice (something like "loose" atoms or "dangling bonds"), which have two alternative positions in space separated by a sufficiently low tunneling barrier. At low temperatures, the dynamics of such a system is described well by a two-level system (TLS) Hamiltonian. If one assumes that the spectral density of these TLSs is flat, one recovers the linear heat capacity. One also finds that the inverse mean free path of a thermal phonon due to resonant scattering off the TLSs is equal to $l_{\mathrm{mfp}}^{-1} \propto T$, which implies thermal conductivity $\kappa \simeq \frac{1}{3}\sum_{\omega} C_{\mathrm{ph}}(\omega)\, l_{\mathrm{mfp}}(\omega)\, c_s \propto T^2$. Here, $C_{\mathrm{ph}}(\omega)$ is the heat capacity of a phonon mode of frequency ω and c_s is the speed of sound (one assumes here that heat is carried primarily by phonons, which was experimentally demonstrated explicitly four years later by Zaitlin and Anderson [10]). Note that the resonant character of phonon scattering implies that the scattering cross section of low-frequency phonons would be independent of the scatterer size, but would scale with the phonon wavelength (squared) itself. Therefore no knowledge of a scatterer's microscopic details are needed. Rather, only a single coupling parameter is needed to estimate the magnitude of scattering at low temperatures. The STM did prove to be very successful [11], as it predicted, among other things, nonlinear sound absorption due to the saturation of the resonant absorption and the phonon echo, both of which were later observed [12, 13]. In spite of these successes, the microscopic nature of these defects had remained unknown, although there later appeared several indications in the literature that the tunneling centers are not single atom entities but rather involve motions within larger groups of atoms [14, 15]. On the experimental front, there had been a growing amount of evidence that the number of these additional excitations and their coupling to the phonons are correlated and also depend on T_g [16–18], which culminated in the observation made by Freeman and Anderson [19] that the heat conductivities of all studied insulating glasses, if scaled by elastic constants, fall onto the same line in two regions, connected by a nonuniversal flat piece corresponding to the so called "plateau." Figure 1 demonstrates this heat conductivity universality. The lower temperature straight line corresponds to the value $\simeq 150$ of the ratio of the

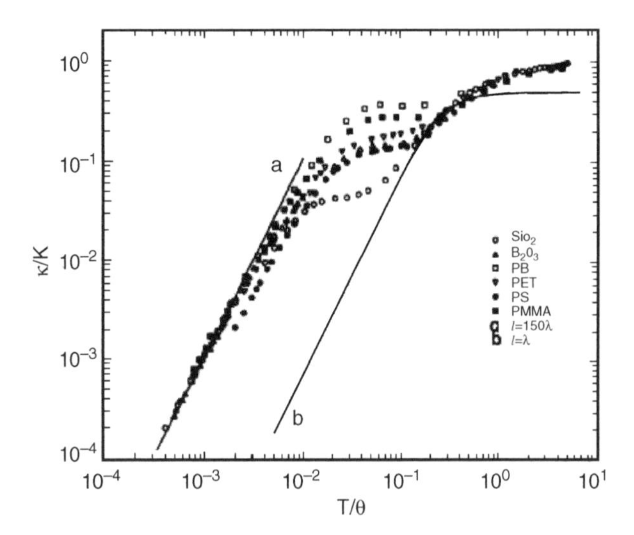

Figure 1. Scaled thermal conductivity (κ) data for several amorphous materials is shown. The horizontal axis is temperature in units of the Debye temperature T_D. The vertical axis scale $K \equiv k_B^3 T_D^2 / \pi \hbar c_s$. The value of T_D is somewhat uncertain, but its choice made by Freeman and Anderson [19] is strongly supported by the resulting universality in the phonon localization region. The solid lines are calculated using $\kappa \simeq \frac{1}{3} \sum_\omega C_{\text{ph}}(\omega) l_{\text{mfp}}(\omega) c_s$ with $l_{\text{mfp}}/\lambda = 150$ and $l_{\text{mfp}}/\lambda = 1$, respectively. Reproduced from Ref. [19] with permission.

thermal phonon mean free path l_{mfp} to the thermal Debye wavelength $\lambda \equiv \hbar c_s / k_B T$. This region spans roughly 1.5 decades in temperature between several mK (lowest T accessed so far for the heat conductivity measurements) to 1–10 K, depending on the substance. The short linear region at higher temperatures (20–60 K) corresponds to $l_{\text{mfp}}/\lambda \simeq 1$, which actually implies complete phonon localization [20] according to the heuristic Ioffe–Riegel criterion. This implies, among other things, that one can no longer use kinetic theory expressions for heat transfer at these temperatures, as a diffusive mechanism must prevail.[1]

The intermediate region ("plateau") is usually observed between 1 and 30 K and does not scale with the Debye temperature and speed of sound. The standard tunneling model of noninteracting two-level systems mentioned previously is normally applied to the region where $l_{\text{mfp}}/\lambda \simeq 150$, that is generically below 1 K. The universality of l_{mfp}/λ can be boiled down [11] to the universality of the following combination of parameters: $\bar{P}g^2/\rho c_s^2$, where $\bar{P}$ is the spectral and spatial density of the TLSs (empirically $\sim 10^{45\pm 1}\text{J}^{-1}\text{m}^{-3}$), g is coupling to the elastic strain on the order of eV, and ρ is the mass density (for reference, $\sim g^2/\rho c_s^2 r^3$ would be the interaction strength between such TLSs at distance r from each other). Now if the defects involved the motion of only a single atom, one would reasonably assume that the value of their spectral density and coupling to the lattice or their combination would be very strongly material dependent. Even though $\bar{P}$ and g^2 vary within almost two orders of magnitude (still surprisingly little), the combination $\bar{P}g^2/\rho c_s^2$ is constant within 50% for different materials (ρ and c_s^2 vary considerably as well). It certainly takes a stretch of imagination to think that this is merely a coincidence, as pointed out by Leggett [22]. In 1988, Yu and Leggett [23] proposed that the density of states of the TLS might itself be a result of dipole–dipole interactions between some original nonrenormalized excitations. In short, this idea is motivated by the observation that for TLS coupled to the phonons with strength g, the coefficient at the dipole–dipole interaction term $g^2/\rho c_s^2$ has dimensions of energy times volume. Therefore the interaction-induced renormalized density of states $\bar{P}$ has to be the inverse of $g^2/\rho c_s^2$ with a coefficient, hence the universality of $\bar{P}g^2/\rho c_s^2$ for different materials. However, it so far has not proved possible to use their approach to justify the value of that coefficient to yield the experimental $l_{\text{mfp}}/\lambda \simeq 150$. This is surprising, since one expects such a simple dimensional argument to be very robust. (Several other studies of the universality [29, 25] were undertaken at the time following the paper

[1]This idea that the heat was transfered by a random walk was used early on by Einstein [21] to calculate the thermal conductance of crystals, but, of course, he obtained numbers much lower than those measured in the experiment. As we now know, crystals at low enough T support well-defined quasiparticles—the phonons—which happen to carry heat at these temperatures. Ironically, Einstein never tried his model on the amorphous solids, where it would be applicable in the $l_{\text{mfp}}/\lambda \sim 1$ regime.

by Freeman and Anderson [19].) There has been subsequent work applying a renormalization group style calculation to a system of interacting TLS [26], but it seems from the results that renormalizations are relevant only at ultralow temperatures (μK and below) [27]. In spite of the difficulties in justifying the strong interaction scenario [28, 29], the works [22, 23] that first challenged the standard TLS paradigm have been a significant motivation behind the present work. In fact, the idea that the observed coupling constant, which is quite small, could be a result of some original "bare" strong interaction is consistent with the microscopic theory if we argue it is the molecular interactions behind the glass transition itself which become "renormalized" in a somewhat unexpected fashion. This microscopic theory suggests [4] that the phenomenological two-level systems are discrete energy levels representing resonantly accessible local degrees of freedom that exist in glasses due to the possibility of collective transitions between alternative structural configurations of compact regions encompassing roughly 200 molecular units. The theory of glassy ergodicity breaking shows the spectrum of these excitations is nearly flat and the density of states scales with the inverse glass transition temperature T_g, echoing the excitation spectrum of a random energy model (REM) with that glass transition temperature. Furthermore, the transitions are an alternative mode of motion that must be in equilibrium with phononic excitations at T_g. This equilibrium requirement makes one realize that TLS–phonon coupling g, T_g, and the material's elastic constants are intrinsically related. The universality of the l_{mfp}/λ ratio is a consequence of this relationship, reflecting the nonequilibrium character of the glassy state. The structural transitions that become tunneling two-level systems at cryogenic temperatures exist because a glassy sample, when it falls out of equilibrium, resides in a metastable configuration chosen from a very high density of states. The sample is broken up into a mosaic of dynamically cooperative regions. Alternatively, the energy landscape is local in nature; that is, rearrangements of compact regions will not change the structural state of the rest of the sample, but only deform the surrounding regions weakly and purely elastically. A (small) fraction of these rearrangements requires overcoming only a very low barrier and can therefore occur even down to sub-Kelvin temperatures. The tunneling occurs by consecutive molecular displacement within the cooperativity length established at T_g. The consecutive motion of atoms is conveniently visualized as a domain wall separating the two alternative local structural states, moving through the local region.

The thermal conductivity plateau has traditionally been considered by most workers as a separate issue from the TLS. In addition to the rapidly growing magnitude of phonon scattering at the plateau, an excess of density of states is observed in the form of the so-called "bump" in the heat capacity temperature dependence divided by T^3. The plateau is interesting from several perspectives. For one thing, it is nonuniversal if scaled by the elastic constants (say, ω_D

and c_s). However, it is located between two universal regions and it is important to understand which *other* scales in the problem determine its location and shape. The excitations that give rise to the dramatically increased phonon absorption at the corresponding frequencies have been circumstantially associated with the excitations observed as the so-called boson peak (BP), directly seen in the inelastic X-ray and neutron scattering experiments, also observed in the optical Brillouin and Raman scattering measurements. These experimental developments in the 1990s became possible, in the neutron spectroscopy case, due to the improved resolution in the neutrons' velocity detection, combined with the ability to generate higher energy incident beams [30]. Similarly, meV resolution was needed to utilize the X-ray scattering techniques to discern the small inelastic wings on the sides of the strong elastic peak [31]. The term "boson peak" comes about because the peak's intensity scales roughly according to the Bose–Einstein statistics. The extraction of the density of states from the spectra is unfortunately model dependent, and those models can roughly be divided [32] into the ones where the boson peak signifies the energy scale on the edge of phonon localization, as promoted in Foret et al. [30], and those following the other school of thought, which asserts that these modes are propagating even well above the frequency of the BP, as supported by the interpretation in Pilla et al. [32]. As far as theoretical interpretation is concerned, it is our impression that most of theories of the boson peak, existing until recently, have postulated a sort of spatial heterogeneity in an otherwise perfectly elastic medium (see a partial list of references in Grigera et al. [33]), with the notable exception of the soft-potential model (SPM) [34, 35]. It is, of course, always possible to recover the observed magnitude of the heat capacity excess at the BP temperatures by a particular choice of parameters. While a contribution of the lattice disorder to the density of states undoubtedly exists and can be very significant (e.g., see simulations of silica's heat capacity by Horbach et al. [36]), we must note that if amorphous lattices were purely harmonic, the phonon absorption at the BP frequencies would be of the Rayleigh type and should be significantly lower than observed in the experiment [37, 38]. There must be internal resonances present in the bulk that scatter phonons inelastically. Although phenomenologically introduced, this feature is present, for example, in the soft-potential model. An analysis of the higher temperature behavior of the tunneling transitions that give rise to the TLS at sub-Kelvin energies was provided in the RFOT approach in Lubchenko and Wolynes [5]. When these transitions occur at high enough temperature, the domain wall separating the two alternative states can have its surface vibrations thermally excited. These vibrational states, characteristic of a two-dimensional membrane, thereby accompany the underlying structural transition. Their large degeneracy is sufficient to account for the enhancement of phonon scattering at the plateau, as compared to the TLS regime. Finally, the superposition of the domain wall vibrations on the underlying tunneling transition leads to an excess of density of states that reproduces well the bump in the heat capacity (we call these compound excitations "ripplons"). We therefore arrive at a rather complete physical picture that allows a unified quantitative explanation of previously seemingly unrelated mysteries in the TLS regime and at the higher, plateau energies.

This chapter is organized as follows: the first section outlines the basics of the RFOT theory and then proceeds in applying that theory to understanding the origin of the tunneling centers in amorphous solids. The spectrum of the two-level systems, their coupling to the phonons, and the origin of the universality of phonon scattering are then discussed. Additionally, we show how details of the derived TLS tunneling amplitude distribution lead to a deviation of T dependence of the heat capacity from a strict linear form. The second section explains how the high-energy vibrational excitations (ripplons) of the tunneling interfaces give rise to an excess of states, which exhibit as the heat capacity bump and yield the rapidly rising phonon scattering at these higher energies. A short discussion of the *relaxational* absorption from these excitations is given and its frequency dependent part is derived. The contents of these first two sections are, for the most part, a detailed account of the calculations underlying two earlier brief works [4, 5] that have reported our explanation of the low-temperature anomalies in glasses within a semiclassical approach. The third and fourth sections are comprised of new results. We establish that, while not altering the main conclusions of the semiclassical picture, a purely quantum phenomenon of level mixing and repulsion has an observable effect on the density of states of the tunneling centers at low T. Finally, the interaction between tunneling centers, mediated by phonons, is estimated and this is argued to make a significant contribution to the negative thermal expansivity (and thus a negative Grüneisen parameter) observed in many amorphous materials.

II. OVERVIEW OF THE CLASSICAL THEORY OF THE STRUCTURAL GLASS TRANSITION

From a physicist's perspective, a theory of the glass transition describes what happens to a liquid when it is cooled down sufficiently but is not observed to crystallize. To a mathematician, this is a generalized problem of packing compact interacting objects of comparable size given a specific constraint on the density distribution (it is not periodic) and total energy of the system. A nearly complete conceptual, microscopic picture of the amorphous state has emerged in the course of the two last decades [1, 2, 4, 5, 39–48]. This framework has led to a unified, quantitative understanding of many seemingly unrelated phenomena in super-cooled liquids above and below the glass transition. The glasses we consider form at temperatures where quantum effects are small, so classical statistical mechanics is used. We review such a classical glass transition in what follows.

First, we make several comments on the phenomenology of supercooled liquids. Strictly speaking, these are nonequilibrium systems: when cooled sufficiently slowly, most simple liquids will crystallize at a temperature just below the melting temperature T_m. Randomly atactic polymers become glassy but presumably never crystallize. The melting point is defined as the temperature at which the liquid and crystal free energies are equal. Cooling the liquid at least a bit below T_m is necessary to create a free-energy driving force so as to make the nucleation barrier finite and to allow the system to equilibrate. The crystal, once formed, is different from the liquid in several ways; for example, it scatters X-rays at precise angles and it is anisotropic. Crucially for us, a crystal supports transverse sound waves, at *all* frequencies (including $\omega = 0$; hence the crystal retains its shape). In contrast, the supercooled liquid is a finite lifetime state since crystallization will eventually occur by nucleation. However, the growth of crystalline nuclei, inside the liquid, is subject to the slowing of all motions in liquids. Owing to this dramatic slowing of liquid motions upon lowering the temperature, one can supercool the liquid substantially below its melting point, which is the key to forming glasses. The extra nucleation barrier ensures there is adequate time to study the properties of the supercooled noncrystalline state. Local structures in supercooled liquids persist for some time—call it $1/\omega_c$. This time is longer than the time it takes to establish a Maxwell distribution of velocity, which is at most a few vibrational periods. Such an amorphous system will support transverse waves at frequencies $\omega > \omega_c$, just as a crystal would, but will in contrast exhibit a liquid-like, equilibrium response to time-dependent perturbations at frequencies $\omega < \omega_c$. As we have said, ω_c drops rapidly upon cooling. If one is intent on observing equilibrium response at *some* frequency range, one must prepare the sample by cooling it more slowly than ω_c. Conversely, for any given cooling rate, no matter how slow, the liquid will fall out of equilibrium on *all* time scales and the sample will appear to be mechanically solid. We say the liquid has undergone the glass transition. (The corresponding ω_c usually ranges between 10^2 and 10^5 seconds, depending on the experimenter's patience.) The liquid just below the glass transition temperature T_g is only subtly different from the liquid just above T_g. Structurally, the two are nearly identical. Even dynamically, both can flow, although the T dependencies of the corresponding transport coefficients are distinct in the two forms of the "equilibrium" supercooled liquid and the nonequilibrium glassy state [48]. The residual dynamics below T_g is referred to as "aging." Aging is at least as slow as the motions just above T_g but can be much slower when the sample is studied well below T_g. This requires a greater amount of the experimenter's patience in studying system properties than even needed for sample preparation. Finally, when the sample falls out of equilibrium at T_g, a jump in the heat capacity is measured by differential calorimetry, thus resembling, crudely, a phase transition.

The dramatic slowing down of molecular motions is seen explicitly in a vast area of different probes of liquid local structures. Slow motion is evident in viscosity, dielectric relaxation, frequency-dependent ionic conductance, and in the speed of crystallization itself. In all cases, the temperature dependence of the generic relaxation time obeys to a reasonable, but not perfect, approximation the empirical Vogel–Fulcher law:

$$\tau_{\mathrm{rlxn}} \propto e^{DT_0/(T-T_0)} \quad (1)$$

For a review, see Angell et al. [49] and Böhmer et al. [50]. A specific example of a $\tau(T)$ dependence is shown in the left-hand side panel of Fig. 7. In Eq. (1), T_0 is a material-dependent temperature at which the relaxation times would presumably diverge, if the experimenter had the patience to equilibrate the liquid at the corresponding temperatures. Needless to say, measurements of equilibrium dynamics near T_0 are essentially nonexistent. The coefficient D is often called "fragility," with larger values of D corresponding to "stronger" substances, while smaller values are associated with "fragile" liquids. This terminology apparently refers to the degree of covalent networking in the material [51], a qualitative trend later rationalized with a density-functional study by Hall and Wolynes [52]. Fragility appears to correlate with the Poisson ratio, at least for nonpolymeric glasses [53]. At any rate, the value of coefficient D is directly related to what glassblowers refer to as "short glasses" and "long glasses" [54]: (molten) glass can be worked or shaped in the range of viscosities 10^4–10^9 poise. If the corresponding temperature range is short, the glass is called "short," and vice versa for the "long" glass. The former and the latter obviously correspond to a small and large value, respectively, of the parameter D.

The nonequilibrium character of a supercooled liquid is exhibited in the entropy of the liquid, which is considerably larger at T_g than that of the corresponding crystal at this temperature. This additional entropy corresponds to all the molecular translations that would have otherwise frozen out at crystallization. In crystallization, this would appear as the latent heat of the liquid-to-crystal transition. In a supercooled liquid, the molecular structure is dense enough to define a lattice locally. Vibrations around lattice sites are small. The excess entropy associated with the locations of these lattice sites has traditionally been designated as the "configurational" entropy. This excess entropy, s_c, is temperature dependent. It refers to all possible liquid configurations that could be surveyed by the liquid if we wait long enough for molecular translations to occur. Experimentally, we determine the configurational entropy by relying on the third law of thermodynamics. Using the third law, we know the total entropy of the liquid at T_m by integrating the crystal's heat capacity (over T) and adding the entropy of melting. Now for the supercooled liquid, we integrate the heat capacity difference between the liquid and the crystal. To do this, of course, we assume the vibrational entropies of the ordered and aperiodic lattices are close. The heat capacity measured by differential calorimetry above the glass transition depends on the rate of the configurational and vibrational entropy decrease with temperature right above T_g. Below T_g the structure of the liquid remains the same as of the moment of vitrification, apart from some (normally insignificant) aging. The vibrational entropy decreases as it did above T_g, but there is no component from configurational change. Thus one observes a nonzero heat capacity jump at T_g. Above T_g, the s_c decreases and the density increases with lowering the temperature. This is expected because there are fewer ways to mutually arrange the molecules at higher densities. When extrapolated past T_g, as was done by Simon [55], and notably by Kauzmann in his review [56], the configurational entropy vanishes at a temperature T_K, which is securely above the absolute zero. This suggests that only a nonextensive number of low-energy aperiodic, liquid arrangements could be found at T_K and the entropy of the liquid is thus equal to the corresponding crystal (correcting for differences in their vibrational spectra). This phenomenon is sometimes referred to as the "entropy crisis," which again would presumably occur only under completely equilibrium cooling. Such an entropy crisis strictly occurs in several mean-field spin glass models with infinite interactions [42, 57, 58]. There are many sound arguments suggesting a strict singular vanishing of configurational entropy at T_K is unlikely for real liquids [59, 60]. Nevertheless, T_K is a useful fiducial point for the analysis. None of the results of the present theory in the experimentally accessible regime depend on the configurational entropy truly vanishing at any point. As we shall see, the configurational entropy is macroscopic but decreases with temperature. s_c is typically $\sim 0.8k_B$ per movable unit at the conventional glass transition temperature corresponding to cooling rate of inverse hour and decreases at a rate proportional to $\Delta c_p/T_g$. For simplicity, we will assume s_c extrapolates so as to scale linearly with the proximity to the entropy crisis (see Ref. [61]): $s_c = \Delta c_p(T - T_K)/T_K$.

Before our formal discussion, let us make several qualitative statements about molecular transport above T_g. The motions of a supercooled liquid are much slower than the local vibrations. The potential felt by an individual molecule comforms to a local "cage." This local "cage" is formed by the neighboring molecules, of course. In order to translate irreversibly a given molecule, as opposed to vibrating about the current position, the cage must be destroyed. In other words, a number of surrounding molecues must be translated as well. Upon lowering the temperature, the density increases and s_c decreases; therefore fewer alternative states are available to any given group of molecules. Thus it is clear that conforming the liquid to an arbitrary translation of a given molecular unit will require readjusting the positions of more and more surrounding molecules at the same time. This leads to a larger cooperative region size, leading in turn to higher barriers for relaxation processes and higher viscosity. At a crude level, this picture underlies the arguments of Adam and Gibbs [62], but those arguments fail to relate the size of the moving regions to the energy landscape itself. In contrast, the random first-order transition (RFOT) theory [1, 2] explicitly shows how these reconfigurational motions occur and thus establishes intrinsic connection between the kinetic properties and the thermodynamics of supercooled liquids. Our account is based on Lubchenko and Wolynes [48], which also discusses the intrinsic connection between cooperative, activated motions in the supercooled liquid above the glass transition and the classical aging dynamics below the glass transition. These arguments also pave the way for understanding the quantum dynamics at cryogenic temperatures.

The main prerequisite of the RFOT theory is the existence of time scale separation between vibrational thermalization and equilibrating structural degrees of freedom that involve crossing saddle points on the free energy surface. This only occurs below a crossover temperature T_A, which is predicted by the theory itself. The existence of local trapping in cages is well established by experiment: there is a long plateau in the time-dependent structure factor as measured by the inelastic neutron scattering [63]. In RFOT, such trapping was first established theoretically using a density functional theory (DFT) in Singh et al. [40]. This paper shows there are aperiodic free energy minima by computing the free energy of an aperiodic variational density distribution function: $\rho(\mathbf{r}) \equiv \rho(\mathbf{r}, \{\mathbf{r}_i\}) = \sum_i (\alpha/\pi)^{3/2} e^{-\alpha(\mathbf{r}-\mathbf{r}_i)^2}$. The set of coordinates $\{\mathbf{r}_i\}$ denotes a particular aperiodic lattice. The typical lattice spacing is a. A zero value for the parameter α would correspond to a completely delocalized, uniform liquid state, such as just below the liquid–vapor transition. $\alpha \rightarrow \infty$ would imply freezing into an infinitely rigid lattice. α can also be interpreted as the spring constant of an equivalent Einstein oscillator forcing each molecule to remain near its proper location in the aperiodic lattice. $F(\alpha)$ develops a metastable minimum, at nonzero $\alpha = \alpha_0 \neq 0$, only below some temperature T_A. This minimum has higher free energy than the lowest minimum at $\alpha = 0$ (see Fig. 2). In the mean-field limit, the appearance of such minimum would lead to a lattice stiffness and would represent a state with a divergent viscosity. This localization transition and the viscosity catastrophe of mode-coupling theories are essentially identical as was established in Kirkpatrick and Wolynes [41]. A single such high-lying free energy minimum would be thermodynamically irrelevant, but one must recall that this $F(\alpha)$ is computed for a *single*, particular aperiodic lattice, which is actually only one of many possibilities. Taking into acount the thermodynamically large number of alternative aperiodic packings increases the entropy of the (set of) localized, aperiodic state(s) and thus lowers the metastable free energy minima just the right amount to make them competitive with the mean-field uniform, delocalized state. The correspondence

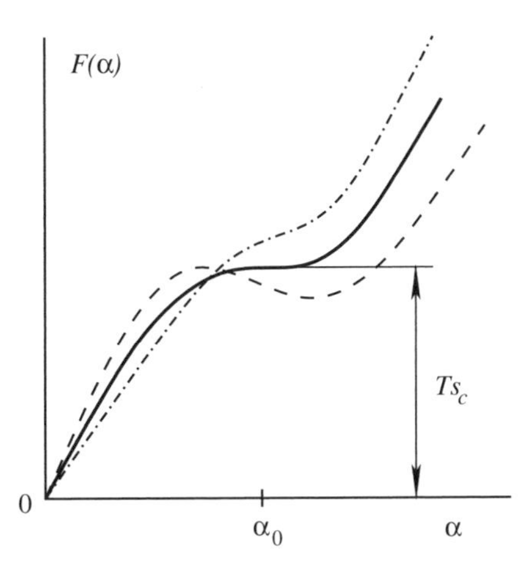

Figure 2. A schematic of the free energy density of an aperiodic lattice as a function of the effective Einstein oscillator force constant α (α is also an inverse square of the localization length used as input in the density functional of the liquid). Specifically, the curves shown characterize the system near the dynamical transition at T_A, when a secondary, metastable minimum in $F(\alpha)$ begins to appear as the temperature is lowered. Taken from Ref. [47] with permission.

between the free energy difference in mean-field theory and the configurational entropy was rigorously shown for the Potts glass by Kirkpatrick and Wolynes [42], who argued such systems have similar symmetry properties to structural glasses. For structural glasses this correspondence may also be shown more formally using a replica formalism [64]. The localization transition at T_A is accompanied by a discontinuous change in the order parameter α. This is why the transition is called "random first order." Although there is a discontinuity in α, the actual structure in which the system freezes is chosen at random out of a multitude of possibilities (given by the configurational entropy). At the same time, such an ordered phase will persist only for finite times; therefore this is a true transition only for high-frequency motions, comparable at first to the vibrational time scale. This transition at T_A only signifies a soft crossover, as far as the *whole* dynamical range is concerned. We emphasize that there are many different "phases" below T_A, all of which are random packings. The number of random packings, thermally available to a region of size N, $e^{s_c N}$, decreases *gradually* with temperature. (This corresponds to a gradual freezing out of the translational degrees of freedom with lowering the temperature, as signified by the decreasing ω_c.) Because the decrease is gradual, the *random* first-order transition does not exhibit a latent heat. In a finite range system, different minima can interconnect by barrier crossing. (Such barriers would be infinite in a mean field.) Even though the transition at T_A is a crossover, the temperature T_A itself is a useful parameter characterizing material properties.

The resulting time scale separation at and below T_A has two important consequences. First, one may perform canonical averaging over the vibrations within a particular structural state. This gives a free energy of a particular structural state: $\Phi = E - TS_{\text{vibr}}$, where S_{vibr} is the vibrational entropy. Note the vibrations are not necessarily harmonic. To define Φ, all that matters is that the local vibrations equilibrate much faster than the structural degrees of freedom. As a consequence, Φ can be termed the bulk, *microcanonical* energy of a given *structural* state. To any value of this energy one may associate a bulk, microcanonical *entropy* $S_c(\Phi)$ counting states with similar contributions from energy and vibrational entropy; both Φ and $S_c(\Phi)$ scale linearly with the size of the system. One may thus have to work with morphologically distinct, globally defined aperiodic phases without actually specifying their precise molecular constitution, as long as we know their spectrum, that is, their number as a function of the microcanonical free energy. These statistics are directly measurable by calorimetry just as in our discussion of the Kauzmann paradox.

Having established the transitory existence of a global aperiodic structure, we may next enquire into how molecular motions allow the system to escape such a phase.[2] This occurs by replacing locally one part of the aperiodic packing by a different local packing. This will be an activated event. The RFOT theory allows one to compute the mean activation barrier and its distribution. Also, the theory determines critical region size and the spatial extent of the excitations corresponding to the cooperative rearrangement. The magnitude of an individual molecular displacement during the transition is determined by α. To estimate the activation free energy, let us make the following construct. Consider a library of possible local aperiodic arrangements at a particular location, as illustrated in Fig. 3. This *local* library of states can be constructed based on the existence of the *global* library of states introduced earlier that we described by the energy variable Φ and the corresponding entropy $S_c(\Phi)$ reflecting the spectrum. Clearly, the energy density $\exp[S_c(\Phi)]$ is extremely high and grows rapidly with Φ. We might perform a full survey of local states by mentally carving out a small region of size N, while freezing in place the lattice sites surrounding the region. One can then heat the local region and allow that region to equilibrate. Unless the new local arrangement is exactly the same as the original one, its energy will likely be significantly higher: a local substitution statistically must cost free energy, stemming from a structural mismatch between two randomly chosen aperiodic packings of a given energy Φ. This mismatch energy corresponds to the usual surface energy, such as that between two different crystal forms or at a liquid–crystal interface. The free

[2]Of course, the issue of producing the aperiodic state in the *laboratory* would also involve estimating whether corresponding quenching rates can be experimentally achieved.

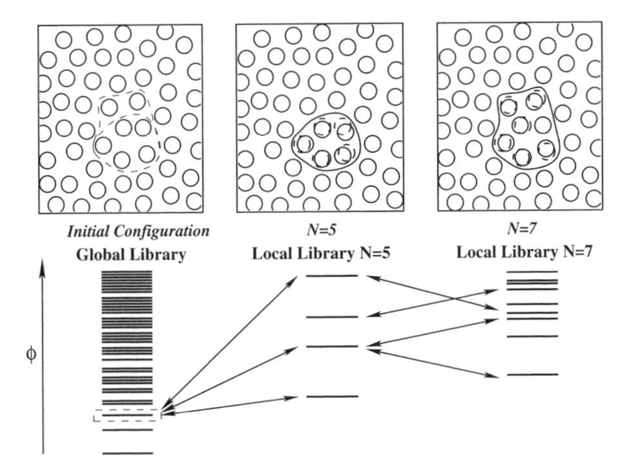

Figure 3. In the upper panel on the left a global configuration is shown, chosen out of a global energy landscape. A region of $N = 5$ particles in this configuration is rearranged in the center illustration. The original particle positions are indicated with dashed lines. A larger rearranged region involving $N = 7$ particles is connected dynamically to these states and is shown on the right. In the lower panel, the leftmost figure shows the huge density of states that is possible initially. The density of states found in the local library originating from a given initial state with 5 particles being allowed to move locally is shown in the center diagram. These energies are generally higher than the original state owing to the mismatch at the borders. The larger density of states where 7 particles are allowed to move is shown in the rightmost part of this panel. As the library grows in size, the states as a whole are still found at higher energies but the width of the distribution grows. Eventually with growing N, a state within thermal reach of the initial state will be found. At this value of N^* we expect a region to be able to equilibrate. Taken from Ref. [48] with permission.

energy cost of locally replacing the initial phase (labeled as "in") by another phase, call it j, can therefore be written

$$\phi_j^{lib}(\mathbf{R}) - \phi_{\rm in}^{lib}(\mathbf{R}) = \Phi_j(N) - \Phi_{\rm in}(N) + \Gamma_{j,\rm in} \tag{2}$$

where $\Gamma_{j,\rm in}$ is the mismatch energy and $\mathbf{R}$ is the location of the local region. As before, Φ denotes the *bulk* energy, corresponding to a distinct aperiodic packing, with the vibrational entropy already included. To compute the likelihood of such a local rearrangement, substitute for the specific surface energy $\Gamma_{j,\rm in}$ its average value, which should scale with size: γN^x. γ depends on the material and on temperature. Naively, the usual surface energy scaling is $N^{(d-1)/d}$, expected in d dimensions. One can argue, however, that x will actually turn out to equal $\frac{1}{2}$. Such a surface tension renormalization was first conceived by Villain [65], in the context of the random field Ising model (RFIM). In RFIM, the Ising spins, in addition to their coupling, are subjected to a random static magnetic field obeying certain fluctuation statistics. A flat interface, or domain wall, between spin-up and spin-down domains will distort so as to conform to the local variation of the field. An RG argument incarnating this distortion on a hierarchy of length scales yields a scale-dependent renormalization of the surface tension, giving a surface free energy exponent $x = \frac{1}{2}$ [65]. The structure–structure interface in a supercooled liquid resembles the RFIM, owing to the fluctuations of local energies of the various aperiodic packings. The statistics of these fluctuating local energies require that $\delta\Phi(\delta N) \sim \Phi_0\sqrt{\delta N}$, where Φ_0 is δN independent, echoing the fluctuation statistics of the frozen random field of the RFIM. Thus, as Kirkpatrick et al. [2] suggest, the originally thin flat interface will become diffuse yielding $x = \frac{1}{2}$. In the liquid case, a vivid interpretation of the surface energy renormalization is possible: since the interface is distorted down to the smallest scale (allowed by the material's discreteness), the region occupied by the now diffuse wall is neither of the two original structures it separates. Instead, it may be interpreted as accommodating *other* structures. These intermediate packings interpolate structurally two randomly chosen, and thus priori energetically disagreeable, packings. In other words, the original thin interface separating two given packings is "wetted" by other packings, thus lowering the overall interface energy. As we shall see, real liquids have only modest size regions of rearrangement, so it is hard to argue about the exact value of the exponent. Nevertheless, we note two felicitous observations: with $x = \frac{1}{2}$, the usual scaling argument will give precisely a discontinuity in Δc_p at any ideal transition, to be seen at T_K. Also, while the RFIM itself remains the subject of discussion, Villain's argument does give a length scale exponent agreeing with the majority of experiments and numerical studies [66, 67].

The role of the interface mismatch energy in the reconfiguration process can be beneficially understood from a statistical point of view, as illustrated in Fig. 3. It costs free energy to reconfigure a small number N of molecules because considering a small region severely limits the number of available liquid configurations. The interface energy grows with N; however, the available density of states also grows with N, both in terms of its absolute value and the distribution's *width*. At some size N^*, which will be computed shortly, all relevant liquid states become available. The rate of escape of a group of N molecules to another structural state can be determined by a canonical type sum accounting for the multiplicity of the final states at energy ϕ_j:

$$\begin{aligned} k &= \tau_{\text{micro}}^{-1} \int (d\phi_j^{lib}/c_\phi) e^{S_c(\Phi_j)/k_B} e^{-(\phi_j^{lib} - \phi_{\text{in}}^{lib})/k_B T} \\ &\simeq \tau_{\text{micro}}^{-1} e^{S_c(\Phi\text{eq})/k_B} e^{-(\phi_{\text{eq}} - \phi_{\text{in}}^{lib})/k_B T} \end{aligned} \tag{3}$$

In the second step, a steepest descent evaluation is made where ϕ_{eq} maximizes the integrand. c_ϕ is some constant of units energy that reflects the local curvatures of the energy landscape. The quantities ϕ_j^{lib} and Φ_j are related through Eq. (2). The time scale τ_{micro} is the time scale of a molecular scale nonactivated process, typically on the order of a picosecond. The value ϕ_{eq} that maximizes the integrand above will be the internal (equilibrium) free energy characteristic of the system at the ambient (i.e., vibrational) temperature T. In other words, the greatest kinetic accessibility of a state, as embodied in the optimization in Eq. (3), implies that the state will be most frequently visited by the system; therefore it must be the *equilibrium* state. The integration in Eq. (3) is similar to a canonical sum; yet it is different in an important way: the summation in Eq. (3) is far more general than the usual expression for the partition function because when relaxation times are continuously distributed, one must *explicitly* weigh the contribution of a state (to the canonical sum) by its kinetic accessibility. The latter, in general, will depend on the spatial extent of the excitation corresponding to a transition between two states; in this regard, the integration variable ϕ_j is, in a sense, a local *microcanonical* energy. Consequently, the energy ϕ_{eq} corresponds to a *canonical* energy. Yet, ϕ_j and ϕ_{eq} would strictly become a microcanonical and canonical energy, in their conventional sense, only in the large N limit, when the boundary effects are small. In contrast, the very thermodynamic relevance of the glassy state is due to the locality of the landscape and nonsmallness of the surface term. Finally, since the bulk entropy $S_c(\Phi_{\text{eq}})$ corresponds to the equilibrium energy, it will be given by the equilibrium configurational entropy $S_c(T)$, measured by calorimetry. Thus given ϕ_{eq}, one can compute the value of the typical escape rate to a structure where N particles have moved. This gives

$$k(N) = \tau_{\text{micro}}^{-1} \exp\left(S_c(N, T) - \frac{\phi_{\text{eq}} - \phi_{\text{in}}^{lib}}{k_B T} \right) \tag{4}$$

The number of particles that must be moved for complete equilibration is determined by the minimum of this expression over N. We thus determine an activation free energy profile

$$\begin{aligned} F^{\ddagger}(N) &= \phi_{\text{eq}} - \phi_{\text{in}}^{lib} - TS_c(N, T) \\ &= \Phi_{\text{eq}}(N) - \Phi_{\text{in}}(N) + \gamma\sqrt{N} - TS_c(N, T) \end{aligned} \tag{5}$$

where we used Eq. (2) in the second equality. The maximum of the $F(N)$ curve defines the bottleneck location. This equation is suitable for finding the rate of structural rearrangement both in the equilibrated supercooled liquid (before it crystallizes!) and in the nonequilibrium glass, which ages below T_g.

Let us first consider equilibrium liquid rearrangements. In this case, typically $\Phi_{eq} = \Phi_{in}$, apart from fluctuations. Thus one arrives at the following simple expression:

$$F(N) = \gamma\sqrt{N} - Ts_c N \tag{6}$$

where we have used the thermodynamic scaling of the configurational entropy, $S_c(N) = s_c N$. In the supercooled equilibrated liquid, molecular transport is driven only by the multiplicity of mutual molecular arrangements. For this reason, the reconfigurations following the activation profile from Eq. (6) have been called "entropic droplets." The graph of the function in Eq. (6) is shown in Fig. 4. The transition state configuration will satisfy $\partial F/\partial N = 0$, corresponding to an unstable saddle point of this free energy. This gives for fixed γ a rearranging region size $N^{\ddagger}$ that grows as s_c diminishes: $N^{\ddagger} = (\gamma/2s_c T)^2$. The resulting barrier also scales inversely proportionally to s_c:

$$F^{\ddagger} = \frac{\gamma^2}{4s_c T} \tag{7}$$

An inverse scaling of the barrier with the configurational entropy was arrived at by Adam and Gibbs [62] in a different (and inequivalent) way. Note that if γ is a smooth function of temperature around T_K and s_c is described by the linear law $s_c \propto (T - T_K)$, the resulting activation barrier is exactly of the Vogel–Fulcher law form (Eq. (1)), which, as we have said, fits data well. Many arguments can lead to increasing relaxation times at low temperatures and, with enough adjustable parameters, can fit data. What is different about the RFOT theory is that it establishes an intrinsic link between the rate law and the entropy crisis. In addition, if the entropy of the equilibrated fluid can be estimated, the density

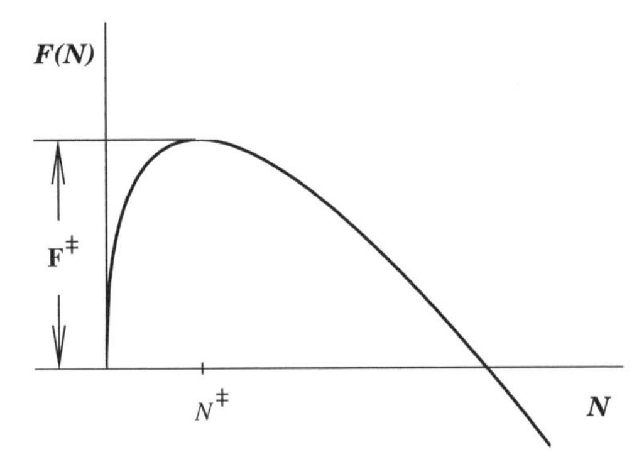

Figure 4. The droplet growth free energy profile from Eq. (6) is shown.

functional theory allows the vibrational entropy and thus, by substraction, the configurational entropy to be determined. Therefore T_K can be estimated from the microscopic force laws. This has been done for simple soft spheres by Mezard and Parisi [64], giving reasonable results. Hall and Wolynes [52] have also calculated T_0 and T_A for a simplified model of a network fluid. Their study is consistent with known chemical trends for T_A and T_K as the network becomes more thoroughly crosslinked.

The idea of the configurational entropy itself driving liquid rearrangements still appears to generate some confusion. One possible reason for this is that s_c is totally unambiguously defined only in the mean-field limit. In the latter limit, rearrangements are infinite so dynamics driven by s_c do not arise. This is a good place to emphasize that the RFOT theory is not mean field! Only the local landscape, within an entropic droplet, is actually well described by a mean-field, random energy model like approximation. We took advantage of this in extracting the energy spectrum of low-energy structural excitations in a frozen glass [4], as explained in detail in the following section. We wish to point the reader to the recent elegant treatment done by Bouchaud and Biroli [68], which re-analyzes the RFOT conclusions for rearrangements in an equilibrated fluid from the viewpoint of Derrida's [69] random energy model (REM).

Now calculations of T_A and T_K are plagued by the usual difficulties of liquid state structure theory and the accuracy of approximations, some of which are hard to control. Still, even in the face of such approximations, such microscopic considerations lead us to expect a universal value of γ/T_g at T_g as we shall discuss next.

The RFOT theory allows the coefficient γ in the mismatch energy to be estimated from a microscopic argument. It turns out to be proportional to T_K and to depend logarithmically on the inverse square of the so-called Lindemann ratio. Early in the 20th century, Lindemann [70] argued that the thermal fluctuations of an atom's position could only be a fraction of the lattice spacing a in a solid, if the packing is to be mechanically stable. Since the threshold value of the vibrational amplitude of an atom in the lattice is finite, the transition in which the lattice disintegrates must be first order. For crystals, the Lindemann ratio of this threshold displacement d_L to the lattice spacing is about $\frac{1}{10}$. For amorphous materials, the d_L/a ratio can be obtained from the plateau in the self-correlation functions measured by neutron scattering experiments [63]. Again, this ratio turns out to be approximately one-tenth (universally!). This number is reproduced in several microscopic calculations consistent with the RFOT theory, such as the self-consistent phonon theory and density functional theories [39, 40], and dynamical mode coupling theory [41, 42, 71, 72], with modest quantitative variations. The meaning of $\alpha \simeq 10$ as a mechanical stability criterion has also been corroborated within the replica formalism [64]. In terms of the DFT calculation discussed earlier, $\alpha_L \equiv a/d_L$ corresponds with the metastable minimum that the free energy $F(\alpha)$ develops below the dynamical transition temperature T_A (see Fig. 2). It has a relatively weak temperature dependence. The logarithmic scaling of the surface tension coefficient with the Lindemann length follows from a detailed calculation by Xia and Wolynes [1] but can be rationalized in a simple way: below T_A, motions span only the length d_L, while in the liquids, they can move a distance a before losing their identity with a neighboring molecule. The entropy of the "caged" fluid is less and thus the free energy cost of confining a molecule within length d_L, as opposed to a, can be assessed by recalling the free energy expression for an ideal monatomic gas: $-f = \frac{3}{2} k_B T \ln[(eV/N)^{2/3} mT/2\pi\hbar^2]$, written deliberately here so as to have a length scale squared in the logarithm.

γ is proportional to T_K and only logarithmically depends on a nearly universal quantity, the Lindemann ratio. If T_g is near T_K (i.e., for slow quenches), γ/T_g is thus nearly material independent and calculable: $\gamma = \frac{3}{2}\sqrt{3\pi} k_B T_g \ln(\alpha_L a^2/\pi e)$. By quantifying the mismatch energy this specifically leads to many predictions about the dynamics near T_g, for a range of substances. The coefficient in the Vogel–Fulcher (VF) law D is predicted to follow from the measured thermodynamics. Using the γ value above, we find not only the VF dependence of the relaxation times on the temperature, $\propto e^{DT_0/(T-T_0)}$, but also a remarkably simple formula relating D and the heat capacity jump [1]: $D = 32R/\Delta c_p$. The coefficient 32 is nearly universal and, as we see, follows numerically from the microscopic theory since the universal value of the Lindemann ratio enters only logarithmically in the localization entropy cost. The numerical relation between D and Δc_p from this simple explicit calculation is in rather remarkable agreement with experiment. In Fig. 5, we plot the so-called fragility index m, as computed from calorimetry and extracted from direct relaxation measurements. m is proportional to the slope of the $\log\tau$ vs. $1/T$ relation at T_g and thus is directly related to D if the VF law is valid. (D values in the literature are obtained from global fits of $\log\tau$ vs. $1/T$ and depend somewhat on the fitting procedure.)

Two other remarkable universalities emerge from the value of γ. First, at a reference laboratory time scale of 1 h $\sim 10^{17}\tau_0$, we have a universal value of $s_c \simeq 0.8k_B$. This implies $s_c(T_g)/s_c(T_m) \simeq 0.7$, where $s_c(T_m)$ is, of course, also the fusion entropy. This relation is independent of the precise identity of the moving subunit and holds very well. A second important universal feature emerges from the universal value of γ/T_g: the cooperative size at T_g is nearly universal.

Let us now consider in greater detail the pattern of cooperative structural rearrangements in a supercooled liquid. These turn out to presage the existence of the residual degrees of freedom in a glass below T_g. Within a period of time shorter than the typical relaxation time τ, the molecular motions within regions of size ξ^3 will be highly correlated and, at the same time, approximately decoupled from the surrounding. That is, the liquid is broken up into a

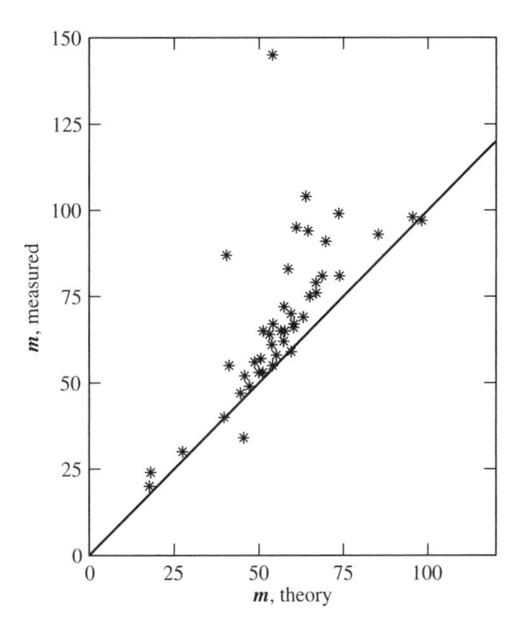

Figure 5. The horizontal axis shows the value of fragility as computed from the thermodynamics by the RFOT theory, and the vertical axis contains the fragility directly measured in kinetics experiments. Here m is the so-called fragility index, defined according to $m = T^{-1}[d\log_{10}\tau(T)/d(1/T)]_{T_g}$. m is somewhat more useful than the fragility D, because deviations from the strict Vogel–Fulcher law, $\tau = \tau_0 e^{DT_K/(T-T_K)}$, are often observed (see text). m essentially gives the apparent activation energy of relaxations at T_g; in units of T_g, it is roughly an inverse of D. In evaluating m theoretically, one needs to know the size of the moving unit, or "bead," in each particular liquid. The latter can be estimated using the entropy loss at crystallization [47], resulting in

$$m_{\text{theor}} \propto \frac{\Delta c_p(T_g)T_g}{\Delta H_m s_c^2(T_g)} \propto \frac{\Delta c_p(T_g)T_g}{\Delta H_m}$$

in view of the near universality of $s_c(T_g)$ (see text). Taken from Ref. [73] with permission.

(flickering) mosaic pattern of cooperative regions. This mosaic structure is directly manifested in the dynamical heterogeneity recently observed in supercooled liquids using single molecule experiments [74], nonlinear relaxation experiments [75], and non linear NMR experiments [76]. (These experimental tools became available only a decade after the RFOT theory was first formulated.) The size of a typical mosaic cell is found from the thermodynamic condition $F(N^*) = 0$. Unlike the regular nucleation of one distinct phase within another (as in crystal growth in the liquid), by crossing the barrier from Eq. (6) the local region arrives at a statistically similar but alternative solution of the free energy functional; thus that solution still represents a typical liquid state! An informal analogy here is that distinct low-energy dense local liquid packings are like the fingerprints of different individuals—different in detail, yet generically equivalent liquid states. Since we have agreed that $F = 0$ is the liquid *equilibrium* free energy at this temperature (the crystalline state is assumed to be hidden behind a high enough crystal nucleation barrier), the condition $F(N^*) = 0$ specifies the size of region to which an arbitrary liquid configuration is available. Therefore a region of size N^* is able to reconfigure on the experimental time scale characterized by $F^{\ddagger}$. In terms of physical length, $F(N^*) = 0$ implies

$$\xi \equiv N^{*1/3} a = a \left[\frac{8}{3\sqrt{3\pi}} \ln\left(\frac{\tau}{\tau_0}\right) / \ln\left(\frac{\alpha_L a^2}{\pi e}\right) \right]^{2/3} \simeq 5.8\, a \quad (N^* \simeq 190)$$

The critical radius $r^{\ddagger}$ at T_g is a multiple of ξ. Droplets of size $N > N^*$ are thermodynamically unstable and will break up into smaller droplets, in contrast to that prescribed by $F(N)$, if used naively beyond size N^*. This is because $N = 0$ and $N = N^*$ represent thermodynamically equivalent states of the liquid in which every packing typical of the temperature T is accessible to the liquid on the experimental time scale, as already mentioned. In view of this "symmetry" between points $N = 0$ and N^*, it may seem somewhat odd that the $F(N)$ profile is not symmetric about $N^{\ddagger}$. Droplet size N, as a one-dimensional order parameter, is not a complete description. The profile $F(N)$ is a projection onto a single coordinate of a transition that must be described by $e^{s_c N^*}$ order parameters—the effective number of distinct aperiodic packings explored by the liquid. At the point N^*, the free energy functional actually has a minimum as a function of the (multicomponent) order parameter. A more detailed discussion of this can be found in Lubchenko and Wolynes [47], where we compute the softening of the barrier $F^{\ddagger}$ near T_A due to order parameter magnitude fluctuations that are important near T_A.

We thus see that the length scale of the mosaic and number density of the mosaic domain walls are determined by the competition between the energy cost of a domain wall and the entropic advantage of using the large number of configurations. We emphasize again, the relative domain size ξ/a depends only on the logarithms of the relaxation rate and the Lindemann ratio, nearly universal parameters themselves, and is therefore the same for different substances. This high-temperature phenomenon of universality at T_g has direct consequences for the universality of the *ultralow* temperature glassy anomalies.

We have seen that the cooperative region, which represents a nominal dynamical unit of liquid, is of rather modest size, resulting in observable fluctuation effects. Xia and Wolynes [45] computed the relaxation barrier distribution. The configurational entropy must fluctuate, with the variance given by the usual expression: [77] $\langle (\delta S_c)^2 \rangle = C_p \propto 1/D$. The barrier height for a particular region is directly related to the local density of states, and hence to

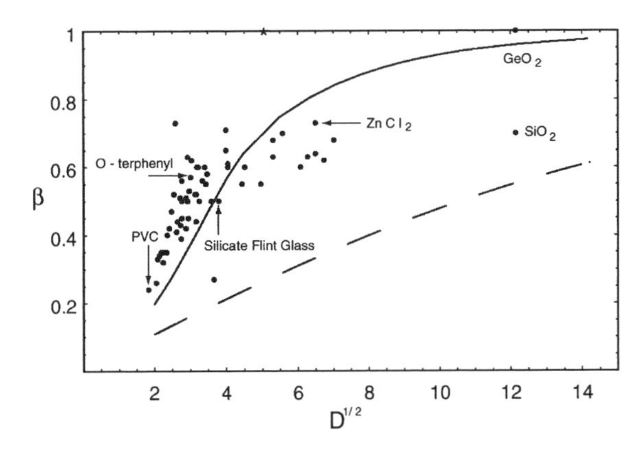

Figure 6. Shown is the correlation between the liquid's fragility and the exponent β of the stretched exponential relaxations, as predicted by the RFOT theory, superimposed on the measured values in many liquids taken from the compilation of Böhmer et al. [50]. The dashed line assumed a simple gaussian distribution with the width mentioned in the text. The solid line takes into account the existence of the highest barrier by replacing the barrier distribution to the right of the most probable value by a narrow peak of the same area; the peak is located at that most probable value. Taken from Ref. [45] with permission.

the configurational entropy itself by Eq. (6), $F^{\ddagger} \propto 1/s_c$. As a result, the barrier distribution width must correlate with the fragility. A gaussian approximation leads to a simple formula [45] $\delta F^{\ddagger}/F^{\ddagger} = 1/2\sqrt{D}$. There are also calculable deviations from gaussianity. The barrier distribution gives rise to nonexponentiality of relaxations. These are well fitted by a *stretched* exponential $e^{-(t/t_0)\beta}$. The measured β correlates with the fragility, in good agreement with the theory; see Fig. 6.

We have so far presented a simplified picture of activated relaxation in liquids, which is more accurate at temperatures close to T_K, and thus sufficiently lower than T_A—the temperature at which activated processes emerge. The transition at T_A where metastable minima emerge, along with a mosaic structure with intermediate tense regions (i.e., domain walls), is in many respects similar to a spinodal for an ordinary first-order transition, except that the number of alternative phases is very large ($e^{s_c N}$ for a region of size N). The proper treatment of this transition must include fluctuations of the order parameter and consequent softening of the droplet surface tension at temperatures close to T_A. As a result of this, closer to T_A the structural relaxation barriers are lowered from what would be expected extrapolating from near T_K: this gives deviations from the VF law. The corresponding length scales $r^{\ddagger}$ and ξ also should be smaller than would be predicted by the "vanilla," T_K-asymptotic version of the RFOT theory. These barrier "softening" effects were quantitatively estimated in Lubchenko and Wolynes [47]. They demonstrated that softening effects do vary between different substances and are more pronounced for fragile liquids. As a result, the value of the configurational entropy at T_g, as predicted by the RFOT theory with softening, varies somewhat, within a factor of 2 or so among different substances. This is in contrast to the universal $s_c(T_g) = 0.82$ of the vanilla RFOT. Nonetheless, the value of ξ at T_g is much less sensitive and seems to be always within 5% of the simple estimate above. This is shown on the right-hand side panel of Fig. 7. Understanding of the softening effect has allowed us to compute the activation barrier for liquid rearrangements in the full temperature range, including the high T part near T_A, where the barriers become low, and the transport is dominated by activationless, collisional phenomena. Consistent with this predicted softening, the T dependence of relaxation times, $\tau = \tau_{\text{micro}} e^{\gamma^2/4T^2 s_c(T)}$, as predicted by the RFOT (see Eq. (6)), fits well the experimental dependencies in the low-frequency range, but underestimates the viscosity near boiling. After softening is included, one can compute the activation component of the molecular transport, with the temperature T_A as a fitting parameter of the theory [47]. Fitting the viscosity was performed using the following obvious constraints: (1) at low temperatures, the order parameter α fluctuations are negligible, the barriers are fully established and high, and the transport is thus fully activated; (2) near boiling, the barrier vanishes, and the viscosity (known to be around a centipoise for all liquids) gives the value of τ_{micro}. The fit, shown in Fig. 7, demonstrates that of the 16–17 orders of the total dynamical range,

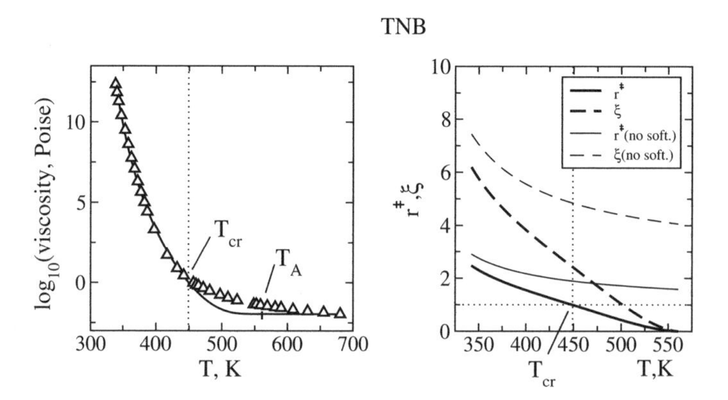

Figure 7. Experimental data (symbols) for TNB's viscosity [78] superimposed on the results of the fitting procedure (line) from Lubchenko and Wolynes [47] are shown. T_A is shown by a tickmark. (TNB = trinaphthyl benzene). The temperature T_{cr} signifies a crossover from activated to collisional viscosity, dominant at the lower and higher temperatures, respectively (see text). The temperature is varied between the boiling point and the glass transition. The right-hand side panel depicts the temperature dependence of the length scales of cooperative motions in the liquid. The thick solid and dashed lines are the critical radius $r^{\ddagger}$ and the cooperativity length ξ, respectively. Taken from Ref. [47] with permission.

about three orders, on the low viscosity side, are dominated by collisions. The experimental curve and activation-only theoretical curve differ from each other above a temperature T_{cr}. The three orders of magnitude time scale separation, arising *internally* in the theory, is indeed consistent with the prerequisite of the transport being fully activated at T_{cr} and below. The earlier discussion indicates that samples quenched (sufficiently fast) from a temperature $T > T_{cr}$ may exhibit somewhat distinct detailed molecular motions, also implying quantitative deviations from the RFOT predictions. At any rate, these samples, being caught in a very high-energy state, are expected to have small cooperative regions and also be very brittle and in general mechanically unstable. Such rapid quenches would be extremely difficult to produce in a lab, because T_{cr} corresponds to relaxation times on the order of 10^{-9} to 10^{-8}s. On the other hand, it is these ultrafast quenches that must be currently employed by simulations owing to the limitations of computer power. We speculate that the thin "amorphous" films made by vapor deposition on a cold substrate also may sometimes correspond to such ultraquenches. While one may expect a number of behaviors in the *bulk* that are qualitatively distinct from what we have discussed here, various *surface* effects are likely to be important too. For one thing, such films are thin, have a large free surface, and strongly interact with the substrate. Furthermore, there is a good reason to believe these films undergo local cracking and spontaneous crystallization [79].

The present chapter deals with phenomena in glasses at temperatures much much lower than the temperatures at which the samples form. If a sample, upon vitrification, is cooled significantly below T_g, its lattice remains practically the same as at the moment of freezing. Indeed, the typical reconfiguration barrier is at least $\ln(10^{15}) \sim 35 k_B T_g$, as already mentioned. If, on the other hand, the sample is maintained at some temperature T close enough to T_g, exceedingly slow structural relaxations take place. These attempts by the sample to equilibrate to a structure characteristic of temperature T can be detected. Achieving quantitative accuracy in such experiments is difficult. Consistent with the notion that the lattice and the barrier distribution freeze-in at the glass transition, the relaxation below T_g obeys approximately the following temperature dependence:

$$k_{\text{n.e.}} = k_0 \exp\left\{ -x_{\text{NMT}} \frac{\Delta E^*}{k_B T} - (1 - x_{\text{NMT}}) \frac{\Delta E^*}{k_B T_g} \right\} \tag{8}$$

where E^* is the equilibrated apparent activation energy at T_g and x_{NMT} lies between 0 and 1. This equation is part of the Nayaranaswany–Moynihan–Tool (NMT) empirical description of aging [80–82]. The difference in the apparent activation energy above and below T_g, as expressed by the parameter x_{NMT}, will depend on how fast the barrier itself was changing, with cooling, above T_g, under "equilibrium" cooling conditions. Since the rate of that change depends on the fragility,

$$m = \frac{1}{T_g} \frac{\partial \log_{10}\tau}{\partial(1/T)}\bigg|_{T_g} = \frac{\Delta E^*}{k_B T_g} \log_{10} e$$

one expects that $x_{\rm NMT}$ and m are correlated. The RFOT-based theory of aging in Lubchenko and Wolynes [98] analyzes structural rearrangements in a nonequilibrium glassy sample by means of Eq. (5), where the initial state is not equilibrium, but instead corresponds to the structure frozen-in at T_g. The predicted correlation between $x_{\rm NMT}$ and m is very simple—$m \simeq 19/x_{\rm NMT}$—and is consistent with experiment; see Fig. 8.

For some of the comparison of theory and experiment it is necessary to be specific about the molecular length scale a (a very detailed discussion of this quantity can be found in Ref. [47]). The molecular scale denotes the lattice

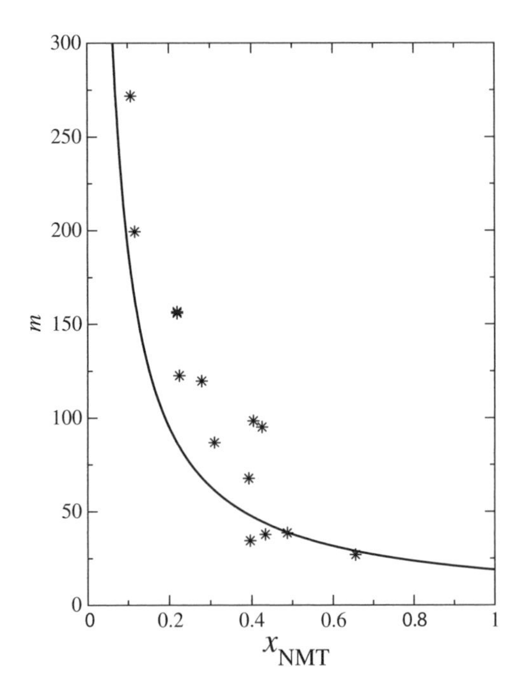

Figure 8. The fragility parameter m is plotted as a function of the NMT nonlinearity parameter $x_{\rm NMT}$. The curve is predicted by the RFOT theory when the temperature variation of γ_0 is neglected. The data are taken from Ref. [49]. The disagreement may reflect a breakdown of phenomenology for the history dependence of sample preparation. The more fragile substances consistently lie above the prediction, which has no adjustable parameters. This discrepancy may be due to softening effects.

spacing between molecular units (or "beads") that act as idealized spherical objects at the ideal glass transition at T_K. The determination of a, though approximate, is rather unambiguous and can be done using the knowledge of chemistry to give values accurate to within 15%. For example, the number of beads in a chain molecule that interacts with the surrounding only weakly is always close to the number of monomers. Highly networked substances, such as amorphous silica, present a more difficult case, because it is not clear how covalent the intermolecular bonds in these substances are. Since melting also involves freeing up molecules, with increased entropy, an independent check on the soundness of a particular bead number assignment can be done by comparing the fusion entropy of the substance (if it exists in crystalline form) with the known entropy of fusion of a hard sphere liquid or Lennard–Jones liquid, equal to $1.16k_B$ and $1.68k_B$, respectively [83]. Note, however, that knowledge of the absolute value of a is not required for most of the numerical predictions the theory will make in the quantum regime.

We thus see that the RFOT theory provides a rather complete picture of vitrification and the microscopics of the molecular motions in glasses. The possibility of having a complete chart of allowed degrees of freedom is very important, because it puts strict limitations on the range of a priori scenarios of structural excitations that can take place in amorphous lattices. This will be of great help in the assessment of the family of strong interaction hypotheses mentioned in the introduction. To summarize, the present theory should apply to all amorphous materials produced by routine quenching, with quantitative deviations expected when the sample is partially crystalline. The presence and amount of crystallinity can be checked independently by X-ray. It is also likely that other classes of disordered materials, such as disordered crystals, will exhibit many similar traits, but of less universal character.

III. THE INTRINSIC EXCITATIONS OF AMORPHOUS SOLIDS

A. The Origin of the Two-Level Systems

In this section we discuss how phenomena near the glass transition temperature, described in the previous subsection, dictate the existence and character of the quantum excitations in glasses at liquid helium temperatures and below. As mentioned earlier, a dynamical pattern of cooperative regions forms in a supercooled liquid below T_A. Each cooperative region is defined by the existence of at least *two* distinct configurations mutually accessible within the time scale τ, which characterizes the lifetime of the local mosaic pattern. Conversely, a molecular transport event is made possible by rearranging molecules within the cooperative length scale. The mosaic pattern "flickers" on the time scale τ; this process slows down dramatically upon vitrification and below T_g is referred to as "aging," as it corresponds to (very slow) structural changes. At T_g, the existing pattern of transitions (with distributed energy changes and reconfiguration barriers) freezes-in because each cell is now surrounded by a rigid lattice (this is because the rearrangements of the neighboring domains were uncorrelated at T_g). Each region of the material can now explore the phase space as prescribed by the environment at the time of freezing. Below T_g, the mosaic is spatially defined by the molecular motions that were not arrested at T_g and is thus strictly speaking only *dynamically* detectable. It is true that the weaker walls will probably be the site of (unstable) instantaneous normal modes in the fluid state with imaginary frequencies. This dynamical correlation pattern does not necessarily imply any easily discernible spatial heterogeneity in the atomic locations. In fact, there has been no direct evidence for any static type of heterogeneity of the appropriate scale in glasses so far, which definitely contributes to the (underappreciated) mystery of glasses.[3] But can the *dynamical* heterogeneity be seen directly? We will claim later that this is done for us by thermal phonons: the magnitude of scattering at the plateau can only be explained by the presence of *dynamical* heterogeneities. The latter are signified by structural transitions that scatter the phonons *inelastically*. Apart from aging (which we will ignore in the rest of the work), a particular pattern of flipping regions, as frozen-in at T_g, will persist down to the lowest temperature. The apparent size of each cell in this mosaic of flippable regions will depend on the observation time. The longer this time is, the more structural relaxation degrees of freedom (from the high barrier tail of the barrier distribution) one should observe. Eventually, in fact, the glass should crystallize.[4] In order to estimate the number of tunneling centers that are thermally active at low temperatures, we will have to find the size of the regions that allow for a rearrangement accompanied by a small energy change and, at the same time, with a low barrier. It seems reasonable that typically such barriers for multiparticle events would be very high. Nevertheless, the lattice is arrested in a high-energy state. We can thus foresee the possibility of stagewise barrier crossing (or tunneling) events, when the width of the barrier for each consecutive atomic movement is only a small fraction of a typical interatomic distance, thus rendering individual atomic movements nearly barrierless. This is as if one could define an instantaneous mode of nearly zero frequency, at each point along the tunneling trajectory. (Yet at no point is the motion harmonic per se!) The presence of such low-frequency modes should be expected given the high

[3]We note, however, that there have been instances of mistaking polycrystalline samples for truly amorphous ones.

[4]Note that there are, in principle, other ways to move molecules in a glass, in addition to the cooperative rearrangements: for example, by creating defects such as vacancies (the corresponding barriers are prohibitively high, of course).

number of configurational states available to the sample at the moment of freezing, as reflected in the high value of the configurational entropy at T_g. After all, the material is unstable, both globally and locally! (Note that the extent of bond deformation during an individual atomic movement is low—within the Lindemann length—actually affording a few "hard" places along the tunneling trajectory, where the "instantaneous" frequencies are not necessarily low.) One may contrast the situation above with, say, tunneling of a substitutional impurity in a crystal, a system which is indeed near its true ground state. Such tunneling would not contribute to the very low T thermal properties owing to a large barrier. Also, we note that multiparticle barrier crossing events *have* been seen in computer simulations of amorphous systems [14], anticipated theoretically [15, 84], and recently inferred from simulations of dislocation motions in copper [85].

We summarize the discussion so far by noting that the preceding section has demonstrated that the possible atomic motions in supercooled liquids are either purely vibrational excitations or structural rearrangements. Any possible motions *below* T_g, in terms of the *classical* basis set, must be a subset of the motions above T_g, although the dynamics of these events become quantum mechanical at low enough temperatures. Even after the system is cooled to an arbitrarily low temperature, it remains essentially in the configuration in which it got stuck at the glass transition. The density of directly accessible states at that high-energy configuration is rather high; the *total* density of states is, of course, exponentially larger, but inaccessible on realistic time scale without other regions of the glass rearranging. Since the *typical* rearrangements near T_g span about a length ξ across, we may make the following, preliminary conclusion: the nonequilibrium character of the glass transition necessarily dictates the existence of intrinsic additional nonelastic degrees of freedom in a glass, tentatively one per region of roughly size ξ, in addition to the usual vibrations of a stable lattice. The universality of ξ, in a sense, is the main clue to the cryogenic universality that is observed. A schematic of a cooperative region is shown in Fig. 9.

Note that showing the existence of low-energy tunneling paths is really a mathematical problem of finding hyperlines, connecting two points of particular latitude on a high-dimensional surface, that meander within a certain latitude range. Visualizing high-dimensional surfaces is prohibitively difficult, while the field of topology, at its present stage of development, is of little help. Yet, a completely general argument is not required here: we only need to consider a very small subset of all surfaces, such that they satisfy the (very severe) constraint on the liquid density distribution above T_g, namely, such that they conform to an "equilibrated" liquid at T_g. Because (and with the help) of this constraint, it is possible to put forth a formal argument showing that there are indeed enough low-energy structural transitions in a frozen glass: this argument will follow (albeit in the reverse order, in a sense) the argument from the

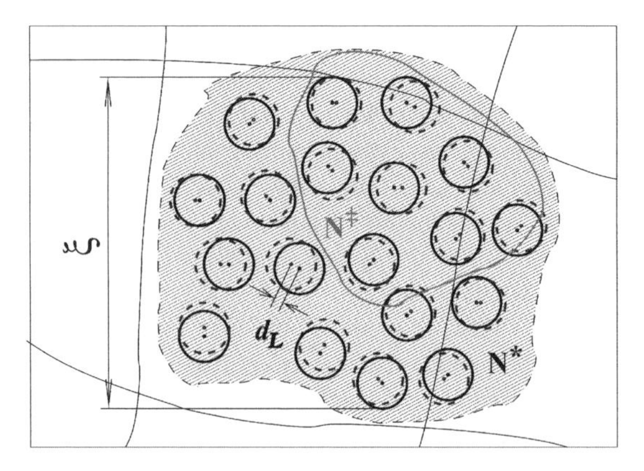

Figure 9. A schematic of a tunneling center is shown. ξ is its typical size. d_L is a typical displacement on the order of the Lindemann distance. The doubled circles symbolize the atomic positions corresponding to the alternative internal states. The internal contour, encompassing N^* beads, illustrates a transition state size, to be explained later in the text.

preceding section, where we found the *typical* trajectory for rearrangement. The key point of the microcanonical-like library construction from Section II is that the distribution width of energies of a region increases with region size. A region is guaranteed to have a state at some low energy, call it $E_{\mathrm{GS}}(N)$, as found by integration in Eq. (3). Past a certain critical size, this energy decreases as N grows larger, giving rise to the existence of a resonant state at a large enough N. One must bear in mind, however, that $E_{\mathrm{GS}}(N)$ reflects the typical freezing energy. It really gives an *upper bound* on the lowest energy level. The *actual* lowest energy state fluctuates and always lies *below* $E_{\mathrm{GS}}(N)$, although most likely not much below. Here we examine in detail the statistics of these energy states below the typical reconfiguration profile, with the aim of finding the probability of a low-energy trajectory for reconfiguring a region of size N.

We will make several preliminary, quite general notions that will guide us in constructing an adequate approximation for the local statistics of the energy landscape of a frozen lattice. First, we give a general argument of the density of frozen-in excitations, valid, as we will see shortly, in the limit of infinitely slow aging: since the atomic arrangement does not change upon freezing, the *classical* density of states of a frozen glass is that of the supercooled liquid at T_g. Those states correspond to configurations in which the system could have frozen at T_g and in principle can explore, provided they are thermally accessible and have a sufficiently low barrier separating them from a given configuration. Take a generic liquid state at T_g as the reference state. Then the Boltzmann probability to switch to a conformation higher in energy by amount ϵ is $\propto e^{-\epsilon/T_g}$. That a configuration with that energy was one of the allowed configurations upon freezing means there must have been $n(\epsilon) = (1/T_g)e^{\epsilon/T_g}$ of them. The factor $1/T_g$ arises because the energy spectrum by construction is continuous, while the actual *local* spectrum is discrete, and $\epsilon = 0$ gives the upper bound on the location of the actual ground state. The latter must be somewhere between 0 and $-\infty$: $\int_{-\infty}^{0} d\epsilon\, n(\epsilon) = 1$. This argument, however, is silent as to what the *spatial* characteristics of such excitations or their time scale are.

This inverse "Boltzmann" density of states has been computed explicitly in frustrated mean-field spin systems [86] but is of more general nature. Indeed, such distributions arise *universally* when describing the statistics of the lowest energy state of a wide class of energy distributions [87], including the random energy model [69] that will be used later on. Kinetic considerations did not explicitly enter our heuristic derivation above (or the mean-field estimates in Refs. [86] and [87]). This is directly seen by differentiating $\partial \log n(\epsilon)/\partial\epsilon = 1/T_g$. Clearly, $n(\epsilon)$ is the microcanonical density of states corresponding to the translational (liquid-like) degrees of freedom, and the system is assumed to be completely ergodic within that set of states. This corresponds to an approximation where we consider all degrees of freedom that are faster than a given time scale as *very* fast, and everything slower than that chosen time scale is regarded to be *much* slower than can be detected in the experiment. By using this same density $n(\epsilon)$, as it was at T_g, also at $T < T_g$, we formally express the fact that this subset of the total density of states no longer thermally equilibrates but stays put where it was at T_g: the subsystem of the translational degrees of freedom has undergone an entropy crisis, a glass transition. Everywhere in the previous discussion, we have been ignoring the contribution of the purely vibrational excitations to the total free energy. We thus assume that the spectrum of those elastic excitations is independent of precisely where on the glassy landscape the liquid is.

We now give the argument, first laid out in Lubchenko and Wolynes [4], that allows one to estimate the classical density of states and also simultaneously yields the size of the region where the excitation takes place. We first address the question of how many structural states are available to a compact fragment of lattice of size N, regardless of the barrier that separates those alternative states from the initial ones. This corresponds to the assumption of time scale separation mentioned earlier. Within this assumption, the low-energy limit of the spectrum must obey e^{E/T_g} so as to give a glass transition at T_g. Next, the spectrum, when integrated, must give $e^{s_c N}$ for the total number of states available to the region. Note further that we expect the reconfiguring regions to be relatively small. The atomic motions within these small regions are directly coupled and so a mean-field, gaussian density of states, that only describes lowest order fluctuations around the mean, should be accurate enough. An energy density satisfying the requirements above actually corresponds to the well-known random energy model (REM) [69], which also describes the pure state free energy in mean-field frustrated spin models:

$$\Omega_N(E) \sim \exp\left(s_c N - \frac{[E - (N\Delta\epsilon + \gamma\sqrt{N})]^2}{2\delta E^2 N} \right) \tag{9}$$

where δE^2 is the variance to be determined shortly. Here, the factor $e^{s_c N}$ gives the correct total number of states; the term $\gamma\sqrt{N}$ takes into account the interface energy cost of considering distinct atomic arrangements with the region. Note the fluctuations in the surface term are expected but are automatically included in the fluctuation of the microcanonical energy E itself. The term $N\,\Delta\epsilon$ is a bulk energy necessary to account for the observed excess energy of the frozen structure relative to the energy of the ideal structure at T_K. It is easy to relate to measured quantities: to do this, recall that the system freezes in its "ground state," with energy E_{GS}, when its entropy becomes nonextensive:

$$\Omega_N(E_{GS}) = 1 \tag{10}$$

We take the energy E_g of the liquid state at T_g as the *reference* energy. Next, note that in the *absence* of the surface energy term $\gamma\sqrt{N}$, the lowest available energy state is that of the liquid at T_K: $(E_K - E_g)/N = -\int_{T_K}^{T_g} dT\,\Delta c_p(T) \simeq -\Delta c_p(T_g - T_K) \simeq -T_g s_c$. (The two latter equalities are accurate for T_g close to T_K. The corrections would be observable [47, 48], but small.) One immediately gets from Eqs. (9) and (10) that $\Delta\epsilon = \sqrt{2\delta E^2 s_c} - T_g s_c$ ($\gamma = 0$ must be used in this estimate, but nowhere else!). Furthermore, using the microcanonical $\partial \ln \Omega_N(E)/\partial E|_{E=E_{GS}} = 1/T_g$ fixes the value of the variance $\delta E^2 = 2T_g^2 s_c$. The resultant density of states is proportional to $e^{(E-E_{GS})/T_g}$ at T_g, as already shown earlier by a general argument. Now that we have determined the thermodynamical quantities entering Eq. (9), we can find how the excess energy of an alternate ground state depends on the size N:

$$E_{GS}(N) = \gamma\sqrt{N} - T_g s_c N \tag{11}$$

where E_{GS} is defined by Eq. (10). Only low-energy excitations will be thermally active at the lowest temperatures. Therefore we are looking for excitations that are nearly isoenergetic with the reference state. This imposes an additional condition $E_{GS}(N) = 0$, thus prescribing the minimal size N_0 of a region such that $\Omega_N(0) \geq 1$ for $N \geq N_0$. A region of this size has at least one alternative structural state at the same energy. One obtains from Eq. (11) that $N_0 = (\gamma/T_g s_c)^2 = N^*$, consistent with our previous argument that any region of size N^* has a spectral density of states equal to $(1/T_g)e^{E/T_g}$. Note that Eq. (11) echoes the free energy profile of droplet growth from Eq. (6), but unlike Eq. (6), it can be used for $N > N^*$ as well. Equation (11) explicitly shows that a droplet of size larger than N^* has an exponentially increasing number of available configurations corresponding to lattices typical of T_g, consistent with the instability of droplets larger than N^* at temperatures above T_g as mentioned earlier.

The earlier microcanonical argument is basically a gedanken experiment in which we had the demon-like ability to browse through all possible atomic arrangements, given the total number of allowed states equal to e^{Ns_c}. The total sample is thus comprised of regions of the type considered in the argument (the interface energy has been taken into account by the term $\gamma\sqrt{N}$; this energy may be viewed as the penalty for considering the states of a given compact region as if this region were totally independent from the rest of the sample; c.f. our earlier comments on the locality of the liquid's energy landscape). Therefore, if the rate of conversion between the alternative glassy states can be ignored, the argument immediately yields the density of residual excitations in a frozen glass: $\simeq (1/N^* a^3 T_g)e^{\epsilon/T_g}$ ($N^* a^3 \equiv \xi^3$, of course). However, even though each of these imaginary regions has an alternative resonant state, there is so far only an undetermined chance to reach it within any particular time. In fact, the typical *classical* barrier for the excitations available to the regions of size $\leq N^* \simeq 190$ is $F^\ddagger \simeq 39 k_B T_g$. Such a barrier would seem to exclude the possibility for tunneling for a typical domain of size N^*. To account for kinetics issues, we should repeat the argument for the critical size, but also simultaneously include the life time of each considered configuration as a selection criterion. In other words, one should compute the *combined* distribution of the excitation energies, their spatial extent, and the corresponding tunneling amplitudes. Later in the chapter, we will discuss one source of correlation between the excitation energy and the tunneling amplitude owing to level repulsion effects. Nonetheless, here we present a simpler argument, given in Lubchenko and Wolynes [4], in which the tunneling rate distribution is assumed to be independent from that of ϵ. Simpler yet, we will look for the density of regions that allow for a rearrangement with a *zero*-height barrier. As vindicated *post factum*, all of these simplifications can lead only to at most a 10% error in the resulting density of states.

Imagine the process of conversion to another state as a stepwise process where the "nucleus" of this new state is increased by adding one atom at a time, as signified by the horizontal axis in Fig. 10. Such addition involves moving the atom a distance on the order of the Lindemann distance d_L. It follows then that the path connecting the two states is likely to encounter a high barrier of the order $F^\ddagger$, which effectively disconnects those two states. However, the possible configurations through which one can pass and therefore the barrier heights are *distributed* and there is a chance even for a region of size N^* to have an

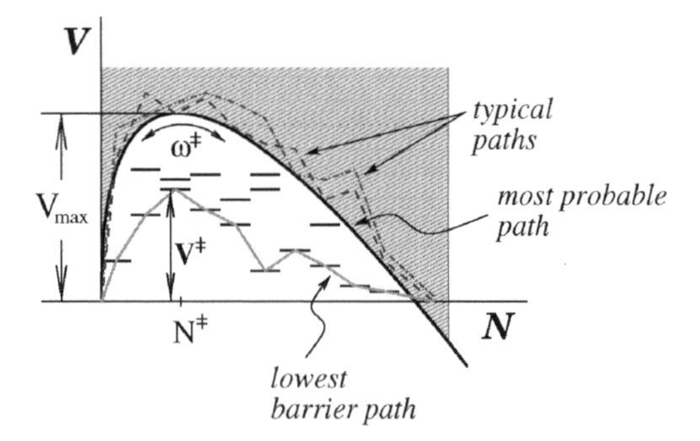

Figure 10. The smooth solid profile denotes the barrier along the most probable path. Thick horizontal lines at low energies and the shaded area at energies above the threshold represent energy levels available at size N. The jagged curves above the most probable path demonstrate generic paths, while the jagged curve beneath the most probable path shows the actual (lowest barrier) path, which would be followed if $\hbar\omega^{\ddagger} > k_B T/2\pi$.

arbitrarily low barrier. What would such a distribution be? We first have to decide whether the tunneling probability is a sum of contributions of many (interfering) paths, or whether it is dominated by a single path that has the lowest barrier. The first scenario would be realized in a highly quantum glass, where Debye temperature rivals or exceeds the glass transition temperatures [88]. Such a highly quantum glass could in fact melt due to quantum fluctuations. In our case, freezing is a completely classical process, which is signified by the fact that the barriers are proportional to a classical energy scale T_g. We now assume more specifically that the contribution of a tunneling path is proportional to $e^{-\pi V^{\ddagger}/\hbar\omega^{\ddagger}}$, where $\omega^{\ddagger}$ is a quantum frequency scale, a multiple of ω_D, and barrier $V^{\ddagger}$ scales with T_g, as mentioned earlier. This would be an accurate assignment in the case of a parabolic barrier. The form of the tunneling amplitude $e^{-\pi V^{\ddagger}/\hbar\omega^{\ddagger}}$ conforms to our expectation that the tunneling trajectory is dominated by a single path with the lowest barrier, as $V^{\ddagger}$ and $\hbar\omega^{\ddagger}$ are taken from distributions characterized by scales $k_B T_g$ and const$\times\hbar\omega_D$, respectively, the former one generally much larger than the latter (the self-consistently will turn out to be less than 1 in Section IVB). Since the energy profile along the tunneling trajectory has a complicated shape formed by *many* intermediate states separated by *small intermediate* barriers (see later discussion), it is fair to say that the state of the system at the highest barrier corresponds to the highest energy intermediate state (the "transition" state). The statistics of energy states have already been found earlier. We therefore use the distribution of Eq. (9) with only one difference: we must double the variance because the barrier height is actually the difference between *two* fluctuating quantities—the final (or initial) energy and the highest energy along the path. As a result,

$$\Omega_N(V) \sim \exp\left\{ s_c N - \frac{[V - (T_g s_c N + \gamma\sqrt{N})]^2}{4\delta E^2 N} \right\} \tag{12}$$

Distribution (12) thus gives the typical value of the barrier for the (quantum) growth of a droplet. It is easy to see from Eq. (12) that the highest barrier corresponds to rearranging a region of size $N^{\ddagger} \simeq 14$ and is equal to

$$V_{\max} = F^{\ddagger}/(2\sqrt{2} - 1) \simeq 26 T_g s_c \tag{13}$$

Since this is the hardest place to get through, we must take it as the transition state. Hence the final distribution of (transition state) barriers is the density of pure states corresponding to Eq. (12) with $N = N^{\ddagger}$ (similar to the e^{ϵ/T_g} obtained earlier). Thus

$$\Omega(V^{\ddagger}) \sim \exp\left\{ \frac{V^{\ddagger} - V_{\max}}{\sqrt{2} T_g} \right\} = \exp\left\{ -18 s_c + \frac{V^{\ddagger}}{\sqrt{2} T_g} \right\} \tag{14}$$

As one can see, the probability of having a small barrier path is exponentially suppressed. Nevertheless, owing to the large value of the energy parameter in distribution (12) the fraction of zero barrier paths per mosaic cell $\sim e^{-18 s_c} \simeq 3 \times 10^{-7}$ is actually not prohibitively small. A region larger by only 18 molecules (less than a single layer) will have $e^{18 s_c}$ more final states (and therefore paths) to go to. We therefore conclude that any region of size $\simeq 200$ molecules will have an accessible alternative state with spectral density $1/T_g$. Finally, we stress a remarkable feature of the tunneling paths statistics in glass. Mark the very rapid (exponential) scaling of the number of paths leading out of a particular local structural state on the size of the respective region. This means that the final estimate of the density of structural transitions that have low enough barriers to be thermally relevant is rather insensitive to the details of correlation between the energy of the transition and its tunneling amplitude. Consequently, even a very simple estimate of this density, such as the one above, is very robust. Finally, note that the tunneling argument above is again a microcanonical argument, such as the one leading to Eq. (9), that also takes into account (in a rather crude manner) the mutual accessibility between alternative energy states.

As we will see later, the tunneling barriers, and hence the relaxation times of the tunneling centers, are distributed. This would lead to a time-dependent heat capacity. Ignoring this complication for now, the *classical*, long-time heat capacity is easy to estimate already (assuming it exists). Since our degrees of freedom span a volume ξ^3 and their spectral density is $1/T_g$ at low energy, one obtains for the low T heat capacity per unit volume: $T/T_g\xi^3$, up to an insignificant coefficient. The coefficient at the linear heat capacity dependence is often denoted $\bar{P}$. For silica, $T_g = 1500$ K and realistic $\xi = 20\,\text{Å}$ yields $\bar{P} \simeq 6 \times 10^{45}\ \text{m}^{-3}\text{J}^{-1}$ in agreement with the experiment. (We took $a = 3.5$Å—a length scale appropriate for a tetrahedron formed by four oxygens around a silicon atom. These tetrahedra appear to be moving units in a-silica [89]). The assumption of the existence of the long-time heat capacity is empirically justified (within logarithmic accuracy) but is also consistent with the present theory; see Sections III C and V.

In conclusion, the main result of this Section is that the nonequilibrium nature of the glass transition results in the existence of residual motional degrees of freedom, a significant fraction of which remain thermally active down to the lowest temperatures. These degrees of freedom are collective highly anharmonic atomic motions within compact regions of size ξ^3, determined mainly by the length scale of the entropic droplet mosaic determined at T_g. The energy scale in the spectrum of these excitations is set by the glass temperature T_g itself. We now turn to the question of what determines the *strength* with which these entropic droplet excitations couple to the phonons. This will explain the universality in the heat conductivity at temperatures below $\sim$1 K.

B. The Universality of Phonon Scattering

First of all, what do we mean by "phonons" in amorphous materials? There is no periodicity; therefore one can only strictly speak of elastic *strain*, even if the structure is completely stable. In the latter case, low gradient strains $\nabla\phi$ are still described by a simple bilinear form:

$$H_{\text{ph}} \equiv \int d^3\mathbf{r} \frac{\rho c_s^2}{2} (\nabla\phi)^2 \tag{15}$$

The $\nabla\phi$ field is defined on an isotropic, translationally invariant flat metric, as in continuum mechanics [90], and so a wavevector k is an operational concept. It is easily seen, by dimensional analysis, that strains arising specifically due to disorder will be of higher order in k than the term in Eq. (15): the corresponding energy terms should scale with some positive power of the lattice inhomogeneity length scale(s), l_{inhom}^{ζ} ($\zeta > 0$), so as to vanish upon "zooming out." The terms will subsequently go as $k^2(l_{\text{inhom}}k)^{\zeta}$. But, as we have already seen, the amorphous lattice is not stable; that is, there are anharmonic transitions with arbitrarily small energy and barrier. Still, the regions encompassing the transitions are quite small, at most 6 lattice spacings across, which is much less than the thermal de Broglie wavelength at 1 K (about $10^3 a$ in a-silica). The unstable regions interact with the strains of the otherwise stable lattice. We conceptualize this interaction by approximating the strain with pure phonons and computing the phonon mean free path. The latter will turn out to be about 150 times longer than the phonon wavelength, so that the phonon approximation is internally consistent in that the phonons are indeed reasonably good quasiparticles, at the plateau temperatures and below. Finally, note in Eq. (15) that we have used only one phonon polarization for simplicity (it will be usually obvious how to account approximately for all three acoustic phononic branches at the end of a calculation; using this "scalar" version of the lattice dynamics, for the purposes of this chapter, boils down to neglecting the difference between the longitudinal and transverse sound, except in the later discussion of the Grüneisen parameter).

The structural transitions interact with the phonons because the energy of the transition changes in the presence of a strain. To see this explicitly, consider the elastic energy within a droplet-sized region, capable of undergoing a *low*-energy transition, as relevant at low T. Since the transition energy is low, the lowest order, quadratic expansion, in terms of *local* stress, suffices. This implies that all individual bonds within the region distort by a very small amount (already shown not to exceed d_L, even at the transition's bottleneck). We have demonstrated that such regions do indeed exist and found their density in the previous section. We separate the total elastic tensor u_{ij} into a contribution ϕ_{ij} due to the elastic stress and d_{ij} due to the tunneling displacement. The *full* elastic energy within the domain includes the energy of the elastic component *and* the anharmonic part that prescribes the tunneling displacements d_{ij}. The latter, anharmonic part of the functional requires specifying the precise configuration of the boundary (call it Ω_b). As a result, the total domain energy can be written

$$\tilde{F} = \frac{1}{2} K_{ij,kl}[(\phi_{ij} + d_{ij})(\phi_{kl} + d_{kl}) - d_{ij}d_{kl}] + \mathcal{H}(\{d_{ij}\}, \Omega_b) \tag{16}$$

where $K_{ij,kl}$ is a compressibility tensor on the order of ρc_s^2, which we are allowed to treat as a constant, with the error contributing in a higher order in k, as already mentioned. The energy functional $\mathcal{H}(\{d_{ij}\}, \Omega_b)$ includes the anharmonic, many-body interactions giving rise to the existence of the many metastable structural minima within the domain. The library construction [48] of the states corresponding to these minima was reviewed in Section II. Of the many metastable minima of the $\mathcal{H}(\{d_{ij}\}, \Omega_b)$, we are interested in the two lowest energy states, relevant at cryogenic temperatures. The size of the domain, as we have seen, is chosen so that one is guaranteed to have at least one alternative structural state of nearly the same energy, and was found to be only slightly larger than the cooperative region size at T_g [4]. By construction, the boundary state Ω_b is independent of the phonon field ϕ_{ik}. The cross term $K_{ij,kl}\phi_{ij}d_{kl}$ in Eq. (16) gives the amount by which the energy of a tunneling transition is modified by the presence of a phonon. This therefore gives the TLS–phonon coupling. For the sake of argument, assume there is no shear deformation and that the stress tensor is isotropic; a similar argument applies to the transverse phonon branches. The stress energy is then $\int_{V_\xi} d^3\mathbf{r}\, K u_{ii}^2/2$, where u_{ii} is the trace of the elastic tensor [90], which has the same meaning as $\Delta\phi$. The $d_{ii}\phi_{jj}$ cross term represents the coupling between the transition and the strain. If the phonon wavelength is larger than ξ, ϕ_{ii} is constant within the integration boundaries and can be taken out of the integral. One consequently arrives at the following *energy difference* for the defect states in the presence of a phonon: $\rho c_s^2 \phi_{jj} \int_{V_\xi} d^3\mathbf{r}\, d_{ii}$. Here d_{ii} corresponds to the transition-induced displacement between two given structural states. We are presently interested in the coupling of the lowest energy transition to the strain; higher energy transitions turn out to be intimately related to the lowest energy transition and are discussed later in Section IV. We therefore conclude that a tunneling transition, active at low temperatures, is linearly coupled to a lattice strain with the strength defined as $g = \rho c_s^2 \int_{V_\xi} d^3\mathbf{r}\, d_{ii}/2$; the corresponding term in the Hamiltonian reads

$$H_{\text{int}} \equiv \mathbf{g}\nabla\phi\sigma_z \tag{17}$$

Note that the present approach is consistent with the early microscopic estimates of the TLS–phonon coupling by Heuer and Silbey [91]. Furthermore, a similiar methodology enables one to compute the coupling of a tunneling center with *photons* as well [92]. One must bear in mind that the atomic constituents of tunneling centers are partially charged. One may thus think, informally, of a tunneling center as a collection of elemental dipoles. These dipoles will rotate during a tunneling transition, leading to an effective *electric* transition dipole. The magnitude of this transition-induced electric dipole has been argued [92] not to exceed a debye or so, consistent with experiment.

We present next two independent ways to estimate the coupling constant g. The first one is based on the realization that, at the glass transition, purely phononic excitations and a frozen-in structural transition must coexist; that is, they are of marginal stability with respect to each other. On the one hand, a local region posed to harbor a structural transition below T_g must not be "crumpled" by a passing phonon. On the other hand, the energy of the transition can only be so high, as to be sustainable by the lattice stiffness. In other words, an atom will be part a of frozen-in transition if that atom is roughly in equilibrium between the transition driving forces and the ambient lattice strain. This stability condition gives, at the molecular scale a, by Eqs. (15) and (17): $\mathbf{g}\sigma_z = -\rho c_s^2 a^3\, \nabla\phi$. The lattice strain will be distributed throughout the material in the usual manner, subject, of course, to the equipartition that fixes the variance of the strain so as to conform to the thermally available energy. We take advantage of this by multiplying the equilibrium condition by $\nabla\phi$ and noting that thermal averaging is also ensemble averaging. Thus for an atom posed to be part of low-energy structural transitions below T_g, it is generically true that

$$|\langle \mathbf{g}\,\nabla\phi\,\sigma_z\rangle| \simeq \rho c_s^2 \langle (\boldsymbol{\nabla}\,\phi)^2\rangle a^3 \simeq k_B T_g \tag{18}$$

Noting that $\langle \mathbf{g}\,\nabla\phi\,\sigma_z\rangle \simeq \langle|\mathbf{g}\,\nabla\phi|\rangle$, one arrives at a simple relation:

$$g = \sqrt{\rho c_s^2 a^3 k_B T_g} \tag{19}$$

which is the main result of this section. It is understood that this estimate is accurate up to a number of order one, and g's are likely distributed. At any rate, we observe that the TLS–phonon coupling is the geometric average between the glass transition temperature and an energy parameter $\rho a^3 c_s^2$ ($\sim 10^1$ eV) related to the cohesion energy of the lattice (note the quantity ρc_s^2 is a multiple of the Young's modulus). We point out the estimate above applies quantitatively to low barrier transitions only. The mechanical stability criterion is a zero frequency, static condition. However, it takes a finite time for a structural transition to respond to an external stress. In other words, a region harboring a *slow* transition will likely appear perfectly elastic to a high-frequency phonon. We thus arrive at the conclusion that TLS–phonon coupling must be frequency dependent; however, deviations from the result obtained above will enter in a higher order in ω, k.

Alternatively, one may attempt to estimate the integral over the derivative of the displacement field that entered in the expression for the coupling constant $g = \rho c_s^2 \int_{V_\xi} d^3\mathbf{r}\, d_{ii}/2$. Since d_{ii} is the divergence of a vector, the integral is reduced to that over a surface within the droplet's boundary: $\int_{S_\xi} ds\, \mathbf{d}(\mathbf{r})$, where $\mathbf{d}(\mathbf{r})$ is the tunneling displacement itself, near the boundary. How near? The Gauss theorem applies as long as the field $\mathbf{d}$ is continuous. This field is roughly d_L in magnitude close to the droplet's center and is zero outside the region. A function defined on a discrete lattice is expressly discontinuous. Requiring that one be able to cast a continuous tunneling displacement field on a discrete manifold, so that the field interpolates smoothly between d_L in the middle and zero outside, imposes constraints on the $\mathbf{d}$ values at the droplet's boundary. We argue this value should generically go as $(a/\xi)d_L$, up to a constant, in order to realize the interpolation and spread evenly the tensile field of the domain wall throughout the droplet. In other words, $\sim(a/\xi)d_L$ is quite obviously the lower bound, while higher values statistically imply a higher stress, and higher transition energy. Now since $\mathbf{d}(\mathbf{r})$ is randomly oriented, the integral over the droplet's border is of the order $a^2\sqrt{N^{*2/3}}(a/\xi)d_L$, where $N^{*2/3}$ is the number of molecules at the boundary. The Lindemann distance at T_g is equal to the magnitude of a thermal fluctuation; hence $d_L/a \simeq |\nabla\phi|$.[5] As a result, the coupling to the extended defects is still about $g \simeq \sqrt{\rho c_s^2 a^3 T_g}$, again within a factor of 2 or so. That the coupling of a tunneling center with the phonons should be expressible through a surface integral over the domain boundary is consistent with the following general notion: the effect of an external *mechanical* stress on the internal displacements within a local, compact region is passed on through the boundary, and so the interaction of the region with the stress can be expressed through a displacement integral over the region surface [90].

Note that, when considering a particular value of lattice distortion $\nabla\phi$ in the previous discussion, we did not specify the wavelength(s) of the phonons that contributed to this distortion; therefore the estimate of g in Eq. (19) is correct as long as the form of the interaction term (Eq. (17)) is adequate. This surely holds for long-wave phonons relevant at the TLS temperatures.

The previous argument, on the g value, allows for a relatively direct experimental verification. Let us briefly return to Eq. (18). In view of $\Delta\phi \simeq d_L/a$, one obtains a very simple expression:

$$g \simeq \left(\frac{a}{d_L}\right) T_g \tag{20}$$

Since the Lindemann ratio $d_L/a \simeq 0.1$ is empirically roughly the same for all substances, one expects the g value, as measured by sound attenuation, to be correlated with the glass transition temperature. Note that this relationship is independent of the details of the bead assignment. Equation (20), if rewritten as $T_g \simeq (d_L/a)\, g$, is almost "obvious," given the interaction of the form in Eq. (17): The typical lattice displacement, at T_g, is roughly $\Delta\phi|_{T_g} \simeq d_L/a$. On the other hand, the typical structural excitations have the energy of about T_g, at the glass transition.

With the knowledge of g, we can estimate the inverse mean free path of a phonon with frequency ω. As done originally within the TLS model, the quantum dynamics of the two lowest energies of each tunneling center are described by the Hamiltonian $H_{\text{TLS}} = \epsilon\sigma_z/2 + \Delta\sigma_x/2$. This expression, together with Eqs. (15) and (17), is a complete (approximate) Hamiltonian of

[5]For reference, $|\nabla\phi| \sim (T_g/\rho c_s^2 a^3)^{1/2}$ at T_g is about 0.05 for SiO_2, 0.06 for B_2O_3 (oxide glasses), and 0.03 for PS and PMMA (polymer glasses), in agreement with the Lindemann criterion. We stress the sensitivity to the value of the molecular size a, which is somewhat arbitrary. Here we have not calculated the bead size based on chemistry, but instead used the values of the speed of sound, as employed in the scaling procedure of Freeman and Anderson [19]. According to the definition of the Debye frequency, $a = (c_s/\omega_D)(6\pi^2)^{1/3}$.

the TLS plus the lattice vibrations. The phonon inverse mean free path is then calculated in a standard fashion [11, 93]:

$$l_{\text{mfp}}^{-1}(\omega) = \pi \frac{\bar{P}g^2}{\rho c_s^3} \omega \tanh\left(\frac{\hbar\omega}{2k_B T}\right) \tag{21}$$

This yields

$$\frac{\lambda_{dB}}{l_{\text{mfp}}} = \frac{2\pi^2}{3} \tanh(1/2) \left(\frac{a}{\xi}\right)^3 \tag{22}$$

where factor $1/3$ comes from the averaging with respect to different orientations of the defects and we used $\bar{P} \simeq 1/T_g \xi^3$. It follows that $l_{\text{mfp}}/\lambda_{dB} \sim (\xi/a)^3 \simeq 200$ up to a constant of order one. Hence the analysis above explains the universality of the combination of parameters $\bar{P}g^2/\rho c_s^2$, and relates it to the geometrical factor $(a/\xi)^3 \simeq 10^{-2}$, which is the relative concentration of cooperative regions in a supercooled liquid, an almost universal number within the random first-order glass transition theory [1], depending only logarithmically on the speed of quenching. Strictly speaking, our argument predicts the universality in l_{mfp}/λ only within 10–20% or so. This is a consequence of indeterminacy of ϵ versus Δ correlation that may be system specific, or could be due to deviations of the ξ/a ratio from the universal 5.8 at T_g. Since the latter ratio depends on the glass preparation time, the corresponding experimental study seems worthwhile.

Numerically, Eq. (22) yields $l_{\text{mfp}}/\lambda \simeq 70$, a factor of 2 less than the empirical 150 [19]. We could not have expected much better accuracy from our estimates, which used no adjustable parameters. Although it may seem that we have slightly over-estimated the number of scatterers, the size of the error is too small to reliably support this suggestion. However, this is a good place to make a few comments on the distribution of the tunneling matrix elements Δ, which will also prove useful for the discussion of the phonon scattering at higher frequencies in Section IV. The estimate for the phonon mean free path in Eq. (21) is not terribly sensitive to the form of tunneling amplitude distribution [11] (within reasonable limits). This is because the contribution of an individual TLS to the total scattering cross section is proportional to Δ^2/E^2, where $E \equiv \sqrt{\Delta^2 + \epsilon^2}$. Two common distributions have been used in the literature. One distribution simply assumed $(\Delta^2/E^2) \sim 1$ in the absence of a more specific knowledge and a flat distribution of the total energy splitting E (this is actually the original TLS model). The earlier standard tunneling model (STM) [8, 9], on the other hand, postulates $P(\Delta) \propto 1/3$ (approximately supported by our own conclusions too), which predicts a nearly flat E distribution as well. In the end, both models differ only in that the TLS model has to postulate the average

(Δ^2/E^2) value when calculating the scattering cross section (this number is absorbed into the TLS phonon coupling constant g). On the other hand, the STM allows for somewhat more closed-form derivations; however, it still has to introduce the cutoff values $\Delta_{\min}$ and $\Delta_{\max}$ as parameters (fortunately, many measured quantities depend on these parameters only logarithmically). (The distinction between the two models is described in detail by Phillips [11].) One point in favor of the STM is that it necessarily predicts time dependence of the specific heat. While a time dependence has been observed, its specific functional form has not been unambiguously established in the experiment (see Refs. [94] and [95]). We also mention, for completeness, that there is a different way to parameterize the two-level system motions within the more general, soft-potential model [34, 35]. At any rate, we see that while a two-level system's contribution to the total phonon scattering depends on the value of its $\Delta \sim E$ ratio, the precise form of the Δ distribution will change the answer quantitatively, but not qualitatively. We note, that in the context of the present calculation, it is preferable to consider the simpler, TLS setup that does not specify the Δ distribution, because our argument has so far been only *semiclassical*. Indeed, so far the tunneling amplitudes have only interested us from the perspective of the volume *density* of *allowed* transitions. We saw that an indeterminacy of the density could not exceed 10% or so due to a weak (logarithmic) dependence of that density on a specific Δ distribution. Therefore we are more confident in the numerical estimates using the TLS model setup, which does not require introducing additional parameters (such as $\Delta_{\min}$) explicitly. Nevertheless, the special role of $\Delta \sim E$ defects in scattering is worth noting. These defects have low classical energy splittings $\epsilon < \Delta$ and their dynamics are mostly determined by the *quantum* energy scale. These are the "fast" or "zero-barrier" excitations discussed earlier in the literature [96, 97], whose tunneling matrix element probably cannot be directly estimated by WKB, but we can still guess that it scales with ω_D. This suggests that using the same distribution function $P(\Delta)$ for *all* thermal defects may not be justified, as circumstantially supported by results of Black and Halperin [96], who found that the density of TLS derived from the heat capacity and conductivity measurements, respectively, are not exactly equal to each other. While this indeterminacy in the exact barrier distribution introduces only an error of order one in quantitative estimates [96] of the density of states and is not of special concern here, we note that the present theory, upon inclusion of the effects beyond the strict semiclassical picture, *does* in fact provide a mechanism for the excess of the "fast" two-level systems, as will be explained in Section V.

Strong Interaction Scenarios. By deriving the density of states of structural transitions, and their coupling to the phonons based on the known properties of the amorphous lattice, we have constructively established the microscopics of glassy excitations in excess to the purely elastic excitations. It follows from the discussion that *no other* excitations are present in glasses at 1 K and below (see also the discussion on the exhaustive classification of excitations in amorphous lattices in Section IVA). Importantly, the density of states (DOS) in excess of the phonons is due to *local* motions. This is in contrast with strong interaction scenarios (SIS) [22–24, 26] that posit that *any* local excitations (other than pure strain) would give rise to a "universal" density of states. Such universal density of states arises in SIS as a consequence of long-range, $1/r^3$, interaction mediated by the phonons, so that the actual observed DOS is a highly renormalized entity. The corresponding excitations are expressly nonlocal, possibly infinite in extent. The idea is very attractive because of its generality but remains a pure abstraction, until those bare excitations are constructively shown to *exist* in the first place. Additionally, even upon assuming some bare excitations are present, demonstrating the quantitative relationship between the effective density of states and the phonon coupling that conforms to the experimental $\bar{P}g^2/\rho c_s^2 \sim 10^{-2}$ has so far proved elusive [28, 29, 98]. On the other hand, we have shown that local structural transitions, that interact with phonons with a particular strength, must indeed take place in amorphous solids. In order to determine where the current theory stands in relation to the SIS, one may inquire whether the phonon-mediated interaction leads to the emergence of some collective density of states. It should be immediately clear that no such additional, collective DOS appears at T_g, because the argument in the previous subsection has already included *all* the effects of the surrounding of a structural transition, which simply amounted to the thermal noise at T_g delivered to the transition by the elastic waves. Of course, it does not matter what the phonon source is. What happens at low temperatures should be considered separately. The effects of interaction turn out to be small in the TLS regime and are discussed in detail in the final section. Here we give several qualitative estimates for the sake of completeness, both at high and low temperatures. The phonon-mediated interaction goes roughly as $(g^2/\rho c_s^2)(1/r^3)$, with a numerical factor less than 1 (see Section VI). Ignoring the factor, the interaction is expressed, with the help of Eq. (19), via the glass transition temperature according to $k_B T_g a^3/r^3$. Two neighboring domains, a distance ξ apart, would thus couple with strength $J_{\text{neigh}} = T_g(a/\xi)^3$. In order to assess the effects of interaction on the effective energies of individual transitions, or whether it even makes sense to talk about on-site energies after the interaction is turned on, one must compare the interaction strength to the width of the distribution of the on-site energies as derived in the absence of interaction, exactly the same way it is done in the context of Anderson localization. According to the previous subsection, the latter width, call it ΔE, is of the order T_g. The ratio $J_{\text{neigh}}/\Delta E \simeq (a/\xi)^3$ is a small number, as expected. There will be no long-range effects, due to resonant interactions, at high temperatures near T_g. At very low temperatures, only tunneling centers (TCs) with energy splitting $\sim k_B T$ or less are thermally active. While the relevant spread of the on-site energies $E_T \sim k_B T$ is now down by a factor T/T_g compared to the glass transition temperature, the *concentration* of active TCs is also down by the same factor, namely, $T/T_g \xi^3$, thus increasing the mutual separation between the regions of mobile transitions. The total dipole–dipole induced static field due to all those thermally active two-level systems at a given spot is simply $(g^2/\rho c_s^2)(T/T_g \xi^3) \simeq k_B T(a/\xi)^3$, again much smaller than the relevant on-site energy range $E_T \sim k_B T$. The motions within the tunneling centers are quantum mechanical at these low temperatures, and so one may consider possible effects of *resonant* interactions between distinct TCs, as in the Burin–Kagan scenario [26]. These effects have been shown to become important only at ultracold temperatures of μK and below [27], as already mentioned in the introduction.

Besides having explained the origin of the universality of combination $\bar{P} g^2/\rho c_s^2$ ubiquitous in the STM, we have also seen why the value of this parameter is different from 1, suggested by the strong defect–defect interaction universality scenario. This value can be traced to the relative concentration of the domains $(a/\xi)^3 < 0.01$, as just mentioned. It is curious that the defect–phonon interaction in the long wavelength limit can be expressed as a surface integral. Besides supporting our picture of the residual excitations as motions of domain walls, it points at a connection with string theories, where the elementary particles exhibit internal structure at high enough energies, which is also true in our case. In fact, this internal structure is ultimately the cause of the phenomena observed in glasses at higher temperatures, namely, the so-called bump in the specific heat and the thermal conductivity plateau, which are dealt with in Section IV.

C. Distribution of Barriers and the Time Dependence of the Heat Capacity

The STM postulated tunneling matrix element distribution $P(\Delta) \propto 1/\Delta$ implies a weakly (logarithmically) time-dependent heat capacity. This was pointed out early on by Anderson et al. [8], while the first specific estimate appeared soon afterwards [93]. The heat capacity did indeed turn out time dependent; however, its experimental measures are indirect, and so a detailed comparison with theory is difficult. Reviews on the subject can be found in Nittke et al. [99] and Pohl [95]. Here we discuss the Δ distribution dictated by the present theory, in the semiclassical limit, and evaluate the resulting time dependence of the specific heat. While this limit is adequate at long times, quantum effects are important at short times (this concerns the heat condictivity as well). The latter are discussed in Section VA.

In the tunneling argument from Section IIIA, we have suggested a WKB-type expression for the tunneling amplitude:

$$\Delta = \Delta_0 e^{-\pi V^\ddagger/\hbar\omega^\ddagger} \tag{23}$$

which would be completely correct in the case of a parabolic barrier with frequency $\omega^\ddagger$ and height $V^\ddagger$ and was motivated by the necessity to maintain the proper scaling with $\hbar$ in the denominator of the exponent, given that the typical barrier height is determined by the *classical* landscape characteristics and should scale with T_g. According to Eq. (14), $P(V^\ddagger) \propto e^{V^\ddagger/\sqrt{2}T_g}$. It follows then that

$$P(\Delta)\, d\Delta = A\left(\frac{\Delta_0}{\Delta}\right)^c \frac{d\Delta}{\Delta} \tag{24}$$

where $c = \hbar\omega^\ddagger/\sqrt{2}T_g$ should be less than 0.1 according to our estimates of $\omega^\ddagger$ (see Section IVB). A is a constant, to be commented on very shortly. The distribution in Eq. (24) becomes $P_{\text{STM}}(\Delta)\, d\Delta \propto d\Delta/\Delta$ postulated in the standard tunneling model, if $c \to 0$. As shown next, the nonzero c gives rise to an anomalous exponent $\alpha = c/2$ in the heat capacity $C \propto T^{1+\alpha}$ and a power law $t^{c/2}$ for the specific heat time dependence at long times, as opposed to a logarithmic one, predicted by the STM. While $c \simeq 0.1$ implies $\alpha \simeq 0.05$, experimentally, α seems to vary from 0.1 to 0.5. This larger value is consistent with quantum mixing effects that go beyond the semiclassical analysis, as we will discuss later.

Scaling $\hbar\omega^\ddagger$ in the denominator of the tunneling exponent implies that $\omega^\ddagger$ must be a quantum energy scale and it is indeed shown in Section IVB that $\omega^\ddagger$ is proportional to the Debye frequency ω_D. While the tunneling argument from Section IIIA only explicitly considered the statistics of the *highest* energy state along the tunneling trajectory, the expression in Eq. (23) actually does not use such a simplified picture but considers a *finite* vicinity of the barrier top. The conclusion of Section IVB that the barrier heights are distributed exponentially, such as in Eq. (14), remains true in either case. The leads to a nonzero value of c, and here we explore what consequences this has for the low-temperature heat capacity and conductivity. As follows from the discussions in Section IIIA, constant A in Eq. (24) is of order 1.

Since the temperatures in question here are so low (1 K and below), we will ignore the energy dependence of $n(\epsilon)$ in this section and take $n(\epsilon) = \bar{P}$. In order to see the time dependence of the heat capacity, we obtain the combined distribution of the TLS energy splittings E and relaxation rates τ^{-1}—$P(E, \tau^{-1})$, much as was done in Jäckle [93]—and then count in only those TLS whose relaxation time τ is shorter than a particular experimental observation time t.

The (phonon irradiation induced) relaxation rate of a TLS is [93]

$$\tau^{-1} \simeq \frac{3g^2\Delta^2 E}{2\pi\rho c_s^5} \coth\left(\frac{\beta E}{2}\right) \tag{25}$$

It follows from Eqs. (24) and (25) that

$$P(E, \tau^{-1}) = \bar{P}A\left(\frac{\tau}{\tau_{\min}(E)}\right)^{c/2}\left(\frac{\Delta_0}{E}\right)^c \frac{\tau}{2\sqrt{1 - \tau_{\min}(E)/\tau}} \tag{26}$$

where

$$\tau_{\min}^{-1}(E) \equiv \frac{3g^2E^3}{2\pi\rho c_s^5} \coth\left(\frac{\beta E}{2}\right) \tag{27}$$

is the fastest relaxation rate of a TLS with energy splitting E, achieved at $\Delta = E$, of course. As follows from Eq. (27), the rate scales roughly as the cube of temperature and is empirically on the order of an inverse millisecond at 10^{-2} K. The resultant sample's heat capacity per unit volume is then

$$C(t) = \int_0^\infty dE\left(\frac{\beta E}{2\cosh(\beta E/2)}\right)^2 \int_{t^{-1}}^{\tau_{\min}^{-1}} d\tau^{-1}\, P(E, \tau^{-1}) \tag{28}$$

where $[\beta E/2\cosh(\beta E/2)]^2$ is the TLS heat capacity. With a change of variables, Eq. (28) reads

$$C(t) = \int_0^\infty dE\left(\frac{\beta E}{2\cosh(\beta E/2)}\right)^2 \int_0^{\log(t/\tau_{\min}(E))} dz\, \frac{A}{2}\left(\frac{\Delta_0}{E}\right)^c \frac{e^{(c/2)z}}{\sqrt{1 - e^{-z}}} \tag{29}$$

At long times Eq. (29) yields a power law for both time and temperature dependence of the specific heat:

$$\lim_{t\to\infty} C(t) \propto t^{c/2} T^{1+c/2} \tag{30}$$

where we note that the temperature dependence also comes from the energy dependence of $\tau_{\min}(E) \propto E^{-3}$ in Eq. (29). The value $c \sim 0.1$ implies the long-time heat capacity should obey $C \propto T^{1+\alpha}$ at low T, where $\alpha \sim 0.05$, a smaller number than observed in amorphous materials. We must bear in mind that the issue of the exact form of the time dependence in the laboratory still appears to be unresolved, as there is no definite agreement between different experiments;

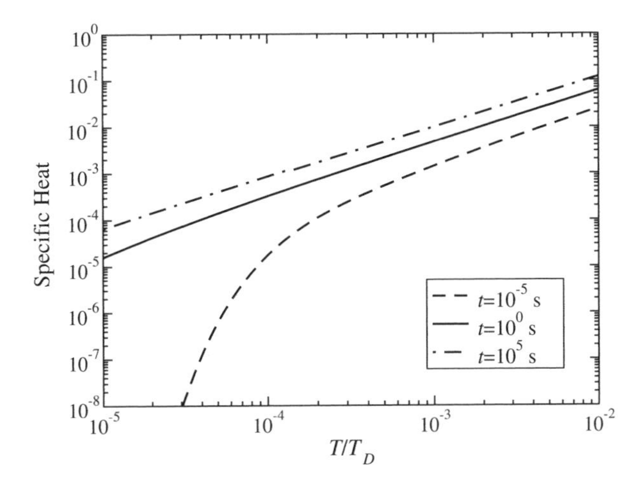

Figure 11. Displayed are the TLS heat capacities as computed from Eq. (29) appropriate to the experiment time scales on the order of a few microseconds, seconds, and hours. A value of $c = 0.1$ was used here. If one makes an assumption on the specific value of Δ_0, it is possible to superimpose the Debye contribution on this graph, which would serve as the lowest bound on the total heat capacity. As checked for $\Delta_0 = \omega_D$, the phonon contribution is negligible at these temperatures.

for references, see Hunklinger and Raychaudhuri [94], Nittke et al. [99], Pohl [95], and Sahling et al. [100]. While there is no doubt that the specific heat *is* time dependent, some experiments agree with the logarithmic time profile, as predicted by the STM; others give a lower or higher speed of variation with time. The present semiclassical prediction with $c = 0.1$ would be hard to distinguish from a logarithmic law. Finally, even though a correction to the linear temperature heat capacity dependence with $c = 0.1$ is most likely smaller than what is experimentally observed, the value of this correction is nonuniversal, consistent with empirical data.

The expression in Eq. (29) can be evaluated numerically for all values of t, and the results for three different waiting times are shown in Fig. 11 for $c = 0.1$. The value of $\tau_{\min} = 2.0\,\mu\text{s}$ at $E/T_D = 5.7 \times 10^{-4}$, derived from the present theory (also consistent with Goubau and Tait [101]) was used. The results for $t = 10\,\mu\text{s}$ demonstrate that, due to a lack of fast relaxing systems at low energies, short-time specific heat measurements can exhibit an apparent gap in the TLS spectrum. Otherwise, it is evident that the power-law asymptotics from Eq. (30) describes well Eq. (29) at the temperatures of a typical experiment.

As is clear from the previous discussion, the long-time power law behavior of the heat capacity is determined by the "slow" two-level systems corresponding to the higher barrier end of the tunneling amplitude distribution, argued to be of the form shown in Eq. (24). If one assumes that this distribution is valid for the zero-barrier tail of the $\log(\Delta)$ distribution as well, one would expect that the heat conductivity would scale as T^{2+c} at the TLS temperatures, in contrast to an observed experimental subquadratic dependence $T^{2-\alpha'}$. As we shall see in Section V, other quantum effects are indeed present in the theory and we will discuss how these contribute both to the deviation of the conductivity from the T^2 law and to the way the heat capacity differs from the strict linear dependence, both contributions being in the direction observed in experiment. Finally, when there is significant time dependence of c_V, the *kinematics* of the thermal conductivity experiments are more complex and in need of attention. When the time-dependent effects are included, both phonons and two-level systems should ideally be treated by coupled kinetic equations. Such kinetic analysis, in the context of the time-dependent heat capacity, has been conducted before by other workers [102].

IV. THE PLATEAU IN THERMAL CONDUCTIVITY AND THE BOSON PEAK

In this section we continue to explore the consequences of the existence of the low temperature excitations in amorphous substances, which, as argued in Section III, are really resonances that arise from residual molecular motions otherwise representative of the molecular rearrangements in the material at the temperature of vitrification. We were able to see why these degrees of freedom should exist in glasses and explain their number density and the nearly flat energy spectrum, as well as the universal nature of phonon scattering off these excitations at low $T(< 1 \text{ K})$.

At higher temperatures ($K_B T \sim 10^{-2.0}$ to $10^{-0.5} \hbar\omega_D$), an apparently different kind of excitation begins to appear, leading to the so-called bump in the heat capacity and plateau in the thermal conductivity, as was discussed in the introduction. We argue in this section that the transitions between the mutually accessible frozen-in minima in the amorphous lattice that give rise to the two-level system behavior at the lowest T also explain the existence of the modes responsible for this "boson peak" and the intense phonon scattering at the corresponding frequencies. This thus removes the need to invoke theoretically any additional mechanisms, although other contributions may well be present to some extent; we will try to assess this possibility in the following section.

A. Classification of Excitations in Glasses

While we believe to have mostly achieved a microscopic understanding of the excitations that are specific to the amorphous solids and are not present in other types of materials, this description is rather new and, naturally, there is a certain lack of established language that could be efficiently used to characterize these excitations. In this section, we will introduce some terminology that will be used in the rest of the chapter. At the same time, we will provide a brief general analysis of what possible qualitatively distinct types of molecular motions can exist in glasses.

Any atomic motions that take place in a frozen glass obviously may also be present in the liquid above T_g. For example, those high T motions that correspond to shear attain stiffness (on realistic time scales) below freezing. The motions in the liquid, apart from pure volume change, corresponding to the longitudinal sound, are molecular translations or, informally speaking, jumps. Above T_A, such jumps are not accompanied by a noticeable volume change and bond stretching, as no metastable structures form in the liquid at these temperatures. The barriers are therefore largely entropic. (It is nice to compare Feynman's discussion [103] of the absence of energy barriers in superfluid He in this regard.) Below T_A, such hopping already involves moving a number of molecules from one local minimum of the free energy functional to another such minimum and thus requires structural rearrangement within a certain cooperative length owing to the formation of metastable local arrangements. Molecular translations do not conserve momentum, which subsequently must be provided by the rest of the bulk. We thus call these degrees of freedom, which are relics of the translational motions in the liquid above T_g, *inelastic* degrees of freedom. They are truly inelastic also in the macroscopic sense of the word, because the existence of alternative configurations in the solid bulk, which are also coupled to the phonons, ultimately leads to irreversible relaxation, if the sample is subjected to mechanical stress thus causing a shift in the thermal population of the alternative internal, structural states. This is the mechanism behind the so-called bulk viscosity [90] (incidentally, it also contributes to the so-called relaxational phonon absorption, which we discuss in Section IVG). The switching from one energy minimum to another is accomplished by moving the domain wall—the interface between the two alternative configurations—through the local region. As mentioned earlier, this domain wall is something of an abstract entity, really a quasiparticle of a sort. Yet it has many ponderable attributes. For one thing, it has a mass (per unit area), which will be obtained in this section. It also has surface tension; therefore it can support surface vibrations, again, of a sort. Although these vibrations are realized through real atomic motions, it is more beneficial to think of them as vibrational modes of an imaginary membrane. In fact, as will be argued later, the oscillations of this membrane correspond to the indeterminacy in the exact boundary of the frozen-in domain that has more than one kinetically accessible internal state. Therefore highly anharmonic atomic motions in the real space correspond to *harmonic* motions in the space where the domain walls are defined. This mental construction does the trick of enabling us to calculate the ripplon spectrum, as demonstrated in Section IVC.

Now since it was shown in Xia and Wolynes [1] that the liquid degrees of freedom below T_A consist of switching to alternative local energy minima, we can claim our assignment of different *inelastic* modes is exhaustive (but not unique, of course!). These are, again, translations and vibrations of the domain walls.

On the other hand, any purely *elastic* motions in the glassy lattice can be thought of as a sum of ordinary, affine, displacements and nonaffine displacements (e.g., see Ref. [104]). The affine component would be the only one present in a perfectly isotropic medium and would follow the stress pattern according to a Poisson equation (the situation with a nonisotropic crystal is conceptually the same). The nonaffine displacements are a consequence of the absence of periodicity. They involve a small number of molecules and are characterized by a nonzero circulation of the displacement field. It is not clear at present whether the size of these nonaffine "islands" could be inferred in present-day computer simulations, since the amorphous structures that can be currently generated on modern computers still correspond to unrealistically rapid quenching rates. The resulting structure corresponds to a sample caught in a very high energy state with extraordinarily low barriers. As is clear by now from the random first-order transition theory, such structures correspond to temperatures close to T_A and will have very small cooperative regions approaching the molecular scale a.

We conclude this subsection by repeating ourselves that one important difference between the elastic and the inelastic modes is in how they absorb the phonons. While any static disorder can only provide Rayleigh scattering with a characteristic length scale equal to the size of the heterogeneity, the inelastic (resonant) absorption's cross section scales as the square of the phonon wavelength; it thus will considerably dominate the Rayleigh mechanism for the longer wavelength phonons (absorption saturation in the TLSs does not occur at the sound intensities typical of heat transfer).

B. The Multilevel Character of the Entropic Droplet Excitations

We hope to have convinced the reader by now that the tunneling centers in glasses are complicated objects that would have to be described using an enormously big Hilbert space, currently beyond our computational capacity. This multilevel character can be anticipated coming from the low-temperature perspective in Lubchenko and Wolynes [4]. Indeed, if a defect has at least two alternative states between which it can tunnel, this system is at least as complex as a double-well potential—clearly a multilevel system, reducing to a TLS at the lowest temperatures. Deviations from a simple two-level behavior have been seen directly in single-molecule experiments [105]. In order to predict the energies at which this multilevel behavior would be exhibited, we first estimate the domain wall mass. Obviously, the total mass of all the atoms in the droplet is so large that the possibility of *simultaneous* tunneling of all atoms is completely excluded. The tunneling, we argue, occurs stagewise; each individual motion encounters a nearly flat potential, implying low-frequency instantaneous modes.

In addition, the effective mass of the domain wall turns out to be low, *also* owing to the collective, barrierless character of the tunneling events. This is because moving a domain wall over a molecular distance a involves displacing, at any one (imaginary!) time, individual atoms only a Lindemann length d_L. Suppose this occurs on the (imaginary) time scale τ. The resulting kinetic energy is $M_w(a/\tau)^2 = N_w m(d_L/\tau)^2$, where $N_w \simeq (\xi/a)^2$ is the number of molecules in the wall and m is the molecular mass. Thus the mass of the wall M_w is only $m(\xi/a)^2(d_L/a)^2$. Using $(\xi/a) \simeq 5.8$ and $(d_L/a)^2 \simeq 0.01$ gives $M_w \simeq m/3$. This implies the mass of the wall per *atom* is very small—about a hundredth of a molecular mass—consistent with the simulations of certain barrierless dislocation motions in copper [85]. Using $(d_L/a)^2 \simeq k_B T_g/\rho c_s^2$, derived earlier, one can express the wall's mass through the material constants as $M_w \simeq (\xi/a)^2 k_B T_g/c_s^2$. The wall mass estimate above, inspired by Feynman's argument on the effective mass in liquid helium [106], is entirely analogous to the well-known estimate of the soliton mass in polyacetylene; for example, see Heeger et al. [107]. In the latter, the soliton moves a large distance, while individual atoms undergo only small displacements leading to a low soliton mass.

We can now use the typical value of the barrier curvature from our tunneling argument in Section IIIA (see Fig. 10) to estimate the typical frequency $\omega^\ddagger$ of motion at the tunneling barrier top. We now express the barrier profile $V(N)$ as a function of the droplet's radius $r \equiv a(3N/4\pi)^{1/3}$ and obtain

$$\omega^\ddagger = -(\partial^2 V/\partial r^2)/M_w \simeq 1.6(a/\xi)\omega_D \tag{31}$$

According to the quantum transition state theory [108], and ignoring damping, at a temperature $T' \simeq \hbar\omega^\ddagger/2\pi k_B \simeq (a/\xi)T_D/2\pi$, the wall motion will typically be classically activated. This temperature lies within the plateau in thermal conductivity [19]. This estimate will be lowered if damping, which becomes considerable also at these temperatures, is included in the treatment. Indeed, as shown later in this section, interaction with phonons results in the usual phenomena of frequency shift and level broadening in an internal resonance. Also, activated motion necessarily implies that the system is multilevel. While a complete characterization of *all* the states does not seem realistic at present, we can extract at least the spectrum of their important subset, namely, those that correspond to the vibrational excitations of the mosaic, whose spectral/spatial density will turn out to be sufficiently high to account for the existence of the boson peak.

C. The Vibrational Spectrum of the Domain Wall Mosaic and the Boson Peak

At low temperatures the two-level system excitations involve tunneling of the mosaic cells typically containing $N^* \simeq 200$ atoms. The tunneling path involves stagewise motion of the wall separating the distinct alternative configurations through the cell until a near-resonant state is found. At higher temperatures, other final states are possible since the exact number and identity of the atoms that tunnel can vary. These new configurations typically will be like the near-resonant level but will also move a few atoms at the boundary, that is, at the interface to another domain. This is schematically shown in Fig. 12. Alternatively, due to the quantum mechanical uncertainty of the exact location of the domain wall, its shape is intrinsically subject to fluctuations (these are zero-point vibrations of the domain wall). It is thus not surprising that the ripplon's frequencies turn out to be proportional to ω_D, the basic quantum energy scale in the system. These fluctuations of the domain boundary shape

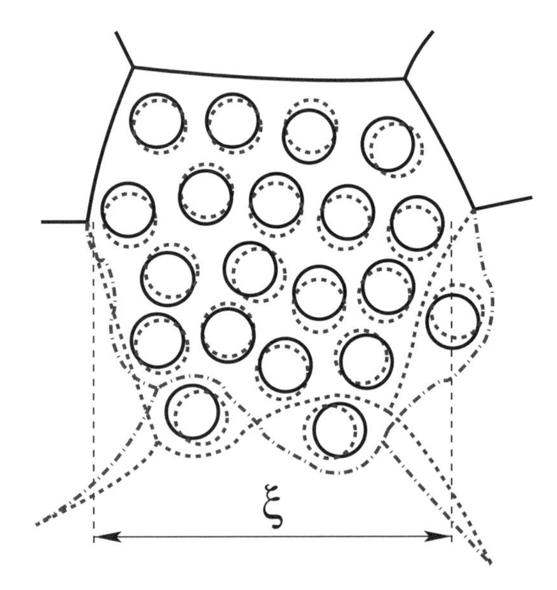

Figure 12. Tunneling to the alternative state at energy ϵ. can be accompanied by a distortion of the domain boundary and thus populating the ripplon states. The doubled circles denote atomic tunneling displacements. The dashed line signifies, say, the lowest energy state of the wall, and the dashed circles correspond to the respective atomic displacements. An alternative wall's state is shown by dash-dotted lines; the corresponding alternative sets of atomic motions are coded by dash-dotted lines. The domain boundary distortion is shown in an exagerated fashion. The boundary does not have to lie *in between* atoms and is drawn this way for the sake of argument; its position in fact is not tied to the atomic locations in an *à priori* obvious fashion.

can be visualized as domain wall surface modes (ripplons). A detailed calculation of the ripplon spectrum would require a considerable knowledge of the mosaic's geometry. At each temperature below T_A, the domain wall foam is an equilibrium structure made up of flat patches of no tension (remember the renormalized $\sigma(r) \propto r^{-1/2}$; however, fluctuations will give rise to finite curvature and tension). To approximate the spectrum we note that the ripples of wavelength larger than the size of a patch will typically sense a roughly spherical surface of radius $R = \xi(3/4\pi)^{1/3}$. The surface tension of the mosaic has been calculated from the classical microscopic theory and is given [1] by $\sigma(R) = \frac{3}{4}(k_B T_g/a^2)\log[(a/d_L)^2/\pi e](a/R)^{1/2}$, where d_L/a is the universal Lindemann ratio. It could appear that such tension could collapse an individual fragment of the mosaic but this tension is, of course, compensated by stretching the frozen-in outside walls. We approximate the effect of this compensation by an isotropic positive pressure of a ghost (i.e., vanishing density) gas on the inside.

The eigenfrequency spectrum of the surface modes of a hollow sphere with gas inside is well known (e.g., see Ref. [109] as well as our Appendix A). If we pretend for a moment that the surface tension coefficient σ is curvature independent, the possible values of the eigenfrequency ω are found by solving the following equation:

$$\cot\left[\alpha_l\left(\frac{\omega R}{c_g}\right)\right] = \left(\frac{\rho_W}{\rho_g R}\right) - \frac{(l-1)(l+2)}{(\omega R)^2}\left(\frac{\sigma}{\rho_g R}\right) \tag{32}$$

where ρ_W is the membrane's mass per unit area, and ρ_g and c_g are the gas's mass density and sound speed, respectively. As stated earlier, Eq. (32) only applies for $l \geq 2$. Finally, function

$$\cot[\alpha_l(z)] \equiv \left(-l + z\frac{j_{l+1}(z)}{j_l(z)}\right)^{-1}$$

where $j_l(z)$ is the spherical Bessel function of lth order, does exhibit behavior similar to that of the regular trigonometric cotangent for arguments on the order of unity and larger, going however to $-1/l$ as $z \to 0$. Its graph for $l = 2$ is shown in Fig. 13. An inspection shows that, for each l, the smallest solution of Eq. (32) gives the frequency of the proper eigenmode of the shell itself (shifted due to the presence of the gas inside), whereas the rest of the solutions represent the standing acoustic waves in the gas. This is especially clear in the $\rho_g \to 0$ limit, when the lowest frequency does not even depend on the gas's sound speed, whereas the rest of the solutions are obviously determined by the inverse time it takes the sound in the gas to traverse the sphere.

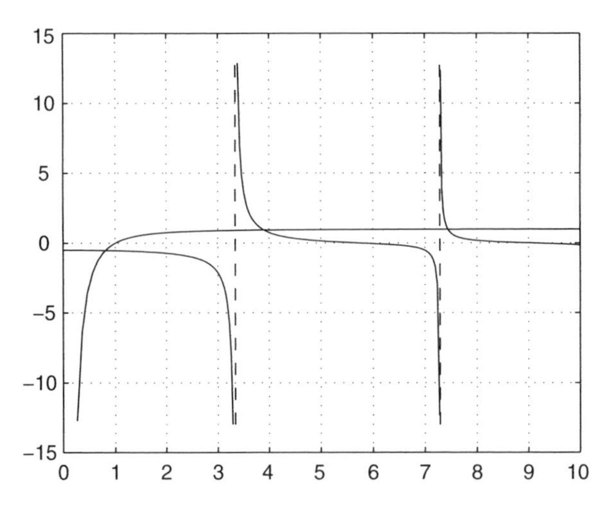

Figure 13. The functions entering Eq. (32) are shown for some arbitrary parameter values. Here, $l = 2$.

Since we are interested only in the wall's *proper* modes in the limit $\rho_g \to 0$, we get unambiguously for the frequency of an lth harmonic

$$\omega_l^2 = (l-1)(l+2)\left(\frac{\sigma}{\rho_W R^2}\right), \quad l \geq 2 \tag{33}$$

Accounting for the unusual r dependence of the surface tension $\sigma(r) \propto r^{-1/2}$ modifies the standard result from Eq. (33) by a factor of $\frac{9}{8}$. The reason is that the peculiar surface energy dependence $F_{\text{surf}}(R) = 4\pi R^2 \sigma = 4\pi\sigma_0 R^{3/2} a^{1/2}$ calls for the following dependence of pressure on the curvature:

$$p = \frac{1}{4\pi R^2} \frac{\partial F_{\text{surf}}(R)}{\partial R} = \frac{3}{2}\frac{\sigma}{R}$$

(as compared to the regular $p = 2\sigma/R$). The eigenfrequencies, in their turn, are determined by calculating the (frequency-dependent) excess pressure due to a variation in curvature. Since now $p \propto R^{-3/2}$, varying p with respect to R brings down another factor of $\frac{3}{2}$, thus giving $\frac{9}{4}$ instead of the 2 of the curvature-independent case: hence the (barely significant, but curious) correction factor of $\frac{9}{8}$ used in Lubchenko and Wolynes [5]. Since we have been assuming that the amplitude is infinitesimally small, this factor is the only consequence of having a curvature-dependent σ, which should have made the membrane oscillations even more nonlinear (as compared to $\sigma =$ const case) in the case of *finite* displacements. Pinpointing this effect, however, is clearly beyond the accuracy attempted by the present model. Finally, one finds a spectrum with

$$\omega_l^2 = \frac{9}{8} \frac{\sigma}{\rho_W R^2} (l-1)(l+2), \quad l \geq 2 \tag{34}$$

where each lth mode of a sphere is $(2l+1)$-fold degenerate. Using $\rho_W = (d_L/a)^2 \rho a$, obtained earlier in the chapter, and $T_g \simeq \rho c_s^2 a^3 (d_L/a)^2$ (Section IIIB), one finds

$$\begin{aligned} \omega_l &\simeq 1.34\, \omega_D (a/\xi)^{5/4} \sqrt{(l-1)(l+2)/4} \\ &\simeq 0.15\, \omega_D \sqrt{(l-1)(l+2)/4} \end{aligned} \tag{35}$$

Because of the universality of the a/ξ ratio [4], ω_l is a multiple of the Debye frequency. Apart from the barely significant $(a/\xi)^{1/4}$ factor, again, due to the R-dependent σ, the ubiquitous scaling $\omega_l \sim (a/\xi)\omega_D$ stresses yet another time the significance of the scale ξ. Such a scale has been previously empirically deduced by interpreting inelastic scattering experiments but has been usually ascribed to the static heterogeneity length scale, in contrast with the dynamical nature of the mosaic in the present theory. We note again that this "static heterogeneity" has never been unambiguously seen in X-ray diffraction. Owing to the material's discreteness, one does not expect harmonics of higher than $\pi[(3/4\pi)N^*]^{1/3}[(R - a/2)/R] \simeq$ 9–10th order, a relatively large number, which justifies the tacitly assumed continuum approximation. The lowest allowed ripplon mode is $l = 2$ (corresponding frequency is ~ 1 THz for silica, in remarkable agreement with the inelastic neutron scatering data [110]).

The requirement $l \geq 2$ can be understood from the symmetry considerations. The case of no restoring force, $l = 1$, corresponds to a domain translation. Within our picture, this mode corresponds to the tunneling transition itself. The "translation" of the defects center of mass violates momentum conservation and thus must be accompanied by absorbing a phonon. Such resonant processes couple *linearly* to the lattice strain and contribute the most to the phonon absorption at the low temperatures, dominated by one-phonon processes. On the other hand, $l = 0$ corresponds to a uniform dilation of the shell. This mode is formally related to the domain growth at $T > T_g$ and is described by the theory in Xia and Wolynes [1]. It is thus possible, in principle, to interpret our formalism as a multipole expansion of the interaction of the domain with the rest of the sample. Harmonics with $l \geq 2$ correspond to pure shape modulations of the membrane.

The existence of the domain wall vibrations explains and allows us to visualize, at least in part, the multilevel character of the tunneling centers as exhibited at temperatures above the TLS regime. Curiously, the existence of TLSs, even though displayed at the lowest T, is basically of classical origin due to the

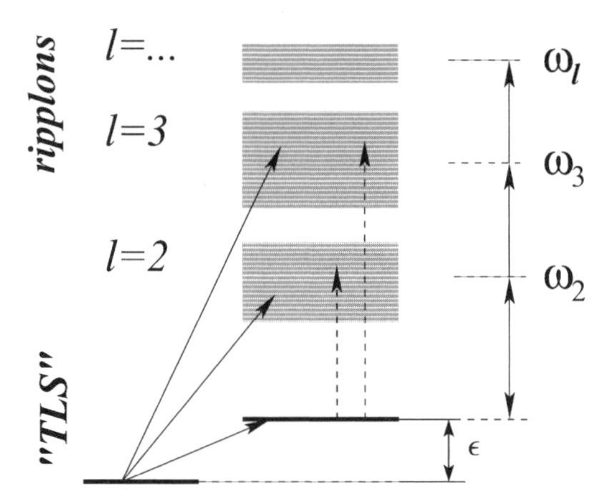

Figure 14. Tunneling to the alternative state at energy ϵ can be accompanied by a distortion of the domain boundary and thus populating the ripplon states. All transitions exemplified by solid lines involve tunneling between the intrinsic states and are coupled linearly to the lattice distortion and contribute the strongest to the phonon scattering. The "vertical" transitions, denoted by the dashed lines, are coupled to the higher order strain (see Appendix A) and contribute only to Rayleigh-type scattering, which is much lower in strength than that due to the resonant transitions.

nonequilibrium nature of the glassy state. Yet the ripplons, even though seen at higher T, are mostly due to quantum effects and would not be predicted by a strictly semiclassical theory, in which $\hbar \to 0$. A schematic of the resultant droplet energy levels is shown in Fig. 14. The arrangement of the combined internal (configurational) and ripplonic density of states, as depicted in Fig. 14, has the following motivation behind it. We include the possibility of distorting the domain wall during the tunneling transition by providing a set of vibrational states on top of the alternative internal state. The arrangement of the energy states as depicted in Fig. 14 ensures that only thermally active tunneling centers have mobile (and thus vibrationally excitable) boundaries. The atomic motions at the *inactive* defects' sites (i.e., that cannot tunnel or cross the barrier) would be indistinguishable from the regular elastic lattice vibrations. Importantly, direct transitions to the ripplonic state can occur from one of the two lowest—"TLS"—energy states of a tunneling center. This inherent assymetry between the two structural states of a tunneling center actually reflects the thermodynamical inequivalence of the two states at the glass transition temperature. While one of the states represents the local structure in (metastable) equilibrium with the current liquid arrangement around it, the other state is a configuration that must only be regarded as one of the structures along the many escape routes from the current equilibrium local state. At T_g, most of those escape routes become too costly energetically.

This is a good place to remind the reader that existing explanations of the large density of states at the BP energies have to do either with purely harmonic excitations of disordered, but perfectly stable lattice (see Introduction), or by empiracally generalizing the low energy inelastic, two-level degrees of freedom to multilevel systems, as was done, for example, by the soft potential model (SPM) [34, 35]. Such generalizations imply a connection between the anomalies seen in the TLS regime and at these higher energies. Such a connection is strongly suggested by experiment, most prominently by the strength of phonon scattering. The latter is *inelastic* at the BP energies, as it was at the TLS energies. We stress, that the rate of increase of the ripplonic density of states is much much higher than that assumed in the purely empirical SPM. Again, there is virtually no freedom to adjust the numbers in our theory.

In order to compute the heat capacity of the ripplons on top of the structural transitions, we will need to consider the (classical) density of the inelastic states in more detail than in the previous section. The density of states $n(\epsilon) = (1/T_g)e^{\epsilon/T_g}$ was derived earlier taking as the reference state the generic global liquid state corresponding to the (high-energy) configuration frozen-in at T_g. It turns out that only transitions to states with $\epsilon < 0$ (relative to the liquid state!) will contribute to the TLS density of states. Indeed, as we have shown, the size of the region that permits a *low-barrier* rearrangement must be slightly (by $\sim$ 18 molecules) larger than the generic cooperative size at T_g. On the other hand, we know from the RFOT theory that larger cooperative regions correspond to lower energy liquid structures. Therefore one of the two alternative states must be lower in energy than the generic liquid state at T_g. As a result, the negative ϵ values correspond to some of the very numerous but mostly unavailable lower lying energy states, now accessed by tunneling. Now if each of those true *local* ground states is taken as the reference one, the spectral density will now be $n(\epsilon) = (1/T_g)e^{-\epsilon/T_g}$ ($\epsilon > 0$). We consequently can let ϵ from Fig. 4 take both positive and negative values by writing

$$n(\epsilon) = \frac{1}{T_g} e^{-|\epsilon|/T_g} \tag{36}$$

We can now calculate each domain's partition function by including all possible ways to excite the system:

$$Z_\epsilon = 1 + \sum_{\{n_{lm}\}} e^{-\beta(\epsilon + \sum_{lm} n_{lm}\omega_{lm})} = 1 + e^{-\beta\epsilon} \prod_l Z_l^{2l+1} \tag{37}$$

where $Z_l \equiv 1/(1 - e^{-\beta\omega_l})$ is the partition function of an lth order ripplon mode and we used $m = -l, \ldots, 1$. Here we assume each ripplon is a harmonic oscillator. Note that since the "harmonic" excitations of frequency ω_l are on top of another (structural) excitation, we must consider the issue of the zero-point energy of these "harmonic" excitations, which is no longer a matter of simply choosing a convenient reference energy. Note that this zero-point energy is actually several orders of magnitude higher than the sub-Kelvin energies that are sufficient to excite *some* of the local structural transitions. Indeed, the energy that comprises the ripplons' ground state energy is not extracted from the thermal fluctuations of the medium but, one may say, is simply "converted" from the zero-point energy of local elastic vibrations of the lattice. At the site of a "slow" (or thermally inactive) structural transition, domain wall vibrations are indistinguishable from the regular lattice phonons, as already mentioned.

The specific heat corresponding to the partition function in Eq. (37) is found by computing $c_\epsilon = \beta^2(\partial^2 \log Z_\epsilon)/\partial^2\beta$.

$$c_\epsilon = \frac{\left(\beta\epsilon + \sum_l (2l+1)\dfrac{\beta\omega_l}{e^{\beta\omega_l}-1}\right)^2}{\left(2\cosh\left(\dfrac{\beta\epsilon + \sum_l (2l+1)\log(1-e^{-\beta\omega_l})}{2}\right)\right)^2} + \frac{\sum_l (2l+1)\left(\dfrac{\beta\omega_l}{2\sinh(\beta\omega_l/2)}\right)^2}{e^{\beta\epsilon + \sum_l (2l+1)\log(1-e^{-\beta\omega_l})}+1} \tag{38}$$

Expression (38) clearly becomes the TLS specific heat $c_{\mathrm{TLS}} = (\beta\epsilon/[2\cosh(\beta\epsilon/2)])^2$ for $T \ll \omega_l$.

In order to obtain the amorphous heat capacity per domain, we (numerically) average c_ϵ with respect to $n(\epsilon)$; the result is shown in Fig. 15 with the thin solid line.

D. The Density of Scatterers and the Plateau

In order to estimate the phonon scattering strength and thus the heat conductivity, we need to know the effective scattering density of states, the transition amplitudes, and the coupling of these transitions to the phonons.

Any transition in the domain accompanied by a change in its internal state is coupled to the gradient of the elastic field with energy $g \sim \rho c_s^2 \int ds\, \mathbf{d}(\mathbf{r})$, where $\mathbf{d}(\mathbf{r})$ is the molecular displacement at the droplet edge due to the transition (see Section IIIB). An additional modulation in the *domain wall* shape due to the current vibrational state cancels out due to the high symmetry ($l \geq 2$), as is easily seen when computing the angular part of the surface integral. We therefore conclude that any transitions between groups marked with solid lines in Fig. 4 are coupled to the phonons *with the same strength as the underlying (TLS-like) transition.* (Note that this also implies inelastic scattering off those transitions!) Incidentally, no selection rules apply for the change in the ripplon quantum numbers, being essentially a consequence of strong anharmonicity of the total transition.

We do not possess detailed information on the transition amplitudes; however, they should be on the order of the transition frequencies themselves, just as is the case for those two-level systems that are primarily responsible for the phonon

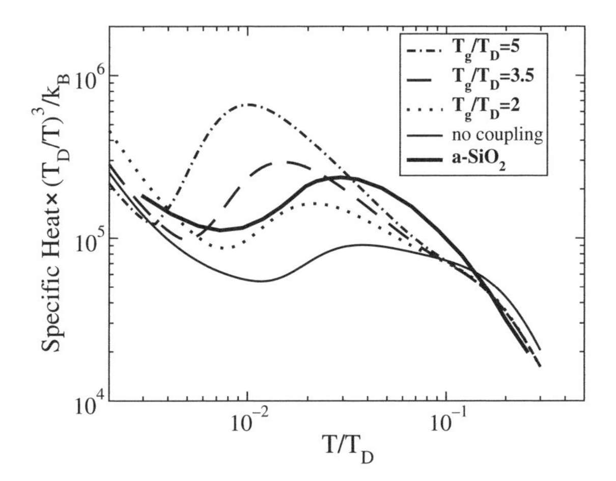

Figure 15. The bump in the amorphous heat capacity, divided by T^3, as follows from the derived TLS + ripplon density of states. The thin curve corresponds to Eq. (38). The thick solid line is experimental data for a-SiO_2 from Pohl [95]. The experimental curve, originally given in J/gK^4, was brought to our scale by being multiplied by $\hbar^3 \rho c_s^3 (6\pi^2)(\xi/a)^3/k_B^4$, where we used $\omega_D = (c_s/a)(6\pi^2)^{1/3}$, $(\xi/a)^3 = 200$, $\rho = 2.2\,\text{g/cm}^3$, $c_s = 4100\,\text{m/s}$, and $T_D = 342\,\text{K}$ [19]. The other curves take into account effects of friction and frequency shift in the ripplon frequencies. They will be explained later in Section IVF. The Debye contribution was included in our estimate of the total specific heat; it was calculated according to [111] (per particle) $c_D = 9(T/T_D)^3 \int_0^{T/T_D} dx[x^4 e^x/(e^x - 1)^2]$. This equals $234(T/T_D)^3$ at the low T. When multiplied by $(\xi/a)^3 \simeq 200$, this gives a value of 4.6×10^4 in good agreement with the experiment (see Fig. 3.10 from Pohl [95]). Note, however, that the amorphous T_D is *lower* than the corresponding crystalline one; still it seems $T_D^{\text{amorph}} > 12 T_D^{\text{cryst}}$. We remind the reader that no adjustable parameters have been used so far.

absorption at the lower T, which also have their transition amplitudes comparable to the total energy splitting. The argument is thus essentially the same as proposed earlier in Section IIIA. It should be noted, however, that the Hilbert spaces corresponding to the *quantum* in nature ripplons and the *classical* inelastic states are quite distinct (although overlapping); it thus should not be surprising that the matrix element beween *superpositions* of these spaces is on the order of the energy differences themselves. In what follows, we circumvent to an extent the question of what the precise distribution of the tunneling amplitudes of the TLS + ripplon transitions is and simply calculate the *enhancement* of the bare TLS-induced scattering due to the presence of the ripplons. This is suggested by an earlier notion that the structural transitions in glasses couple to the phonons with the same strength even if accompanied by exciting vibrational modes of the mosaic.

We now calculate the density of the phonon scattering states. Since we have effectively isolated the transition amplitude issue, the fact of equally strong coupling of all transitions to the lattice means that the scattering density should directly follow from the partition function of a domain via the inverse Laplace transform. We will not proceed this way for purely technical reasons. In addition, we will separate the cases of positive and negative ϵ (see Fig. 14), corresponding to absorption from ground and excited states, respectively.

The phonon–ripplon interaction exhibits itself most explicitly through the phonon scattering, which becomes so strong by the end of the plateau as to cause complete phonon localization. This interaction also results in other observable consequences, such as dispersion (or frequency shift) of the ripplon frequencies, as well as rendering resonances of finite width. Furthermore, we will argue that this interaction suffices to account for the nonuniversality of the plateau. First, however, we consider a simpler situation, where we assume the ripplon spectrum itself is unaffected by coupling to the phonons.

E. Phonon Scattering Off Frictionless Ripplons

If $\epsilon > 0$, the phonon absorbing transition occurs from the ground state. The total number of ways to admit energy ω into the system is

$$\begin{aligned}\rho(\omega) &= \int_0^\infty d\epsilon\, n(\epsilon) \sum_{\{n_{lm}\}} \delta\left(\omega - \left[\epsilon + \sum_{lm} n_{lm}\omega_{lm}\right]\right) \\ &= \frac{1}{T_g} \sum_{\{n_{lm}\}} \theta\left(\omega - \sum_{lm} n_{lm}\omega_{lm}\right) e^{-\beta_g(\omega - \sum_{lm} n_{lm}\omega_{lm})}\end{aligned} \tag{39}$$

where we sum over all occupation numbers of the ripplons with quantum numbers l, m ($m = -l, \ldots, l$). Using an integral representation of the step function θ, this can be rewritten

$$\rho(\omega) = \frac{1}{T_g} \sum_{\{n_{lm}\}} \lim_{\epsilon_1 \to 0^+} \int_{-\infty}^{\infty} \frac{dk}{2\pi(ik + \epsilon_1)} e^{ik(\omega - \sum_{lm} n_{lm}\omega_{lm})} e^{-\beta_g(\omega - \sum_{lm} n_{lm}\omega_{lm})} \tag{40}$$

The integral in Eq. (40) will be taken by the steepest descent method (SDM). The reason why we do not apply an analogous technique directly to the δ-function in Eq. (39) is not only because we want to get rid of the ϵ integration, but also because the SDM proved more forgiving in terms of accuracy when used to approximate the θ-function, rather than the δ-function.

For each k on the real axis, the sum over the occupation numbers n_{lm} diverges, so each integral should be taken before the summation. However, in the vicinity of the point that will turn out to be the saddle point k_0 ($\Im k_0 < -\beta_g$), all the sums are finite, so we reverse the order of summation and integration. The integration contour is shifted as shown in Fig. 16.

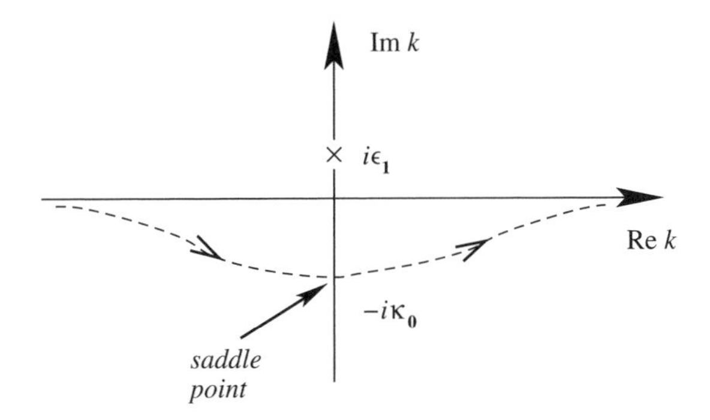

Figure 16. The integration contour from Eq. (40) is distorted as explained in text. The pole $k = i\epsilon_1$ is shown by a cross. Note the contour does not cross the pole when being shifted off the real axis.

Hence the saddle point approximation yields, for the value of the integral in Eq. (40) ($\kappa \equiv ik$),

$$\rho(\omega) = \frac{1}{T_g} \frac{1}{\sqrt{2\pi |f''(\kappa_0)|}} \frac{1}{\kappa_0} \exp\{(\kappa_0 - \beta_g)\omega - \sum_{lm} \log[1 - e^{-(\kappa_0 - \beta_g)\omega_{lm}}]\} \quad (41)$$

where the saddle point κ_0 is determined from

$$\omega = \sum_{lm} \frac{\omega_{lm}}{e^{(\kappa_0 - \beta_g)\omega_{lm}} - 1} + \frac{1}{\kappa_0} \quad (42)$$

and the curvature at the saddle point is equal to

$$|f''(\kappa_0)| = \sum_{lm} \frac{\omega_{lm}^2}{4 \sinh^2[(\kappa_0 - \beta_g)\omega_{lm}/2]} + \frac{1}{\kappa_0^2} \quad (43)$$

As is clear from Eq. (42), the approximation amounts to finding the effective temperature so as to populate the ripplonic states to match the excitation energy ω. The expression for the curvature, Eq. (43), appropriately involves the corresponding heat capacity of the excitations.

The $\omega \to 0$ and the barely relevant $\omega \to \infty$ asymptotics are easily found. As luck has it, the $\omega \to 0$ limit of Eq. (41), apart from the $1/T_g$ factor, gives $\exp(1)/\sqrt{2\pi} \simeq 1.08$, only 8% away from the correct 1. The $\omega \to \infty$ yields, on the other hand, $\rho(\omega) \propto \prod_{lm}(\omega/\omega_{lm}) \propto \omega^{96}$, as expected ($\sum_{l=2}^{9}(2l+1) = 96$). The SDM is thus reasonably accurate in this case, which could be at least somewhat evaluated by computing the value of the fourth-order term under the

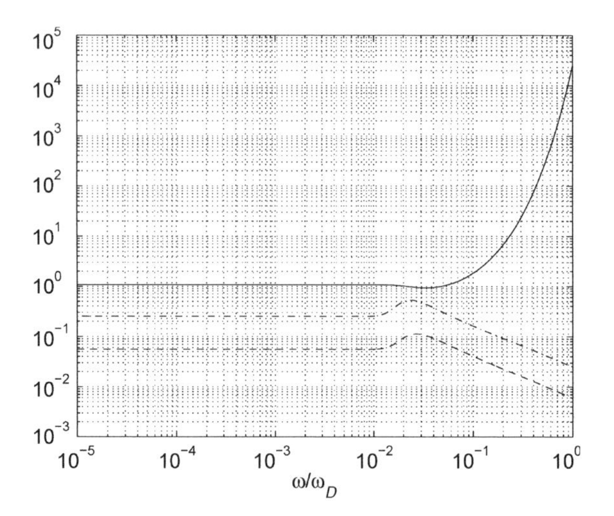

Figure 17. The solid line shows $\rho(\omega)T_g$ from Eq. (41). The dash-dotted line shows the value of the fourth-order term. The third-order term, being purely imaginary, contributes only in the sixth order; it is shown as the dashed line. However, there are other contributions to the sixth order.

exponent at "one sigma" distance from the extremal action point. This turns out to be satisfactorily small, as demonstrated in Fig. 17, along with the density of states itself as a function of ω.

When estimating absorption from the ground state, we totally ignore the depletion of ground state population at finite temperatures, when the system spends some time in an excited state. This is fine because by the relevant temperatures, the excited state absorption dominates anyway (see Fig. 14 and note that $\omega_l - |\epsilon| < \omega_l + |\epsilon|$). This case (i.e., $\epsilon < 0$) is somewhat less straightforward. Let us calculate

$$N_E(\omega) \equiv \int_0^E d\epsilon\, n(\epsilon) \sum_{\{n_{lm}\}} \delta\left(\omega - \left[\sum_{lm} n_{lm}\omega_{lm} - \epsilon\right]\right) \tag{44}$$

This expression gives the *cumulative* density of absorbing states between energies 0 and E (note the change of sign in front of ϵ). This expression can be used to estimate the total excited state absorption by computing

$$\rho_{\text{exc}}(\omega, T) \equiv \int_0^\infty dE\, f(E,T)\, \frac{\partial N_E(\omega)}{\partial E} \tag{45}$$

where $f(E,T) \equiv 2/(e^{\beta E} + 1)$ gives the appropriate Boltzmann weights. The factor of 2 is used in order to calibrate the excited state absorption relative to the ground state case: $f(0,T) = 1$. We now have two θ-functions and consequently two integrations. The SDM value for $N_E(\omega)$ is given by

$$N_E(\omega) = \frac{1}{T_g} \frac{1}{2\pi|\text{“Det”}|^{1/2}} \frac{1}{\lambda_0 \mu_0} \times \exp\left((\beta_g + \lambda_0 - \mu_0)\omega + \lambda_0 E - \sum_{lm} \log[1 - e^{-(\beta_g+\lambda_0-\mu_0)\omega_{lm}}] \right) \tag{46}$$

The corresponding saddle points are determined from

$$\omega + E = \sum_{lm} \frac{\omega_{lm}}{e^{(\beta_g+\lambda_0-\mu_0)\omega_{lm}} - 1} + \frac{1}{\lambda_0} \tag{47}$$

and

$$\omega = \sum_{lm} \frac{\omega_{lm}}{e^{(\beta_g+\lambda_0-\mu_0)\omega_{lm}} - 1} - \frac{1}{\mu_0} \tag{48}$$

Here

$$|\text{“Det”}| \equiv \sum_{lm} \frac{\omega_{lm}^2}{4\sinh^2[(\beta_g + \lambda_0 - \mu_0)\omega_{lm}/2]} \left(\frac{1}{\lambda_0^2} + \frac{1}{\mu_0^2} \right) + \frac{1}{\lambda_0^2 \mu_0^2} \tag{49}$$

is the determinant of the curvature tensor in the direction (i.e., 2D subset) of the fastest descent in the four-dimensional (complex) λ, μ space. The steepest descent approximation turns out to perform well, except at very low frequencies ($\omega < 10^{-2}\omega_D$). However, even though it overestimates the answer, it is still very small compared to the $\rho(\omega)$ calculated earlier at these frequencies, much as the complete result would be. The appropriate graph is shown in Fig. 18.

An accurate calculation of the heat conductivity requires solving a kinetic equation for the phonons coupled with the multilevel systems, which would account for thermal saturation effects and so on. We encountered one example of such saturation in the expression (21) for the scattering strength by a two-level system, where the factor of $\tanh(\beta\omega/2)$ reflected the difference between thermal populations of the two states. Neglecting these effects should lead to an error on the order of unity for the thermal frequencies. Within this single relaxation time approximation for each phonon frequency, the Fermi golden rule yields, for the scattering rate of a phonon with $\hbar\omega \sim k_B T$,

$$\tau_\omega^{-1} \sim \omega \frac{\pi g^2}{\rho c_s^2} [\rho(\omega) + \rho_{\text{exc}}(\omega, T)] \tag{50}$$

The heat conductivity then equals $\kappa = \frac{1}{3}\sum_\omega l_{\text{mfp}}(\omega) C_\omega c_s$. The mean free path cannot be less than the phonon's wavelength λ (which occurs at the Ioffe–Riegel

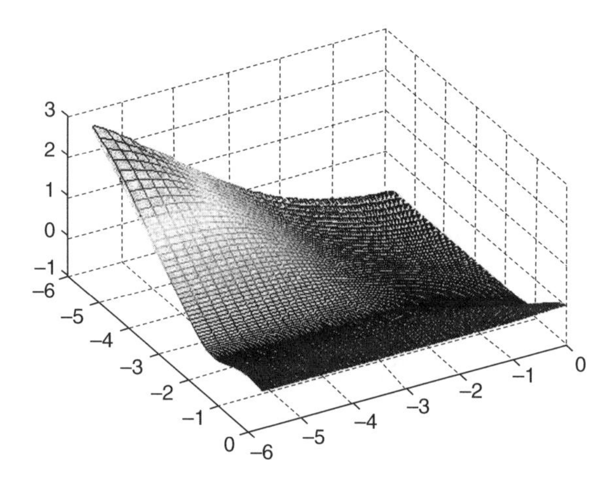

Figure 18. The value of the fourth-order correction in the exponent at the distance of 1σ from the saddle point, used to evaluate the adequacy of the SDM, is shown here. Each axis is shown in the base 10 logarithm scale. The error at the low energies ($< 10^{-2}\omega_D$) is extremely large, which is a consequence of the fact that the expression in Eq. (46) gives an incorrect asymptotics as $\omega, E \to 0$. However, it still gives a density of states that is less than 1, which is all that matters to us. The performance at the plateau energies, which is essential here, is good. Actually, in order to save space, we have presented the data for the case with the phonon coupling effects on the ripplon spectrum are taken into account ($T_g/\omega_D = 5$). This case is more interesting anyway, and the error here is slightly larger (but still tolerable!); we thus have covered all the cases.

condition). Since our theory does not cover the phonon localization regime, we account for multiple scattering effects by simply putting $l_{\text{mfp}} = c_s \tau_\omega + \lambda$. At high T, the heat is not carried by "ballistic" phonons, but rather is transferred by a random walk from site to site, as originally anticipated by Einstein [21] for homogeneous isotropic solids. The resultant heat conductivity is shown in Fig. 19. We postpone further discussion of the results above until we include the effects of coupling of the resonant transitions to the phonons on the transitions' spectrum.

F. The Effects of Friction and Dispersion

A transition linearly coupled to the phonon field gradient will experience, from the perturbation theory perspective, a frequency shift and a drag force owing to phonon emission/absorption. Here we resort to the simplest way to model these effects by assuming that our degree of freedom behaves like a localized boson with frequency ω_l. The corresponding Hamiltonian reads

$$H = \omega_l a^\dagger a + \sum_{\mathbf{k}} \omega_k b_{\mathbf{k}}^\dagger b_{\mathbf{k}} + \sum_{\mathbf{k}} \frac{(\mathbf{gk})}{\sqrt{2\omega_k V \rho}} (a^\dagger b_{\mathbf{k}} + b_{\mathbf{k}}^\dagger a) \tag{51}$$

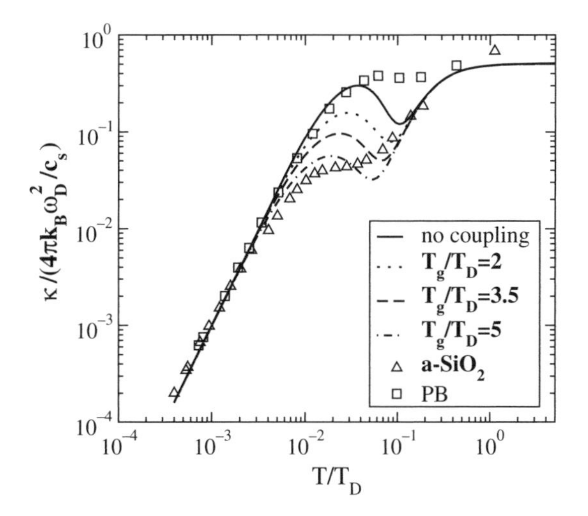

Figure 19. The predicted low T heat conductivity. The "no coupling" case neglects phonon coupling effects on the ripplon spectrum. The (scaled) experimental data are taken from Smith [112] for a-SiO$_2$ ($k_B T_g/\hbar\omega_D \simeq 4.4$) and from Freeman and Anderson [19] for polybutadiene ($k_B T_g/\hbar\omega_D \simeq 2.5$). The empirical universal lower T ratio $l_{\mathrm{mfp}}/l \simeq 150$ [19], used explicitly here to superimpose our results on the experiment, was predicted by the present theory earlier within a factor of order unity, as explained in Section IIIB. The effects of "nonuniversality" due to the phonon coupling are explained in Section IVF.

The ensuing equations of motion are

$$\begin{aligned} \dot{a} &= -i\left[\omega_l a + \sum_{\mathbf{k}} \frac{(\mathbf{gk})}{\sqrt{2\omega_k V\rho}} b_{\mathbf{k}}\right] \\ \dot{b}_{\mathbf{k}} &= -i\left[\omega_k b + \frac{(\mathbf{gk})}{\sqrt{2\omega_k V\rho}} a\right] \end{aligned} \tag{52}$$

We next introduce the following (retarded) Green's functions $A(t) \equiv -i\theta(t)\langle[a(t), a^\dagger(0)]\rangle$ and $B(t) \equiv -i\theta(t)\langle[b(t), a^\dagger(0)]\rangle$. The Fourier transforms of these Green's functions will consequently obey

$$\begin{aligned} (\omega - \omega_l)\tilde{A} &= \frac{1}{2\pi} + \sum_{\mathbf{k}} \frac{(\mathbf{gk})}{\sqrt{2\omega_k V\rho}} \tilde{B}_{\mathbf{k}} \\ (\omega - \omega_k)\tilde{B}_{\mathbf{k}} &= \frac{(\mathbf{gk})}{\sqrt{2\omega_k V\rho}} \tilde{A} \end{aligned} \tag{53}$$

From Eqs. (53), one determines the real and imaginary parts of the Green's functions self-consistently. However, we can disregard the phonons' dispersion and damping, which introduces an error in a higher order, in so far as the shifted frequencies ω_l are concerned. This yields

$$\tilde{A} = \frac{1}{2\pi}\left(\omega - \omega_l - \frac{1}{3}\frac{g^2}{4\pi^2\rho c_s^2}\lim_{\epsilon_1\to 0^+}\int_0^{k_c}\frac{k^3\,dk}{\omega/c_s + i\epsilon_1 - k}\right)^{-1} \tag{54}$$

where k_c is the cutoff wavevector whose value will be discussed shortly (we have also replaced $\sum_{\mathbf{k}} \to V\int[d^3\mathbf{k}/(2\pi)^3]$). Equation (54) immediately gives for the inverse lifetime of the internal resonance

$$\tau_{\omega_l}^{-1} = \frac{g^2}{4\pi\rho c_s^2}\left(\frac{\omega}{c_s}\right)^3 \simeq \frac{3\pi}{2\hbar}T_g\left(\frac{\omega}{\omega_D}\right)^3, \quad \omega \le \omega_c \tag{55}$$

and its frequency shift

$$\begin{aligned}\omega_l(\omega) &= \omega_l - \frac{g^2}{4\pi^2\rho c_s^2}\int_0^{\omega_c}\frac{d\omega'(\omega'/\omega_c)^3}{\omega' - \omega} \\ &\simeq \omega_l - \frac{3}{2\hbar}T_g\left(\frac{\omega_c}{\omega_D}\right)^3\int_0^{\omega_c}\frac{d\omega'(\omega'/\omega_c)^3}{\omega' - \omega}\end{aligned} \tag{56}$$

where the factor of $1/3$ has disappeared because we have accounted for the three phonon polarizations and also ignored the distinction between the longitudinal and transverse branches. The singularity in Eq. (56) at $\omega \to \omega_c$ is completely artificial, as the cutoff is not supposed to be sharp. In our numerical estimates, we use a cutoff smeared by $\delta\omega_c = \omega_c/\sqrt{D}$, where D is the glass's fragility (see Appendix A); this is, however, totally unimportant as the divergence is only logarithmic. According to Eq. (56), the frequency shift scales roughly with ω_c^3 and is thus rather sensitive to its value. Due to the dispersion, the resonance in Eq. (56) is effectively broadened because the value of the integral in Eq. (56) is positive for sufficiently small ω, but turns negative at a frequency that is a multiple of ω_c.

We approximate the phonon coupling effects by replacing in our spectral sums in Eqs. (41)–(43) and (46)–(49) the discrete summation over different ripplon harmonics by integration over "Lorentzian" profiles:

$$\sum_l \int d\omega\,\delta(\omega - \omega_l) \to \sum_l \int d\omega\,\frac{\gamma_\omega/\pi}{[\omega - \omega_l(\omega)]^2 + \gamma_\omega^2} \tag{57}$$

where $\gamma_\omega \equiv \tau_\omega^{-1}$ is a (frequency-dependent) friction coefficient and $\omega_l(\omega)$ is the ripplon frequency shifted due to the dispersion effects. This approximation amounts to having the total inverse lifetime of a transition involving more than one mode being the sum of the inverse lifetimes of the participating modes. This would be correct in the case of a frequency-independent γ but should still be adequate at the low T end of the plateau, where the absorption is mostly due to single ripplon mode processes.

The value of the cutoff frequency ω_c is close to but larger than $(a/\xi)\omega_D$ (see Appendix B), as the phonons whose wavelength is shorter than ξ cause an increasingly smaller effective gradient of the phonon field as sensed by a region of size ξ. These shorter wavelength phonons will still strongly interact with the droplets; however, at this point we could only emulate that to some extent by increasing ω_c. This also brings us back to the radiation lifetime's frequency dependence. It is now clear that for $\omega_l(\omega) > \omega_c$, γ_ω will not follow the simple cubic dependence cited earlier, the latter being probably still a safe lower estimate. We will thus use the above expression as it makes little difference computationally in the region of such intense damping. At the corresponding temperatures, the scattering is probably better formally described by the stochastic resonance [113] methodology anyway.

We are now ready to discuss the nonuniversality of the plateau. It is evident from Eqs. (55) and (56) that even though the absorbers' frequencies are determined by the quantum energy scale ω_D, the overall effective frequency *shifts* scale with T_g. The ratio T_g/ω_D seems to vary within the range of 2 to 5 among different glasses, and the nonuniversality in this number could have a substantial effect subject to the value of ω_c. As argued in Appendix B, a value of $\omega_c < 2.5(a/\xi)\omega_D$ is justified. $\omega_c = 1.8(a/\xi)\omega_D$ seems to yield the experimentally observed spread in the plateau's position. Our results for three values of T_g/ω_D are shown in Fig. 19. Since ω_c should be regarded as an adjustable parameter, we can claim to possess only circumstantial evidence that the plateau's nonuniversality is caused by the spread in the value of the ratio of the two main energy scales in the problem: the classical T_g and the quantum ω_D. On a speculative note, this phenomenon may be a sign of strong mixing (and thus level repulsion) between the ripplons and the phonons, as implicitly confirmed by a phonon localization transition at frequencies just above those at the plateau. Indeed, the self-energy of an internal resonance of dimensions ξ coupled with strength g to an elastic medium scales (within perturbation theory) as $g^2/\rho c_s^2 \xi^3 \propto T_g$. This can be viewed as lowering of an impurity band edge due to the interaction with the phonons, yet another way to express the existence of mixing between the resonant transitions and the elastic waves. Within our theory, the nonuniversality of the plateau is an internally consistent proof that the degrees of freedom causing the boson peak are *inelastic* ones, whose coupling with the phonons then must be equal to g related to the value of T_g through the stability requirement explained in Section III.

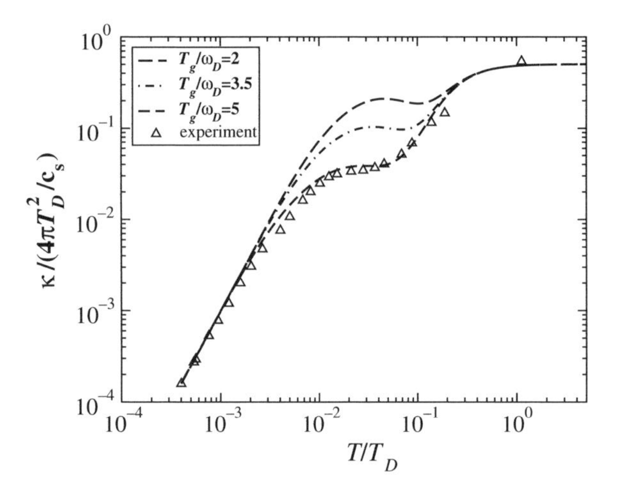

Figure 20. The predicted low T heat conductivity for three different values of $T_g/\omega_D = 5, 3.5, 2$ for a simpler model of the scattering density as explained in text. The a-SiO_2 is the same as in Fig. 19.

We now comment on the plateau slopes in Fig. 19 being noticeably more negative than the experimental value. The explaination is that we did not solve the full kinetic equation for the interacting system; instead we used a simplistic single lifetime approximation. We demonstrate this issue by briefly presenting a slightly different way to estimate $\rho_{\text{exc}}(\omega, T)$ from Eq. (45). Here we imagine we do not exactly know the thermal weight function $f(E, T)$ due to the lack of knowledge of the lifetimes in the multilevel system. On general grounds, however, this function should decrease rapidly for $\omega > \alpha T$, where the α is of order unity. This yields $\rho_{\text{exc}}(\omega, T) \simeq N_{\alpha T}(\omega)$ (where $N_E(\omega)$ was defined in Eq. (44)). In Fig. 20, we show the result of this approximation for reasonable $\alpha = 1$ and $\omega_c = 2(a/\xi)\omega_D$. Even though these curves resemble the experimental data better than in the previous figure, they do not really provide more material support for the theory than the earlier method. This discussion simply demonstrates that the basic estimates are robust enough to "survive" different levels of treatment. Also, curiously, these curves reflect the experimental tendency that the higher T plateaus seem to have a more negative slope as compared to the low T ones (see Fig. 1), which was less obvious in Fig. 19.

Finally, we return to the specific heat. The effects of the phonon coupling on the ripplon spectrum can be taken into account in the same fashion as in the conductivity case. Here we replace the discrete summation in Eq. (38) by integration over the broadened resonances, as prescribed by Eq. (57). The bump, as shown in Fig. 15, is also predicted to be nonuniversal depending on T_g/ω_D. The predicted bump for $T_g/\omega_D = 2$ seems to match well the available data for a-SiO_2, whereas the more appropriate $T_g/\omega_D \sim 4$ is about a factor of 3 lower in temperature. It is somewhat unsatisfying that the plateau's and the bump's positions cannot be thus both made to exactly match the experiment at the same time, say, by adjusting ω_l, which is certainly allowed given the qualitative character of some of the estimates. However, since we had to employ an approximation when calculating the scattering density of states, the discrepancy does not warrant too much concern, in our opinion.

G. The Relaxational Absorption

In addition to the resonant absorption, an internal resonance will also provide a so-called "relaxational" scattering mechanism. Since a crossover to the multilevel behavior of the tunneling centers leads to an increased resonant scattering, we must check whether the relaxation mechanism is enhanced as well. This latter mechanism arises because a passing phonon modifies the energy bias of a particular pair of internal states. This causes irreversible thermal equilibration processes within each pair, resulting in energy dissipation [114, 115]. This phenomenon is sometimes referred to as the bulk viscosity [116]. One important difference between the relaxational and resonant absorption is that the former does not saturate and can easily exceed the latter at low enough temperature and high enough sound intensity, which is what is usually observed in ultrasonic experiments unless special care is taken [94] (this saturation is not an issue in heat conductance, owing to the rather low sound intensities in these experiments). Applying the notion of the relaxational absorption to the two-level systems explained well the shape of the maximum in the temperature dependence of the sound speed at very low frequencies at $\sim$1 K [94], which is one of the impressive achievements of the TLS model. In Hunklinger and Raychaudhuri [94] the relation between the slopes of the logarithmic temperature profiles around the maximum was explained. At higher T, the logarithmic decrease in c_s is followed by what has been viewed by others as a mysterious linear law [117]. At higher frequencies still, the logarithmic decrease is outweighed by the just mentioned linear T dependence. We have argued that the increase in the density of the scattering states is due to thermal activation of the vibrational states of the domain walls, or matching of the thermal phonon frequency with that of a ripplon on a mobile domain wall. Does the existence of the vibrational modes modify the relaxational scattering as compared to a bare underlying two-level system? The answer is: not significantly, for the following reason. The magnitude of the dissipation due to the bulk viscosity depends on the number of local distinct molecular configurations, populated according to the Boltzmann statistics. A shift in this population results in relaxational dissipation. While having a domain wall excited may modify the energy scale in the Boltzmann distribution, which may produce some effect, it does not change the number of the intrinsic ("inelastic") glassy states, and thus will not on average enhance the relaxational scattering. This is to be compared to the resonant scattering, which depends on the degeneracy of the ripplon states and will thus intensify at higher T, subject to the degree of the ripplon's linearity. While the relaxational mechanism thus seems to play only a minor role in the phonon absorption at the plateau temperatures, its effects are observable and can explain, as we will argue later, the temperature-independent $\log \omega$ part in the sound speed variation as measured in Belessa [117]. According to Jäckle et al. [114], the variation in the speed of sound due to a collection of two-level systems is

$$\left.\frac{\delta c_s}{c_s}\right|_{\omega} = \left\langle\left\langle \frac{g^2}{2\rho c_s^2}\left(\frac{\epsilon_i}{E_i}\right)^2 \frac{\beta}{\cosh^2 \beta E_i} \frac{1}{1+\omega^2\tau_i^2} \right\rangle\right\rangle \tag{58}$$

where

$$\tau_i \simeq \frac{3g^2\Delta_i^2 E_i}{2\pi c_s^5} \coth\left(\frac{\beta E_i}{2}\right) \tag{59}$$

is the radiative lifetime of the ith TLS [93] (see also Eq. (55)), and the double angular brackets denote averaging with respect to E_i, Δ_i, and τ_i. While it would seem that detailed information on the relevant parameters' distribution is necessary to use Eq. (58), some qualitative conclusions can be made on general grounds. First, for small ω the average is dominated by the long-lifetime systems, that is, those with $\Delta \ll E$ and thus $\epsilon \sim E$. As a result, the averaging over these systems is not very sensitive to the possible correlation between E_i and τ_i, and thus the summation over the two-level system (nearly flat!) spectral density $\sim \int d\epsilon (1/T_g \xi^3)$ introduces, within order unity, only a numerical factor proportional to T/T_g (and eliminates the explicit temperature dependence). As just argued, the $(\epsilon/E)^2$ factor should only give a correction factor of order unity, and we are left with averaging expression $1/(1+\omega^2\tau_i^2)$ with respect to the lifetime distribution. At low frequencies ω, this averaging will be dominated by the TLS with the long lifetimes. Quite generally, for large τ, $P(\tau)d\tau \propto d\tau/\tau$ because τ^{-1} scales algebraically with Δ, and the distribution of $\log \Delta$ is flat (at least for small Δ), or almost flat, up to a weak power law, as argued earlier. More specifically, for a two-level system coupled linearly to the elastic strain, $\tau^{-1} \propto \Delta^2 E$ (Eq. (59)). Therefore at each E (which is incidentally only weakly dependent on Δ in the relevant long-lifetime case $\Delta/E \ll 1$), obviously $d(\log \Delta) = \text{const} \Rightarrow d(\log \tau) = \text{const}$. Thus the averaging w.r.t. τ will produce a term of the order $\sim (g^2/\rho c_s^2)(1/T_g \xi^3) \log \omega$, which is of the right order of magnitude (and sign!). Since the dimensionless factor in front of the $\log \omega$ term has been shown to be universal ($\propto (a/\xi)^3$), the present theory predicts that it should not vary significantly among the insulating glasses; in fact, according to our argument, it is proportional to the coefficient α at the logarithmic temperature dependence of the sound speed variance in the TLS regime, a rather universal quantity indeed [22]. We stress, however, that the just predicted TLS-like property should be observed in the *plateau* regime. A deviation would be a sign of more than two inelastic states playing a role in the transition. We finally mention that the lower limit in the integral over the lifetime distribution should produce a $\log T$ term, which would be masked, however, by the stronger linear dependence.

V. QUANTUM EFFECTS BEYOND THE STRICT SEMICLASSICAL PICTURE

A. Quantum Mixing of a Tunneling Center and the Black–Halperin Paradox

The preceding sections have shown that structural transitions, accompanied at high enough temperatures by vibrational excitations of the mosaic, account for the most conspicuous departures of the low-temperature behavior of glasses from the prescriptions of a standard harmonic lattice theory—namely, the existence of multilevel intrinsic resonances in an amorphous sample made by quenching a supercooled liquid. At the lowest temperatures, these resonances behave for the most part as if they were two-level systems, while at higher T the density of states of these intrinsic excitations grows considerably and leads to the boson peak phenomena. While we have computed the density of states accessible by *tunneling* even at the lowest temperatures, we have assumed, within a semiclassical approach, that having a small tunneling barrier between alternative local structural states does not affect significantly the corresponding spectrum $n(\epsilon)$ of the lowest energy transitions from its classically defined value. Likewise, we have assumed that the vibrational spectrum of moving domain walls is unaffected by the presence of tunneling, that would in principle mix those vibrations quantum mechanically. Clearly, the transitions that are active at low T must have some significant (even if small) overlap between the wavefunctions corresponding to the alternative structural states. This overlap would lead to the familiar effects of repulsion between the semiclassically determined energy levels. This could be described as partial quantum melting of some tunneling centers, but it is probably better to use the term "quantum mixing."

In this section we estimate the magnitude of these quantum mixing effects. Even though the strictly semiclassical theory agrees well with experiment as is, making such estimates that go beyond it is useful for two distinct reasons. First, we must check to what extent the semiclassical picture, tacitly assumed earlier, is a consistent zeroth order approximation to a more complete treatment.

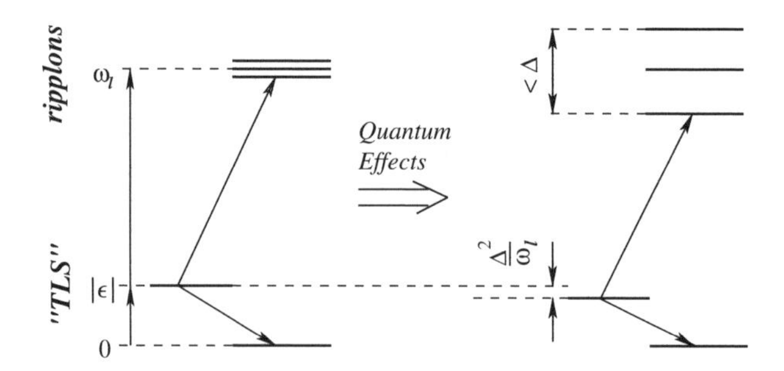

Figure 21. A low-energy portion of the energy level structure of a tunneling center is shown. Here $\epsilon < 0$, which means that the reference, liquid, state structure is *higher* in energy than the alternative configuration available to this local region. A transition to the latter configuration may be accompanied by a distortion of the domain wall, as reflected by the band of higher energy states, denoted as "ripplon" states.

Second, it is important to ask whether the expected corrections to the strict semiclassical theory lead to observable consequences. In what follows, we provide approximate arguments that indeed such corrections are discernible and may even potentially answer some longstanding puzzles in this field.

As the starting point in the discusion, we consider a simplified version of the diagram of a tunneling center's energy states from Fig. 14 with $\epsilon < 0$, as shown on the left hand side of Fig. 21. We remind the reader that the $\epsilon < 0$ situation, explicitly depicted in Fig. 21, implies lower transition energies than when the semiclassical energy difference $\epsilon > 0$ and thus dominates the low-temperature onset of the boson peak and the plateau.

Accounting for tunneling, the low-energy portion of the Hamiltonian that corresponds to Fig. 21 is as follows:

$$H_i = \begin{pmatrix} 0 & \Delta/2 & 0 & 0 \\ \Delta/2 & |\epsilon| & \Delta_{i_1} & \Delta_{i_2} \\ 0 & \Delta_{i_1} & \hbar\omega_{i_1} & 0 \\ 0 & \Delta_{i_2} & 0 & \hbar\omega_{i_2} \end{pmatrix} \qquad (60)$$

where the semiclassical values of the ripplonic energies are denoted as $\hbar\omega_i$ and the transition amplitudes to those levels are Δ_i, respectively (only the two lowest of those ripplonic states are shown in Eq. (60)). As argued in detail in Section IV, only one of the lowest two energy levels in a tunneling center (the top one in this case) is directly coupled to the higher, ripplonic, energy states. Obviously, virtual transitions to those high-energy states will result in lowering the energy of the higher level. There are no direct transitions from the bottom state of energy 0, as explained in the previous section, and therefore its position is unaffected by the presence of the ripplons. Consequently, the effective energy splitting of the two-level system (with $\epsilon < 0$) will be lower than the classical value obtained earlier, and the smaller the original value of ϵ was, the more pronounced the effect will be. In what follows we estimate the consequences of this effect on the apparent energy spectrum of the lower excitations, that is, the empirical two-level systems. In the limit of infinitely small tunneling amplitude Δ, the decrease in ϵ could be estimated using a perturbative expansion:

$$|\tilde{\epsilon}| = |\epsilon| - \sum_i \frac{\Delta_i^2}{\hbar\omega_i - |\tilde{\epsilon}|} \tag{61}$$

Here $\tilde{\epsilon}$ is the new value of the energy splitting, the ω_i are the ripplon frequencies, and the Δ_i are tunneling amplitudes of transitions that excite the corresponding vibrational mode of the domain wall. Those amplitudes will be discussed in due time; for now, we repeat, the expression above will be correct in the limit $\Delta_i/\hbar\omega_i \to 0$. Finally, the renormalized value $\tilde{\epsilon}$ was used in the denominator. While, according to Feenberg's expansion [118], including $\tilde{\epsilon}$ in the resolvent is actually more accurate, we do it here mostly for convenience.

Given that the semiclassical values of eigenvalues $\hbar\omega_i$ are known, the low-energy portion of the energy level structure of the tunneling center, as shown in Fig. 21, gives a quantitative idea of the eigenenergies of the full Hamiltonian only in the limit of a very small tunneling splitting Δ. In a complete treatment, all transition amplitudes must be included and the Hamiltonian diagonalized. In general, such diagonalization (and, in our case, the system's "quantization") is difficult; however, it could still be conducted approximately in some cases of interest. Consider, for the sake of argument, the following situation, where Δ is not necessarily smaller than ϵ but $\sum_i \Delta_i^2/\hbar\omega_i$ is. In this arrangement, the energy shift due to the higher lying states can be computed using perturbation theory and yields a "renormalized" value of the classical energy difference that we have called $\tilde{\epsilon}$. This procedure also modifies the tunneling amplitude Δ of the underlying TLS by a *multiplicative* factor according to

$$\tilde{\Delta} = \Delta\left[1 - \frac{1}{2}\sum_i \frac{\Delta_i^2}{(\hbar\omega_i)^2}\right] \tag{62}$$

Following this, the full energy splitting of the TLS tunneling transition is computed using $E = \sqrt{\Delta^2 + \tilde{\epsilon}^2}$. The important feature of the argument is that Δ (or $\tilde{\Delta}$) is allowed to take arbitrarily large values relative to ϵ and the ratio of the two parameters is not treated perturbatively. The lowering of ϵ due to virtual transitions among the higher energy states changes somewhat the effective density of transition energies E that directly enters into the heat capacity and

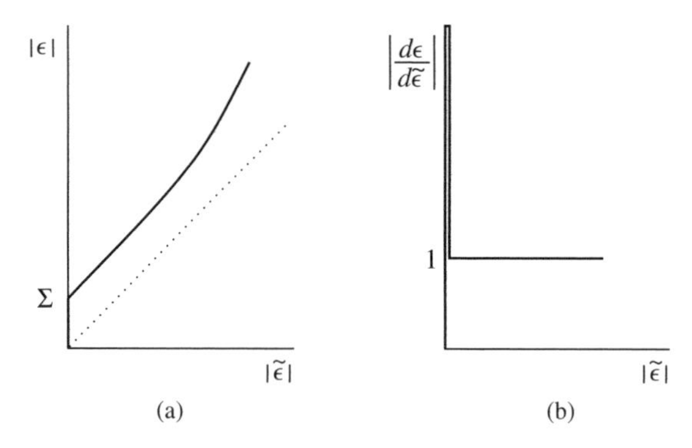

Figure 22. Shown in panel (a) is the relation between the bare energy difference ϵ between frozen-in structural states in a glass and the effective splitting $\tilde{\epsilon}$ that is smaller due the level repulsion in the tunneling center. Panel (b) depicts schematically the derivative of ϵ with respect to $\tilde{\epsilon}$, which is used to compute the new effective distribution $P(\tilde{\epsilon})$ of the transition energies.

conductivity calculations. While Eq. (61) is perturbative, it should accurately give finite effects in the mean-field limit of infinitely many transitions $\hbar\omega_i$ coupled infinitely weakly to one of the two bottom states of the tunneling center. We will analyze the physical consequences assuming the accuracy of Eq. (61). We must, of course, bear in mind that while a tunneling center is nearly a mean-field entity, owing to the strong correlations, it is actually of finite, albeit molecularly large, size. Let us plot $|\epsilon|$ as a function of $|\tilde{\epsilon}|$ (see Fig. 22a). Clearly, the smallest allowed value of the effective classical splitting $|\tilde{\epsilon}|$ of zero corresponds to a finite value of $|\epsilon|$ equal to

$$\Sigma \equiv \sum_i \frac{\Delta_i^2}{\hbar\omega_i} \tag{63}$$

Therefore smaller values of $|\epsilon|$ do not correspond to physically realizable systems, according to Eq. (61). The overall excitation spectrum of the structural transitions with those (small) values of $\epsilon < \Sigma$ is strongly affected by the ease of tunneling. As a consequence, the ϵ and Δ distributions become correlated. The quantity $\sum \Delta_i^2/\hbar\omega_i$ is of central importance for this section of the chapter; therefore we will discuss it now in some detail. The wavefunctions of the highly quantum tunneling centers are heavily mixed combinations of the classical states corresponding to potential energy minima. Transitions between such states are strongly coupled to lattice vibrations and, among other things, would strongly scatter phonons. This result is expected on rather general grounds and was exploited earlier when we noted that the eigenstates of the classical potential energy are largely unrelated to the eigenstates of the vibrational modes of the domain wall; hence one expects that transition amplitudes of exciting a ripplon (which, loosely speaking, is a highly anharmonic combination of both structural and vibrational modes of the lattice) are expected to be comparable to the ripplon energy itself. If this is the case, then number Σ is actually very large relative to ϵ and, strictly speaking, one should not use a *perturbative* expression, such as in Eq. (61), in order to assess the lowering of ϵ due to quantum effects. We note, however, that it is more likely that the seeming discrepancy simply stems from our "quantization" procedure being so far rather naive, so let us slow down a bit and attempt to outline briefly a more careful way to quantize the tunneling centers' dynamics. First of all, recall that we actually know the values ω_i in a strongly quantum regime, because they were computed assuming a freely moving membrane. On the other hand, we know that in the classical limit domain wall vibrations are indistinguishable from the lattice vibrations. Remarkably, the vibrational eigenfrequencies of a "box" of dimensions $\xi - \omega_D$ times a number from a/ξ to 1 span roughly the same range of energies that the ω_i do. Therefore, even though the quantization procedure will "reshuffle" all the ripplonic states, it will not significantly shift their position as a whole. Next, since static lattice inhomogeneities scatter phonons only elastically, the coupling of the ripplonic excitations to the phonons must scale with a positive power of Δ so as to vanish in the classical limit. Consequently, in the limit of small tunneling matrix element Δ, the transition amplitudes Δ_i must scale with a positive power of Δ. On the other hand, as mentioned many times, in the quantum regime, the Δ_i are of the order $\hbar\omega_i$ and are not directly related to Δ. Therefore a careful quantization procedure of introducing quantum tunneling in the system must combine and rationalize both of those seemingly conflicting notions. It will turn out that, in the quantum regime, the Δ_i are related to the strength of the underlying tunneling transition Δ only in a certain renormalized sense. We will now outline this renormalization/quantization procedure. This procedure imposes certain restrictions on how adding the possibility of tunneling to a local structural transition can be performed, so that the structure of the energy levels of the transition, that we know from general arguments, is *preserved*. Recall that "switching on" tunneling to the higher energy states $\epsilon + \hbar\omega_i$ not only lowered ϵ, but also made Δ smaller by a factor of $[1 - \sum_i \Delta_i^2/(\hbar\omega_i)^2]$. This expression is only valid if the sum is a small number, so that the whole correction factor is necessarily positive and only changes the effective *magnitude* of the matrix tunneling element, but not its sign. We must require that Δ not change its sign in the course of the "quantization," but only its absolute value, because the (ordinarily small) value Δ only reflects the (ordinarily small) *configurational* overlap between two local structural states, while the sign (or complex phase, in general) bears no special meaning here because no particular *spatial* symmetry is involved in the problem. (Such spatial symmetry is important, for instance, when computing overlaps between eigenstates, or near-eigenstates, of orbital momentum centered around close locations in space.) Thus the final answer should depend only on $|\Delta|^2$. This becomes especially evident after the following realization. Note that the expression in the brackets in Eq. (62) is a small coupling limit of what can be considered a Franck–Condon factor. The appearance of such a factor after the introduction of nonzero transition amplitudes is natural: the degrees of freedom that used to be classical and static can now follow to some extent a selected motion in the system. The Franck–Condon (FC) factor is the overlap between the the initial and final wavefunctions of these other ("ripplonic") degrees of freedom, corresponding to the initial and final configuration of that selected motion. (A well-known example of a FC factor arising in an analogous, dynamical fashion is the tunneling matrix element renormalization in the spin–boson problem [119].) Let us suppose, in a simplified manner, that the effective renormalization of *all* of the newly introduced tunneling amplitudes occurs in a similar fashion. This allows one to self-consistently close Eq. (62) and rewrite it for a representative amplitude Δ_Q:

$$\tilde{\Delta}_Q = \Delta_Q \left[1 - \frac{B}{2} \frac{\tilde{\Delta}_Q^2}{(\hbar\omega_D)^2} \right] \tag{64}$$

where replacing Δ_Q by $\tilde{\Delta}_Q$ inside the brackets preserves the approximation's order. B is a numerical constant, reflecting the sum over ripplon states with their vibrational frequencies ω_i, and we have replaced ω_i by the Debye frequency ω_D. The two must be related at the end of the renormalization, as we already know. Identifying the expression in brackets with a Franck–Condon factor reminds us that Δ_Q (or $\tilde{\Delta}_Q$) is not only a (generic) tunneling amplitude but also can be considered a coupling constant; therefore only its absolute value is physically relevant in the present context, not its sign. The smallness of Δ_Q and $\tilde{\Delta}_Q$ lets us recast Eq. (64) as

$$d \log\left(\frac{\tilde{\Delta}_Q}{\Delta_Q}\right) = -\frac{B}{2} d\left[\frac{\tilde{\Delta}_Q^2}{(\hbar\omega_D)^2}\right] \tag{65}$$

where the reference state is $\tilde{\Delta}_Q = \Delta_Q = 0$, $\lim_{\Delta_Q \to 0}(\tilde{\Delta}_Q/\Delta_Q) = 1$ (i.e., "no tunneling" $\Rightarrow$ "no renormalization"). Note that the right-hand side of Eq. (65) depends explicitly only on the effective tunneling element $\tilde{\Delta}_Q$, but not on the original (tunable) perturbation strength Δ_Q. We therefore can use the differential relation in Eq. (65) to extend the perturbative construction from Eq. (64) into the region of arbitrarily large values of the bare coupling Δ_Q by using the outcome of the previous (infinitesimal) change in Δ_Q as the initial input in the subsequent increment $d\Delta_Q$. Each of these increments is a small perturbation around a new, self-consistently determined, value of $\tilde{\Delta}_Q$. One gets the following self-consistent equation for the effective tunneling amplitude as a result:

$$\tilde{\Delta}_Q = \Delta_Q e^{-(B/2)[\tilde{\Delta}_Q^2/(\hbar\omega)^2]} \tag{66}$$

Our renormalization procedure is internally consistent in that the physical value of the tunneling amplitude depends on the scaling variable—the bare coupling Δ_Q—only logarithmically. This bare coupling must scale with the only quantum scale in the problem—the Debye frequency, as pointed out in the first section.

In a more complete treatment, the Δ renormalization would not be characterized by a single "Franck–Condon" parameter, but by a distribution of Franck–Condon factors. Therefore the exponential form in Eq. (66) might possibly be replaced by a different, perhaps polynomial, expression. In fact, one may think minimally of Eq. (66) as one of the possible Padé extensions of the perturbative formula (64). At any rate, such a Padé approximant will retain the main feature of Eq. (66) in that the value of the observable tunneling matrix element $\tilde{\Delta}_Q$ is bounded from above and depends strongly on the (semi)classical energies $\hbar\omega_i$. According to the discussion above, this restriction stems from a self-consistency condition, namely, that the leading term in the exponent in Eq. (66) must scale with $\tilde{\Delta}_k^2$, if this same number $\widetilde{\Delta}_k$ is on the left-hand side in that equation. An important corollary of this is that the perturbative term inside the brackets of Eq. (62) must scale with $\tilde{\Delta}^2$ itself; hence the perturbation correction Σ will scale with $\tilde{\Delta}^2$ too. (From now on, we will drop tildes from the symbols denoting the physical tunneling amplitudes, but retain them for the effective ϵ's.) That is, with our definition of B, it is roughly true that

$$\Sigma = B\frac{\Delta^2}{\hbar\omega_D} \tag{67}$$

We have thus demonstrated explicitly that the magnitude of quantum effects on the classical energy splitting ϵ on a particular site should depend on the facility of tunneling at that same site. We have therefore established that the fact of quantum Δ_i being close in value to a rather large energy scale ω_i is consistent with a relatively small value of the correction in Eq. (61) and its scaling with Δ^2. As another dividend from the argument, we obtain a ballpark estimate of the constant B. A structural transition that is thermally active at a plateau temperature $k_B T \sim \hbar\omega_i$ and is an efficient phonon scatterer will have $\Delta \sim \hbar\omega_i$. Therefore $B = \sum_i \omega_D/\omega_i$ will be a number on the order of several hundreds, since the total number of the ripplonic modes (at the laboratory glass transition, at which $\xi/a \simeq 5$–6) is approximately a hundred, and ω_i is proportional to, but somewhat smaller than, the Debye frequency ω_D.

Equation (67) implies that while the distributions of the clasical energy splittings ϵ and the bare semiclassical tunneling amplitude Δ may be uncorrelated, quantum corrections require that $\tilde{\epsilon}$ and Δ be correlated for systems with sufficiently low barriers and which simultaneously have small energy difference between the initial and final structural state. Conversely, the independence approximation is valid when, roughly, $|\epsilon| > B\Delta^2/\hbar\omega_D$. Since Δ is proportional to $\hbar\omega_D$, this criterion is a formal restatement of an earlier comment that the theory is strictly valid in the classical limit. (Note that there is also a (much stronger) $\hbar$ dependence in the exponent of the tunneling element Δ (see Eq. (23).) While, obviously, only a negligible fraction of the *total* number of structural rearrangements in the liquid at T_g would not satisfy the classicality criterion, these particularly facile transitions do actually comprise a significant portion of those transitions that are thermally active at *cryogenic* temperatures. We will now indicate what the observable consequences of this deviation for the strict semiclassical limit are. In order to do this, let us discuss first the difference between the strongly quantum and the bare "classical" structural transitions.

According to Eq. (61), for all transitions whose diagonal energy difference would be $|\epsilon| < \Sigma$ in the classical limit, the effective diagonal splitting $\tilde{\epsilon}$ is actually zero, meaning that the full energy splitting E is entirely comprised of the originally *off-diagonal* energy scale Δ. This implies that the energy eigenstates of such highly mixed tunneling centers are heavy superpositions of the original classical structural states and would not be easily interpreted in terms of the atomic coordinates of the potential minima alone, but must include the kinetic energy term as well. This is directly related to the well-known ambiguity in separating the energy of such systems into potential and kinetic components even at conditions that are entirely classical, such as at a high temperature. Of course, in such cases *free energy* formulations must be employed that allow one to count the number of configurational states unambiguously, while using "inherent" structures based on *potential* energy stationary points alone is of limited utility. The strongly quantum case can be loosely understood by transcribing the complex multiparticle rearrangements onto a single collective "reaction" coordinate (as in the soft potential model [120]). In fact, this analogy to a single coordinate soft potential model is quite loose because of the much higher density of states of the ripplons (that give rise to the boson peak and correspond to the vibrations of the membrane) compared with the density of states of the soft potential model, which is one dimensional so that only one coordinate is vibrationally excited. Nevertheless, following this analogy, consider a two-well potential (with very steep outer walls) with a barrier high enough so that the physical coordinate eigenstates corresponding to the particle being in the left or the right well are unambiguously definable and the diagonal component of the transition's energy is equal to the difference in the potential energy of the two wells with high accuracy. Imagine next lowering the barrier. In the limit of zero barrier the system is simply a particle in a square box, whose energy scale is determined by the quantum energy scale in the problem—that is, the particle's kinetic energy alone. This analogy reminds us that, just like the transition from a largely classical to quantum behavior in a double-well potential, the transition at $|\epsilon| = \Sigma$ is not sharp (note, however, that unlike in a one-dimensional soft potential model, the density of excited states of a tunneling center is very high, thus possibly leading to a sharper crossover). Put another way, this "phase transition" clearly corresponds to term-crossing and therefore would be gradual in a finite system. From a mean-field perspective, the transition at $\epsilon = \Sigma$ resembles a delocalization phase transition (e.g. see Ref. [121]), which we may think of as quantum depinning of the domain wall. Alternatively, one could say that the local structure of classical energy levels melts out locally in that the energy variation on the mostly classical landscape (determined by T_g) happens locally to be smaller than the confinement kinetic energy of the domain wall motion. Of course, this is occuring for only small parts of an otherwise rigid matrix. Again, since the system is finite, one expects a soft crossover rather than a sharp transition when such "melting" occurs. Both ways of interpreting the quantum mixing/melting described above are consistent with our view of the tunneling process leading to the expression (23) for the tunneling amplitude Δ. The action exponent in Eq. (23) scales as the height of the barrier relative to the under-barrier frequency. The former quantity, while distributed, scales with the classical energy scale in the problem—T_g; while the latter is proportional to the Debye frequency (and, most likely, is somewhat distributed too). The quantum limit of large $\hbar\omega_D$ corresponds to a narrow barrier and a short tunneling path. This would imply the relative unimportance of the classical energy landscape modulation during the tunneling process. Finally, in order to avoid ambiguity, we stress that the structural transitions of both types of tunneling centers, that we have called "classical" (in that the wavefunctions are well localized near minima and are well defined structurally, i.e., in position space) and "quantum" (i.e., in a superposition of structural states), at low temperature occur in a purely quantum mechanical fashion, that is, by tunneling.

We now show that the presence of a somewhat distinct class of such low barrier, or "fast," two-level systems, whose effective diagonal splitting is zero, leads to additional phonon scattering in comparison with the strictly semiclassical analysis, which neglects the renormalization from quantum mixing effects. This additional scattering at low energy is consistent with the apparent subquadratic temperature dependence of the heat conductivity in the TLS regime. The mixing also leads to a superlinear addition to the heat capacity at sub-Kelvin temperatures. These highly quantum tunneling centers in strongly mixed superpositions of structural states therefore give a mechanism to resolve a quantitative deviation from the standard tunneling model, which was brought up by Black and Halperin [96] in 1977. They noted that the short-time heat capacity of a-SiO_2 is larger than would be predicted by the logarithmic dependence obtained in the STM, if one uses the TLS parameters extracted from ultrasonic measurements. The quantitative mismatch appears to be as if there were two kinds of two-level systems: one set obeying the distribution postulated in STM, and another set of "fast" tunneling centers responsible for the short-time value of the heat capacity. We can see our analysis of mode mixing leading to the existence of a finite number of two-level systems with $\tilde{\epsilon}$ very nearly 0, as suggested by Eq. (61), is quite consistent with this empirical notion.[6]

To see this more explicitly, we note that Eq. (61) allows one to formulate the effects of quantum mode mixing as a change in the apparent distribution of the diagonal energy splitting. Whatever the old distribution of classical energy difference $n(\epsilon)$, the new distribution of the effective classical component of the transition energy can be found using $n(\tilde{\epsilon})|d\tilde{\epsilon}| = n(\epsilon)|d\epsilon|$. For $\tilde{\epsilon}$ values not too close to $\hbar\omega_i$ (case $\tilde{\epsilon} \sim \hbar\omega_i$ will be discussed later), which is appropriate in the TLS regime, the function $|\partial\epsilon/\partial\tilde{\epsilon}|$, which describes the relative probability distribution of the two quantities, is given by

$$\left|\frac{\partial\epsilon}{\partial\tilde{\epsilon}}\right| = \Sigma\delta(\tilde{\epsilon}) + 1 \tag{68}$$

where the δ-function is positioned to the right of the origin: $\int_0^{0^+} d\epsilon\, \delta(\epsilon) = 1$ (see also Fig. 22b). Consequently, the distribution of the effective diagonal splitting is

$$n_\Delta(\tilde{\epsilon}) = \frac{1}{T_g \xi^3}\left[B\frac{\Delta^2}{\hbar\omega_D}\delta(\tilde{\epsilon}) + e^{-|\tilde{\epsilon}|/T_g}\right] \tag{69}$$

The coefficient of the δ-function reflects the "pile-up" of the two-level systems that would have had a value of $|\epsilon| < \Sigma$ were it not for quantum effects. These fast two-level systems will contribute to the short-time value of the heat capacity in glasses. The precise distribution in Eq. (69) was only derived within perturbation theory and so is expected to provide only a crude description of the interplay of clasical and quantum effects in forming low-barrier TLS. Quantitative discrepancies from the simple perturbative distribution may be expected owing to the finite size of a tunneling mosaic cell, as mentioned earlier,

[6]We must stress, however, that the Black–Halperin analysis has been conducted for only a single substance, namely, amorphous silica, and systematic studies on other materials should be done. The discovered numerical inconsistency may well turn out to be within the deviations of the heat capacity and conductivity from the strict linear and quadratic laws, repsectively. Finally, a controllable kinetic treatment of a time-dependent experiment would be necessary.

and the finite lifetimes of each energy state due to phonon emission. These effects would also smooth the local quantum melting transition as $\tilde{\epsilon} \to 0$. While various improvements of the functional form of $n(\tilde{\epsilon})$ might be suggested, it seems unwarranted, at present, to use any more complicated expressions for this function. Thus to see the main consequences of the quantum mixing effect, we will proceed with the perturbative expression. Assuming a particular value of the coefficient B allows one to derive the contribution of the fast two-level systems to the heat capacity and scattering of the thermal phonons. Before we start, let us note that since we now have to deal with a specific coupled distribution of ϵ and Δ, the generic two-level system model that only specifies the distribution of the total splitting E is not sufficient. We must use the full tunneling model, where the tunneling elements Δ are distributed according to Eq. (24). The exact value of constant A in Eq. (24) depends (weakly!) on the (possibly ϵ-dependent) cutoff value of the $P(\Delta)$ distribution. *Both* the heat capacity and the phonon scattering strength depend on the coefficient A; therefore it is possible to check the *relative* contribution of the "quantum" centers to both of those quantities, regardless of A's value. The $n(\epsilon, \Delta)$ distribution obtained in this way is now a product of the $P(\Delta)$ distribution from Eq. (24) and the density of states from Eq. (69). The new normalization coefficient A_1 is found from the requirement that $\int d\epsilon \, d\Delta \, n(\epsilon, \Delta) = 1/\xi^3$. This gives

$$A_1 = \left[\frac{B}{T_g \hbar\omega_D} \Delta^{2-c} + \frac{1}{c} \left(\frac{1}{\Delta_{\min}^c} - \frac{1}{\Delta_{\max}^c} \right) \right]^{-1}$$

In order to compute the lifetime of a phonon of energy E, one averages the Golden Rule scattering rate $(\pi g^2 \Delta^2 / \rho c_s^2 E) \tanh(\beta E/2)$ with respect to $n(\epsilon, \Delta)$, subject to the resonance condition $E = \sqrt{\epsilon^2 + \Delta^2}$ [8, 11, 93]. This yields two contributions to the decay rate:

$$\tau_E^{-1} = \frac{\pi}{3} A_1 \left(\frac{a}{\xi} \right)^3 E \left(\frac{\Delta_{\max}}{E} \right)^c \times \left[\frac{BE}{\hbar\omega_D} + \int_{\Delta_{\min}/E}^{1} dx \frac{x^{1-c}}{\sqrt{1-x^2}} \right] \tag{70}$$

The first term in the square brackets is the contribution owing to the fast, or highly quantum, two-level systems. Note that this term scales faster with E than the other term. Provided the magnitude of this first term is comparable to the other term, the fast modes will somewhat modify the overall scaling of the heat conductivity κ. Without the first term, κ scales superquadratically according to T^{2+c} (recall that the heat conductivity is *inversely* proportional to the scattering rate from Eq. (70)). If we use a numerical value of B on the order of 100, this leads to a *subquadratic* T dependence of κ: experimentally, $\kappa(T)$ scales like $T^{1.9\pm0.1}$ as extracted from a decade and a half of data (see Fig. 1). Without the fast TLS, one again would have $\kappa \propto T^{2+c}$. Using the theoretical approximation for c, this differs from the empirically observed value at least by a factor of $(10^{1.5})^{c+0.1} \sim 2$ at $T \simeq 10^{-2}T_D$. Obviously, this is a very crude estimate because (1) we do not know how far down in temperature the power law scaling of κ goes; and (2) our correction, while going in the right direction, summed with the older result, is not strictly a power law. Since the integral in the square brackets of Eq. (70) varies between 1 and $\pi/2$ for $0 < c < 1$ ($\Delta_{\min}/E \ll 1$, surely at $E \simeq 10^{-2}T_D$), we conclude that the first term must be between 10^0 and 10^1 in order to make a sizable contribution to the phonon scattering and modify its functional form. Since $E \simeq 10^{-2}T_D$, this shows that B indeed must be on the order of several hundreds, consistent with our expectations based on the number of vibrational modes in the boson peak.

Does this mixing-induced correction to the density of states with the value of B around a hundred make an appreciable contribution to the time-dependent heat capacity? Following the calculation from Section IIIC, but now using the new distribution $n(\epsilon, \Delta)$, one finds

$$C(t) = \frac{A_1}{T_g \xi^3} \int_0^\infty dE \left(\frac{\beta E}{2\cosh(\beta E/2)} \right)^2 \left(\frac{\Delta_{\max}}{E} \right)^c \times \left[\frac{BE}{\hbar\omega_D} \theta(t - \tau_{\min}) + \int_0^{\log(t/\tau_{\min}(E))} dz \, \frac{e^{(c/2)z}}{2\sqrt{1 - e^{-z}}} \right] \quad (71)$$

where $\theta(t)$ is the usual step-function and $\tau_{\min}$ is the fastest possible relaxation time of a TLS with the total energy splitting E, defined in Eq. (27). Again, the first term in the square brackets gives the contribution of the "fast" TLS. Using the same numbers as given in Section IIIC, it is straightforward to show that the second, regular, term is on the order of a hundred at temperatures $T \sim 10^{-2}T_D$ when measured on the time scale of minutes. At the same time, the first term is at most on the order of ten. Note that at the shortest times $t \sim \tau_{\min}$, when the regular two-level system only begins to contribute to the heat capacity, the theory with quantum corrections says that the actual heat capacity is *finite* and is at most one-tenth of the long-time value. At the same time, the fast tunneling centers do not seem to contribute significantly to the long-time heat capacity. We note, however, that the result obtained, $c(T) \propto T^{1+c/2}$ with $c = 0.1$, gives a somewhat slower rise with temperature than seen in experiment. The quantum correction again goes in the right direction of *increasing* the rate of the heat capacity growth with temperature relative to the $T^{1+c/2}$ law.

We have established that effects beyond the strict semiclassical analysis give rise to a subset of tunneling centers that undergo faster tunneling than the rest. Nevertheless, there are some quantitative issues in the heat capacity magnitude that remain to be understood, namely, that the computed contribution of the "fast" centers seems somewhat lower than what is necessary to explain the deviation of the experimental T dependence from the supelinear dependence $T^{1+c/2}$ predicted by the present (approximate) argument. It is posssible that ultimately a broader view of the time dependence of the heat capacity needs to be taken. Since, in fact, the system will clearly be aging by tunneling at those low temperatures, the notion of fixed frozen-in "defects" may no longer be adequate—essentially interactions between defects play a role. "Aging" by definition implies irreversible *structural* changes. More work on understanding the long-time evolution of the tunneling centers is necessary.

We have concentrated on the quantum corrections to the low-lying tunneling states with low barriers. Quantum mixing applies to the higher energy states too. Energy shifts and quantum melting occur within subbands of the ripplonic states of order l and respective degeneracy $(2l + 1)$, thus mixing these states. As tunneling can take place on a given time scale and the vibrationally excited levels become observable, their apparent energies cannot be degenerate because the levels are coupled through those same tunneling transitions. The magnitude of energy level repulsion from the quantum mixing can be assessed qualitatively. In the limit of weak coupling, the deviation of a ripplonic frequency from its classical value scales as $\sum_i \Delta_i^2/\hbar\omega_i$. The width of the ripplonic band of order l is probably limited from above by the tunneling amplitude Δ_i itself. Does this band broadening affect our previous results on the boson peak phenomena? Not very much. Since the observables depend mostly on the number of new excitations and the *number* of the ripplonic modes is not changed by these mixing effects, the essential core of our conclusions from Section IV remains intact. Nevertheless, some quantitative modifications are to be expected. For example, the lowest ripplonic energies may be lowered so as to cause a crossover to a multilevel behavior in some of the internal resonances, thus possibly modifying the derived magnitude of the heat capacity and phonon scattering at subplateau temperatures. This effect will further contribute to the phonon interaction-induced broadening of the ripplonic transitions, as estimated in Section IV.

B. Mosaic Stiffening and Temperature Evolution of the Boson Peak

Equation (61) raises another interesting point. According to that equation, the values of both the bare and the effective classical energy bias of a transition—ϵ and $\tilde{\epsilon}$, respectively—are limited from above by the lowest ripplon frequency (ω_2). (Note that this is only realized in the $\epsilon < 0$ case, discussed in this section.) This is unimportant at low temperatures. But what happens at higher T, near this limit? Unlike in the low-energy situation just discussed, one simply cannot ignore that all the energy states have a rather short lifetime. Therefore the singularity in Eq. (61) does not occur, but will be rounded. This observation does not completely answer the question that one should have asked in the first place on general grounds alone: What happens to the structure of the energy spectrum of a tunneling center, when the energy of the transition becomes comparable to a vibrational eigenfrequency of the domain wall?[7]

When attempting to answer this question, a general multilevel perspective on each tunneling center is somewhat easier to use than the very mechanical view of the wall's excitations that we have mostly employed so far, in which the ripplonic energy states are obtained by quantizing vibrations of a freely moving classical mambrane. The "singularity" at $|\tilde{\epsilon}| \sim \hbar\omega_i$ is actually a term-crossing phenomenon that, again, would not take place in the strict classical limit. Let us go back to our argument on the density of states, but consider a case when ϵ is larger than a ripplonic frequency. As mentioned many times already, vibrational excitations of a domain wall can be defined meaningfully only when a structural *transition* takes place in a given region of the material. The energy of the transition must be the *lowest* excited state of a mosaic cell. On the other hand, the values of the ripplon frequencies are determined by a (fixed) surface tension coefficient and the wall's mass density. They have fixed values. The necessary conclusion from this is that the tunneling centers will not have ripplons whose frequency is lower than the transition frequency. We provide a cartoon illustrating this idea in Fig. 23. We see the quantum mixing reduces the number of the lower frequency vibrational modes. The mosaic appears stiffer than expected. This effect may contribute to the temperature evolution of the boson peak as observed in inelastic scattering experiments. Wischnewski et al. [110] find that, at temperatures between 51 K (numerically close to silica's $\hbar\omega_2$) and above the glass transition, the left-hand side of the boson peak decreases in size as the temperature was raised. At the same time, the high-frequency side remained relatively unchanged. Note that, as temperature is raised, the total area of the peak in Fig. 2 of Ref. [110] does not increase. In this temperature range mosaic cell motion loses oscillating character and becomes a rather featureless activated relaxation process.

To summarize this section, we have seen that the possibility of quantum tunneling between structurally close states in glass does have a predictable effect on the spectrum and must be taken into account when computing the density of low (and not so low) energy structural excitations in these materials.

[7]We remind the reader that the tunneling transition energy could also be thought of as an eigenenergy of the wall's motion, but of a lower, $l = 1$, order, associated with the translational motion of the shell's center of mass.

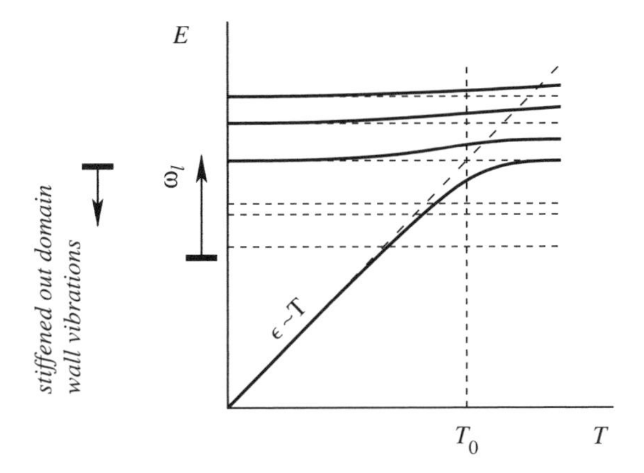

Figure 23. This caricature demonstrates the predicted phenomena of energy level crossing in domains whose energy bias is comparable or larger than the vibronic frequency of the domain wall distortions. The vertical axis is the energy measured from the bottom state; the horizontal axis denotes temperature. The diagonal dashed line denotes roughly the thermal energies. A tunneling center that would become thermally active at some temperature T_0 will not possess ripplons whose frequency is less than T_0.

At the same time, the main conclusions of the original semiclassical argument remain valid: each structural transition may be thought of as a rearrangement of about 200 molecules accompanied by distortion of the domain wall that separates the two alternative local atomic arragements.

VI. THE NEGATIVE GRÜNEISEN PARAMETER: AN ELASTIC CASIMIR EFFECT?

With the exception of the plateau's position and the quantum mixing effects, we have so far dealt with those anomalies in low-temperature glasses that are more or less universal. These universal patterns are of particular interest because they cannot be easily blamed on chemical peculiarities of each substance. Indeed, given the flatness of the low-energy excitation spectrum in glasses, the apparent universal ratio $l_{\mathrm{mfp}}/\lambda \simeq 150$ is the dimensionless quantity that seems to express the general, intrinsic character of those low-energy excitations, as arising from the nonequilibrium nature of the glass transition. The number 150 reflects the size of nearly independent fragments into which a supercooled liquid is broken up at the laboratory glass transition. Yet, there is another dimensionless quantity, namely, the Grüneisen parameter γ, that also reflects the necessity of going beyond a harmonic picture for amorphous solids. This parameter is always a positive number of order one for simple cubic crystals (at low enough T), but varies wildly among amorphous materials [122] (see also a discussion in Ref. [22]). γ in glasses has been reported to be as large as several tens and often negative in sign! A negative γ implies a negative thermal expansion coefficient $(1/V)(\partial V/\partial T)_p$ (the *linear* expansion coefficient $\alpha \equiv (1/L)(\partial L/\partial T)_p = \frac{1}{3}(1/V)(\partial V/\partial T)_p$ is a commonly used quantity chracterizing anharmonicity too). Contraction with heating is observed in some crystals at *not too low* temperatures, owing to the details of the anharmonic couplings in a specific substance that may result in the negativity of the Grüneisen parameter of a lattice mode of a finite frequency (e.g., see Ref. [123]). Thermal contraction along a single direction in anisotropic materials is even more common. Nonetheless, as the temperature is lowered, the thermal expansion coefficient in an insulating crystal eventually becomes positive and approaches the cubic T dependence predicted by standard thermodynamics. In contrast, an isotropic negative thermal expansivity is observed in many amorphous substances even at the lowest temperatures. In addition, the expansivity is not cubic in T. The most widely known example of a substance with a negative α is rubber. Rubber owes this property to the largely entropic nature of its elasticity. Here, we will see that a distinct mechanism of thermal contraction in glasses in the TLS temperature range arises, which is a direct consequence of the existence of the spatially extended tunneling centers that give rise to the universal phenomena considered earlier.

As shown previously, the excitation spectrum of the tunneling centers may be represented as a combination of the two lowest energy levels, corresponding to the structural transition and a set of higher energy states involving vibrations of the moving domain wall. By the exchange of phonons, these local (quantum) fluctuations in the elastic stress will be attracted to each other much like in the van der Waals interaction between neutral molecules. The elastic Casimir effect seems a more appropriate name for this phenomenon, since the moving domain walls are not point-like but, instead, resemble fluctuating membranes. While we do not claim this attraction is solely responsible for the negative expansion coefficient, it turns out to provide a large contribution to the thermal contraction in glasses. We will see how this effect arising from interaction of amorphous state excitations depends on the material constants and the preparation speed of the glass and, therefore, is not universal!

We note first that not all amorphous substances actually exhibit a negative α in the experimentally probed temperature range. In such cases, it is likely that the contraction coming from those interactions in these materials is simply weaker than the regular, anharmonic lattice thermal expansion. Other contributions to the Grüneisen parameter will be discussed later as well.

Coupling the motion of the mosaic cell (TLS and boson peak) to phonons is necesssary to explain thermal conductivity; therefore the interaction effects discussed later follow from our identification of the origin of amorphous state excitations. The emission of a phonon followed by its absorption by another cell will give an effective interaction, in the same way that photon exchange leads to interparticle interactions in QED. The longest range coupling between local degrees of freedom coupled linearly to the elastic stress has the form of a dipole–dipole interaction. Since the structural transitions are of finite size, the dipole assumption is only approximate for the closer centers. For the time being, we take for granted that there is no *first* order, static, interaction between the vibrating domain walls, which, if nonzero, could be priori of either sign. The next, second-order interaction is always negative in sign and is proportional to

$$-\sum_{ij} \frac{1}{r_{ij}^6} \propto -\left(1 - \frac{\delta V}{V}\right)^2 \simeq -1 + 2\frac{\delta V}{V}$$

This favors a sample's contraction (V is the volume). This attractive force, which will be temperature dependent, is balanced by the regular temperature-independent elastic energy of the lattice: $F_{\text{elast}}/V = (K/2)(\delta V/V)^2$. Calculating the equilibrium volume from this balance allows us to estimate the thermal expansion coefficient α. More specifically, the simplest Hamiltonian describing two local resonances that interact off-diagonally is

$$H = \frac{\omega_i}{2}\sigma_x^i + \frac{\omega_j}{2}\sigma_x^j + J_{ij}\sigma_z^i\sigma_z^j$$

where ω_i and ω_j would be the frequencies of ripplons on sites i and j and

$$J_{ij} \equiv \frac{3}{4\pi\rho c_s^2} \frac{(\mathbf{g}_i\mathbf{g}_j) - 3(\mathbf{g}_i\mathbf{r}_{ij})(\mathbf{g}_j\mathbf{r}_{ij})/r_{ij}^2}{r_{ij}^3} \tag{72}$$

is the dipole–dipole interaction following from Eqs. (15) and (17). (Having the interaction be off-diagonal automatically removes the first-order term in J_{ij}.) The factor 3 accounts in our usual simplistic way for all three acoustic phonon branches. This ignores a distinction between the longitudinal and transverse speed of sound. This simplification, however, is accurate enough for our purposes. Since $g \simeq \sqrt{\rho c_s^2 a^3 k_B T_g}$, the J_{ij} turn out to scale in a very simple way with the glass transition temperature and the molecular size a, giving $J_{ij} \sim k_B T_g (a/r)^3$.

Since only mobile domain walls give rise to local dynamic heterogeneities, one may conclude intuitively that only the sites of thermally active structural transitions can contribute to α. Therefore one expects that as temperature is increased, more tunneling centers will contribute to the van der Waals attraction, thus leading to negative expansivity. As already mentioned, the excitations of tunneling centers are conveniently subdivided into a low-energy TLS-like pair of states and higher energy "ripplonic" excitations corresponding to distortions of an active center's domain wall. Hence we may view the total van der Waals attraction as having three somewhat distinct contributions: "TLS–TLS", "ripplon–ripplon," and "TLS–ripplon" attractions. In this section, we focus on the relatively low, subplateau temperature regime, for reasons that will be explained later. At these low temperatures, transitions to the ripplonic states are only virtual, whereas the TLS structural may well be thermally active. This, in addition to the differences in the respective spectra of these excitations, will lead to some difference in the dependence of the mutual interactions between those excitations on temperature and other parameters. In order to assess the magnitude of those interactions, let us consider the following, very simple, three-level Hamiltonian that is designed to model a transition of energy ϵ_i between two different structures that may also be accompanied by a wall vibration of frequency ω_i:

$$H_i = \begin{pmatrix} 0 & 0 & 0 \\ 0 & \epsilon_i & 0 \\ 0 & 0 & \epsilon_i + \omega_i \end{pmatrix} \tag{73}$$

Note that, even though, for simplicity's sake, we use the semiclassical energy ϵ in the Hamiltonian above, the latter is meant as (the lowest energy portion of) the full, diagonalized Hamiltonian with quantum corrections included. This corresponds to the plain two-level system formalism that does not specify a distribution of the tunneling matrix element Δ. Also, in comparison with the general case of Eq. (37), we only include an excitation by a *single* quantum of a *single* ripplon. The latter simplification is obviously justified in the lowest perturbation order, where all pairs of excitations contribute to the total in an additive fashion. Considering only single-quantum excitations is a low-temperature approximation, made mostly to avoid adopting extra modeling assumptions necessary to embody the mixed spin/boson statistics on each site. This simplification is nevertheless adequate, as will become clear later in the discussion.

Since the contributions of the three constituents of the van der Waals attraction are additive, one can consider each contribution separately. This indeed proves to be convenient not only because all the contributions exhibit distinct scaling with the parameters, but each contribution comes to dominate the expansivity at somewhat distinct temperatures. We consider first the ripplon–ripplon attraction. This contribution appears to dominate the most studied region around 1 K. The off-diagonal (flip-flop) interaction between the ripplons has the form

$$H_{ij}^{\text{int}} = J_{ij} |2_i 3_j\rangle\langle 3_i 2_j| + H.C. \tag{74}$$

where the rows and columns in the unperturbed Hamiltonian from Eq. (73) are numbered in the conventional way from the upper left corner. The "ripplon–ripplon" case appears the simplest of the three because here the issue of how many tunneling centers contribute to the effect is more or less separate from the strength of the interaction. The former is (qualitatively) determined by the number of thermally active two-level systems, which scales roughly with the heat capacity. The latter is nothing but the ground state lowering of a pair of resonances after interaction is switched on, which scales as $-J_{ij}^2/(\omega_i + \omega_j)$ and is T independent at these low temperatures. This contribution to the negative thermal expansion is therefore expected to be roughly quadratic in temperature (this corresponds to linear *expansivity*), which is similar to, if not somewhat slower than, that observed in amorphous silica around 1 K.

Calculating the correction to the system's free energy in the lowest order in J_{ij}, which corresponds to the interaction term from Eq. (74), is entirely straightforward and yields

$$\delta F_{\mathrm{rr}} = -\sum_{ij} J_{ij}^2 \frac{e^{-\beta(\epsilon_i+\epsilon_j)}(1+e^{-\beta\omega_i})(1+e^{-\beta\omega_j})}{Z_i Z_j} \times \frac{\omega_i \tanh(\beta\omega_i/2) - \omega_j \tanh(\beta\omega_j/2)}{\omega_i^2 - \omega_j^2} \tag{75}$$

Here $Z_i \equiv 1 + e^{-\beta\epsilon_i} + e^{-\beta(\epsilon_i+\omega_i)}$ is the unperturbed on-site partition function, corresponding to Eq. (73). Here subscript "rr" signifies the "ripplon–ripplon" contribution.

At low subplateau temperatures $T < \omega_i$, which we are primarily interested in here, the expression above reduces to the following van der Waals energy:

$$\delta F_{\mathrm{rr}} = -\sum_{ij}\sum_{l_1 l_2} \frac{J_{ij}^2}{\omega_{l_1}^i + \omega_{l_2}^j} \frac{1}{(1+e^{\beta\epsilon_i})(1+e^{\beta\epsilon_j})} \tag{76}$$

where we have explicitly written out summation over distinct ripplon harmonics l_1 and l_2 at sites i and j.

A few intermediate calculations are needed to compute the sum in Eq. (76). First, averaging of J_{ij}^2 with respect to different mutual orientations of $\mathbf{g}_i$, $\mathbf{g}_j$, and $\mathbf{r}_{ij}$ yields an effective isotropic attractive interaction $\frac{2}{3}(3/4\pi)^2 T_g^2 (a/r)^6$. Second, the sum over all harmonics amounts to $\sum_{l_1,l_2=2}^{l_{\max}} [(2l_1+1)(2l_2+1)/(\omega_{l_1}+\omega_{l_2})]$, where ω_l is found using the dispersion relation from Eq. (35). Here we assume that the ω_i are not correlated with J_{ij} and ϵ_i. As we already know from the discussion in the previous section, the latter assumption is adequate for values ϵ smaller than the boson peak frequency. Now recall that $l_{\max}$ actually depends on the droplet's perimeter, thus introducing an additional (cubic!) scaling with ξ/a. In the end, the sum over the l's is equal, within sufficient accuracy, to $1.5\omega_D^{-1}(3/4\pi)\pi^3(\xi/a)^{5/4}(\xi/a)^3$. Finally, assuming the J_{ij} and ϵ's to be uncorrelated enables one to present the double sum over ϵ_i as a product of two identical sums: $(\sum_i(1+\beta\epsilon_i)^{-1})^2$. Each sum is the effective concentration of thermally active tunneling centers: $k_B(\ln 2)T/T_g\xi^3$ as computed by integrating $1/(1+e^{\beta\epsilon})$ with the density of states from Eq. (36). Note that here we use the simple $1/T_g\xi^3$ expression for density of the tunneling transitions, in keeping with the assumption $E \sim \epsilon_i$ of the plain two-level system model adopted in this section. This is reasonable, given the qualitative character of this calculation. Finally, the summation over the ripplon sites can now be reduced to an integration with the lower limit equal to $\xi(3/4\pi)^{1/3}$.

As a result of the previous discussion, one recovers the following expression for the energy gain (per volume) due to a volume change δV:

$$\delta F_{\mathrm{rr}}/V \simeq 1.5(\ln 2)^2\pi^2 \frac{k_B T^2}{\xi^3 T_D}\left(\frac{a}{\xi}\right)^{7/4}\left(\frac{\delta V}{V}\right)$$

This works against the regular elastic energy $\delta F_{\mathrm{elast}}/V = (K/2)(\delta V/V)^2$, introduced earlier. The equilibrium relative change $\delta V/V$ as a function of T is obtained by setting $\partial F/\partial V = 0$. Differentiating the equilibrium value of δV with respect to temperature yields the following estimate for the thermal (volume) expansion coefficient:

$$\frac{1}{V}\left(\frac{\partial V}{\partial T}\right)_p \simeq -3.0(\ln 2)^2\pi^2 \frac{1}{K}\frac{k_B T}{\xi^3 T_D}\left(\frac{a}{\xi}\right)^{7/4} \tag{77}$$

This can already be used to estimate the magnitude of the ripplon–ripplon contribution to the "Casimir" effect numerically. One can do it in several ways. The simplest thing to do that does not require knowing K is simply to use Eq. (77) to calculate the Grüneisen parameter γ itself according to $\gamma = (\partial p/\partial T)_V/c_V$ [111], also using $(\partial p/\partial T)_V = -(\partial p/\partial V)_T/(\partial T/\partial V)_p$. This yields a temperature-independent Grüneisen parameter:

$$\gamma_{\mathrm{rr}} \simeq -3.0(\ln 2)^2\pi^2 \frac{T_g}{T_D}\left(\frac{a}{\xi}\right)^{7/4} \tag{78}$$

Using $(\xi/a)^3 \simeq 200$ and silica's $T_g/T_D \simeq 1500/350$, one obtains $\gamma \simeq -3$, within an order of magnitude of what is observed in amorphous silica at low temperatures (that experimental number varies between -5 and -20 among different kinds of silica at 1 K and seems to grow larger with lowering the temperature; see Fig. 3 from Ref. [122]). We will argue shortly that this growth may be explained by other contributions to the attraction between local resonances.

We can also directly compare the contribution in Eq. (77) to the *linear* thermal expansion coefficient $\alpha = (1/3V)(\partial V/\partial T)_p$ for silica as measured in Ref. [122]. According to Fig. 2 from Ackerman et al. [122], the α of silica is linear (possibly slightly sublinear) in temperature and equals $-1.0 \times 10^{-9} K^{-1}$ at 1 K. The compressibility K was obtained by Ackerman et al. [122] from the measured speed of sound and density. For internal consistency, we use the scalar elasticity to estimate K in this way. Summing up three single polarization phonon Hamiltonians from Eq. (15) yields $K \simeq \rho c_s^2/3$ (remember, $\Delta V/V = 3\Delta\phi$). Using silica's constants, given in Fig. 15 and the earlier obtained $\xi = 20$ A and recalling that $\delta l/l = \delta V/3V$, Eq. (77) gives linear expansion coefficient $\alpha \simeq -0.4 \times 10^{-9} K^{-1}$ at 1 K, indeed strongly suggesting that attraction between the tunneling centers is a significant contributor to the negativity of the expansion coefficient. The numbers just obtained are also a convenient benchmark in assessing other contributions to the negative thermal expansivity.

Next, we estimate the magnitude of the attraction between virtual transition and the direct, lowest energy transitions on different sites. The corresponding coupling term—$J_{ij}|2_i 2_j\rangle\langle 3_i 1_j| + H.C.$—leads to the following contribution to the free energy in the lowest order:

$$\delta F_{\mathrm{rT}} = -\sum_{ij} J_{ij}^2 \frac{e^{-\beta\epsilon_i}(1+e^{-\beta\omega_i})}{Z_i} \times \frac{\omega_i \tanh(\beta\omega_i/2) - \epsilon_j \tanh(\beta\epsilon_j/2)}{\omega_i^2 - \epsilon_j^2} \tag{79}$$

At subplateau temperatures, when $\beta\omega_i \gg 1$, $\tanh(\beta\omega_i/2)$ can be replaced by unity. Furthemore, the summation with respect to ϵ_j is no longer cut off by the temperature and the respective integral (weighted by $n(\epsilon) = (1/T_g)e^{-|\epsilon|/T_g}$) picks up most of its value at $\epsilon \gg T$. Therefore $\tanh(\beta\epsilon_j/2)$ may be replaced by unity as well. (Actually, both of those replacements must be made simultaneously, otherwise the sum becomes potentially ill-behaved when $\omega_i \sim \epsilon$.) As a result, the expression in Eq. (79) simplifies to

$$\delta F_{\mathrm{rT}} = -\sum_{ij} J_{ij}^2 \frac{1}{1+e^{\beta\epsilon_i}} \frac{1}{\omega_i + \epsilon_j} \tag{80}$$

The ϵ_j integral is related to an exponential integral E_1 and yields, in the two lowest orders: $(\ln(T_g/\omega_i) - \gamma_E)$, where $\gamma_E = 0.577\ldots$ is the Euler constant. As in the previous calculation, we regard ϵ_i, ω_i, and J_{ij} as uncorrelated. The summation over ω_i can be approximately represented as a continuous integral between 0 and $l_{\max}$ and leads to a quantity that scales as the area of the domain wall with a logarithmic correction. The final result is

$$\frac{\delta F_{\mathrm{rT}}}{V} = 0.5 \frac{T}{\xi^3} \left(\frac{a}{\xi}\right)^4 \ln\left[2.0 \frac{T_g}{\omega_D} \left(\frac{\xi}{a}\right)^{1/4}\right] \left(-1 + 2\frac{\delta V}{V}\right)$$

Up to a logarithmic correction, the expression is independent of the energy parameters in the problem and thus must scale linearly in T. Note that we have written out the full expression of $\delta F_{\mathrm{rT}}/V$ that includes the bigger, δV independent term "-1", for the following reason. This larger negative term is linear in temperature, which apparently would lead to a nonzero (positive) entropy at $T = 0$. This observation signals a breakdown of a perturbative picture of largely noninteracting two-level systems. For the sake of argument, let us estimate at what temperature this breakdown occurs. We compare the magnitude of the $\delta F_{\mathrm{rT}}/V$ term, *assuming* it is correct, to the free energy of noninteracting two-level systems per unit volume: $\int (d\epsilon/T_g\xi^3) e^{-\epsilon/T_g}[-T\ln(1+e^{-\beta\epsilon})]$, where we have appropriately chosen $E = 0$ as the reference energy. The latter expression is equal to $(\pi^2/12)T^2/T_g\xi^3$ and becomes smaller (in absolute value) than the $\delta F_{\mathrm{rT}}/V$ term at temperatures below $10^{-3}T_g$. This temperature is actually less, but still within an order of magnitude from the lower end of the plateau, which is well within the empirical validity of the noninteracting two-level systems regime. Let us recall, however, that a perturbative expansion is an asymptotic one and therefore always *over*estimates the magnitude of a correction (we suspect that most of the error comes from the low ϵ two-level systems). Therefore a more accurate estimate would probably yield a breakdown temperature lower in value than the estimate above. There is a reason to believe the "break-down" temperature is just at the edge of the lowest temperatures routinely accessed in the experiments. This is suggested by several experiments such as on internal friction, where deviations from the standard noninteracting two-level system picture have been seen (e.g., see a recent review by Pohl et al. [124]). In general, the effect of interaction between two-level systems could exhibit itself under several guises. One of those is an apparent gap in the excitation spectrum of the effective individual TLS. Such effects may have in fact been observed [125, 126]. The estimates above show these effects are more likely to be observed in substances with a higher glass transition temperature, such as amorphous silica or germania (GeO_2). Note, however, that the effects of interaction on the apparent TLS spectrum must be separated from quantum effects of level repulsion on each site, which we have considered in Section V.

At any rate, the volume expansion coefficient, corresponding to the computed value of the ripplon–TLS term, is approximately equal to

$$\frac{1}{V}\left(\frac{\partial V}{\partial T}\right)_p \simeq -1.0\frac{1}{\xi^3 K}\left(\frac{a}{\xi}\right)^4 \ln\left[2.0\frac{T_g}{\omega_D}\left(\frac{\xi}{a}\right)^{1/4}\right] \tag{81}$$

Substituting the numerical values for a-SiO_2 in Eqs. (81) and (77) shows that, at 1 K, the ratio of the ripplon–TLS contribution to the ripplon–ripplon term is about 1.2; that is, they contribute comparably to the "contraction" free energy at this temperature. However, since the ripplon–TLS α is temperature independent, it will dominate at *sub*-Kelvin temperatures. The Grüneisen parameter's value corresponding to Eq. (81) is

$$\gamma_{\mathrm{rT}} \simeq -1.0\frac{T_g}{T}\left(\frac{a}{\xi}\right)^4 \ln\left[2.0\frac{T_g}{\omega_D}\left(\frac{\xi}{a}\right)^{1/4}\right] \tag{82}$$

The ripplon–TLS term, as estimated here, therefore seems somewhat larger relative to the ripplon–ripplon term than seen in experiment, consistent with our earlier notion that it is somewhat overestimated. Still, qualitatively our estimates are consistent with the observed tendency of γ to increase in magnitude, when the temperature is lowered. We point out that the results obtained above disregard possible effects of a specific distribution of Δ that will influence the precise value of the coupling between phonons and tunneling centers.

Note that the heat capacity-like expression reflecting the number of thermally active sites i enters into the expressions from Eqs. (81) and (82) in a linear fashion. Therefore, in contrast to Eq. (78), the temperature dependence of expression (82) is expected to be largely independent of the exact T-scaling of the heat capacity. Thus, according to Eq. (82), the Grüneisen parameter should eventually scale as $1/T$ at low enough temperatures in all substances (however, unrealistically long observation times may be required to verify this prediction; see the discussion at the end of this section). And again, the apparent density of states of the tunneling centers may be modified at those low temperatures due to interaction effects (such as the Burin–Kagan [26] effect).

According to Eqs. (78) and (82), the *dimensionless* contribution of the attractive forces between the tunneling centers can be expressed in a simple manner through the T_g/T_D and T_g/T ratios, as well as the relative size of the mosaic. Note that the effects of varying the quenching speed of the liquid on the number in Eqs. (78) and (82) add up. For instance, making the quenching faster will increase T_g and decrease ξ. The ripplon–TLS term is especially convenient with regard to testing our results, because it is nearly insensitive to changes in the Debye temperature potentially induced by altering the speed of glass preparation.

Finally, we show that the second-order coupling between direct tunneling transitions is subdominant to the already computed quantities. Consider an interaction of the form $J_{ij}|1_i 2_j\rangle\langle 2_i 1_j| + H.C.$ If one repeats simple-mindedly the steps leading to Eq. (75), one obtains the following simple expression for the free energy correction due to interaction between the underlying structural transitions:

$$\delta F_{\mathrm{TT}} = -\sum_{ij} J_{ij}^2 \frac{\epsilon_i \tanh(\beta\epsilon_i/2) - \epsilon_j \tanh(\beta\epsilon_j/2)}{\epsilon_i^2 - \epsilon_j^2} \tag{83}$$

Assuming, again, that the J_{ij} and ϵ_i are uncorrelated, the ϵ summation can be performed via averaging with respect to the distribution from Eq. (36). One can show that the low-temperature expansion of the above expression yields, within two leading terms, $\delta F_{\mathrm{TT}}/V = -(2T_g/3\xi^3)(a/\xi)^6[1 + (\pi T/T_g)^2 \ln(T_g/T)/3]$. The T-independent term in itself is curious in that it is a contribution to the "vacuum energy" of the lattice that is of purely glassy origin and is entirely due to the locality of the free energy landscape of a liquid. Indeed, as attested by its scaling with T_g/ξ^3, this "vacuum energy" contribution would disappear at the *ideal* glass transition at which the whole space is occupied by a *nonextensive* number of distinct aperiodic solutions of the free energy functional. However, this constant term will have no effect on the *thermal* expansion. The lowest order T-dependent term—$T^2 \ln T$—actually has a slightly stronger temperature dependence than the ripplon–ripplon contribution; however, the latter is larger by at least three orders of magnitude, mostly owing to the large number of ripplon modes. Apropos, we would like to stress again that the presence of vibrational modes of the (extended) mosaic walls is essential to the existence of the negative thermal expansivity effect that we just estimated. Therefore, while the present theory predicts that many (and most conspicuous) effects that distinguish amorphous lattices from crystals should be described well by a set of noninteracting two-level-like entities at cryogenic temperatures, the intrinsic *multilevel* character of the structural transitions, which follows from the present theory, in glasses exhibits itself even at these low energies in higher order perturbation theory.

To complete the discussion of the second-order interaction between tunneling centers, we note that the corresponding contribution to the heat capacity in the leading low T term comes from the "ripplon–TLS" term and scales as $T^{1+2\alpha}$, where α is the anomalous exponent of the specific law. Within the approximation adopted in this section, $\alpha = 0$. However, it is easily seen that the magnitude of the interaction-induced specific heat is down from the two-level system value by a factor of $10(a/\xi)^5(d_L/a)^2 \sim 10^{-5}$–$10^{-4}$ and therefore may be safely neglected.

We have so far considered the second-order part of the induced interactions (square in J_{ij}^2, but fourth order in g). There could also be priori lower order contributions—first order in g, and first order in J_{ij}. First, let us consider the term linear in J_{ij}, which *also* has to do with interaction, mediated by the phonons. If nonzero, it could be of either sign. In our case, it is identically zero for the following reason. It is known [23, 27] that the apparent TLSs are only weakly interacting. (One could also infer this implicitly from the smallness of the second-order term that we have already estimated. The first-order term, if nonzero, is comparable to the second-order term in a mean-field disordered system. The dipole–dipole $1/r^3$ interaction is long range and is indeed well described by the mean field.) But we are dealing here with a nonpolarized state, for which the first-order term, linear in the average on-site magnetization, vanishes. In any case, even if the system were in a "ferromagnetic" state, the first-order term would still be only very weakly temperature dependent and thus would not contribute to the thermal contraction. Whether to consider such first-order term nonzero or not is, to some degree, a matter of choice. If nonzero, it must simply be thought of as the effective Weiss-like field that is part of the molecular field at each site. That field implies a *hard* gap on the order of T_g and indeed is negligible at low T. Yet, at low enough temperatures (microkelvins or so) [27], the *phonon-mediated* first-order interaction between the tunneling centers may become important and one can no longer use the bare frozen-in values of the on-site TLS energies, but must use those determined by the interaction. In this regime an independent two-level system picture breaks down and more complicated renormalized excitations may begin to play a role [26].

On the other hand, the other possible contribution to α, a term linear in g, does not have to do with interactions between the anharmonic amorphous solid excitations but is due to the direct coupling of the tunneling centers with the phonons. This direct TLS–phonon interaction has so far been the main suspect [127–129] behind the anomalous thermal expansion properties of the glasses. This mechanism requires, however, the existence of a correlation [129] (in our notation) between the on-site values g and ϵ, or else between Δ and $\partial\Delta/\partial\phi_{ii}$. In other words, the value of either classical or quantum splitting of a two-level system must be correlated with the way its energy changes when elastic stress is applied locally. The Δ with $\partial\Delta/\partial\phi_{ii}$ correlation has been argued to make a small contribution relative to the g versus ϵ correlation because of the smallness of the Δ values for the majority of the thermally active TLS [129]. On the other hand, a correlation between g and ϵ could produce, in principle, both a negative or positive Grüneisen parameter and therefore could explain, by itself, the observed variety of expansion anomalies in the low T glasses. However, the degree of correlation between g and ϵ and its temperature dependence is not really known and has to be parameterized. The soft potential model offers enough richness in behavior to accommodate two possible contributions—one dilating and the other contracting—to the sample's volume. In fact, Galperin et al. [127] suggest that those two types of the TLS may well be the two types of tunneling centers that were postulated early on by Black and Halperin [96] in order to resolve the apparent discrepancy between the value of the TLS density $\bar{P}$ as deduced from the phonon scattering experiments and the equilibrium and time-dependent heat capacity measurements. This, of course, could be checked experimentally by comparing the degree of the discrepancy in $\bar{P}$ and the sign of the thermal expansion coefficient in different substances. (We have shown in the previous section how the Black–Halperin paradox is, at least partially, explained by quantum corrections to the semiclassical landscape picture of structural transitions in glass.) With regard to the linear in g effect, we suggest here a modification to the original argument of Phillips [129]. According to Phillips (note some notational differences), $|\gamma| = 12g\alpha_0 \ln(2)/\pi^2 k_B T$ (he *also* assumed a heat capacity in T). Here $|\alpha_0| \leq 1$ is an (unknown) coefficient that reflects the degree of correlation between g and ϵ: $\langle \epsilon_i g_i(\epsilon_i) \rangle = \alpha_0 \epsilon_i g$. $\alpha_0 = \pm 1$ means complete correlation and $\alpha_0 = 0$ means no correlation. Now, due to symmetry, α_0 must be odd power in ϵ, the dominant term being therefore linear (see the form of $\alpha_0(\epsilon)$, somewhat cryptically mentioned as a remark of B. Halperin, at the end of Phillips' article). We must note that, although we have pretended, within our one-polarization phonon theory, that g is a vector quantity, it is in reality a tensor, if the phonons are treated properly. The off-diagonal terms, corresponding to interaction with shear, will indeed be uncorrelated with ϵ due to symmetry. However, the *trace* of the tensor, corresponding to coupling of the TLS to a uniform volume change, could be correlated, in principle, with the energy of the transition. For example, it may happen that when the sample is locally dilated, the structural transitions in that region will require less energy to occur. At present, we do not have an argument in favor of or against such a correlation. Note, however, that at the glass transition temperature, when the current arrangement of the defects freezes in, most structural transitions involve a thermal energy around T_g. On the other hand, the energy splitting ϵ of the tunneling centers relevant at the cryogenic temperatures is significantly smaller. Informally speaking, relative to the thermal energy scale at T_g, all two-level systems with low splitting will feel the same as the phonons. Therefore, qualitatively, the correlation factor α_0 should be at least a factor of ϵ/T_g down from the largest value of one. Note that this coincides with the form $\alpha_0(\epsilon) \propto \epsilon$ suggested by Halperin. Therefore the contribution of TLS–phonon coupling to the thermal expansivity of Phillips (who left the issue of the degree of correlation open at the time) should be multiplied by a factor of T/T_g. This takes into account, in a very naive way, both the symmetry and our knowledge of the energy scales relevant at the moment of the tunneling centers' formation.

This modifies Phillips's result to yield

$$|\gamma| < \frac{12g \ln 2}{\pi^2 k_B T_g} = \frac{12 \ln 2}{\pi^2} \sqrt{\frac{\rho c_s^2 a^3}{T_g}} = \frac{12 \ln 2}{\pi^2} \frac{a}{d_L} \tag{84}$$

The temperature independence of this contribution to the Grüneisen constant is the main difference between Eq. (84) and the original calculation by Phillips. The numerical value of the expression should be nearly the same for all substances and is about 8. This suggests that the direct coupling to phonons is a potential contributor to the elastic Casimir effect at temperatures around 1 K. Remember, however, the sign of the expression in Eq. (84) is unknown and its numerical value of 10^1 only provides an estimate from the above.

From the qualitative analysis in this section, we tentatively conclude that there are several contributions of comparable magnitude to the thermal expansion at low temperatures. Higher order effects may also be present. In this case, it may be more straightforward to estimate the interaction between ripplons as extended membranes without using a multipole expansion, as indeed is done when computing the regular Casimir force between extended plates.

The earlier qualitative treatment of the second-order interaction between the ripplons on different sites can be extended to higher temperatures as well. It is easily seen from Eq. (75) that an excitation of energy ω_l will contribute only βJ_{ij}^2 at temperatures comparable to ω_l and above. Therefore one might expect that at the temperatures near the end of the plateau the ripplonic transitions become thermally saturated and this attractive mechanism becomes increasingly less important. The expression in Eq. (79), in contrast, is subject to thermal saturation to a lesser degree. Still, we have seen that its scaling with temperature is subdominant to the ripplon–ripplon term at temperatures above 1 K. Finally, we remind the reader about the effect of mosaic stiffening explained in the previous sections. This should also diminish the attraction between the tunneling centers, owing to a smaller number of resonant modes at the sites of centers thermally active at these higher temperatures. On the other hand, the usual anharmonic effects also become more significant at a higher T leading to a turnover in the temperature dependence of α, as circumstantially supported by the old data on several materials cited in Krause and Kurkjian [130]. However, in order to assess this "crossover" temperature, one needs to know the magnitude of the regular thermal expansion due to the nonlinearities of the lattice. This is something that would be extremely difficult to measure independently, because even a crystal with the same stoichiometry as the respective glass is not guaranteed to have the same nonlinearity. Direct computer simulation estimates of the Grüneisen parameter, on the other hand, may be problematic due to the current difficulty of generating amorphous structures corresponding to realistic quenching rates. This is the main reason we have confined ourselves here to subplateau temperatures.

We have already mentioned, in Section IIIB, that in many substances an *electric* dipole will be generated during a tunneling transition. This will lead to yet another source of interaction between tunneling centers (i.e., via exchanging *photons*). The magnitude of the electric dipole–dipole interaction depends quadratically on the partial charge on the bead. According to recent estimates in Lubchenko et al. [92], the photon-mediated interaction is, in some cases, comparable in magnitude to the elastic, phonon-mediated interaction analyzed previously. In such cases, one may argue that the electric dipole–dipole interaction will contribute to the nonuniversality in the Grüneisen parameter, because the partial charge magnitude is only weakly (if at all) correlated with the elastic constants or the glass transition temperature.

Finally, we note again that even at the low temperatures we have been discussing, not all glasses have been shown to exhibit a negative α. According to our theory, however, the "Casimir" contribution to α is negative and sublinear in T, whereas the regular nonlinear expansion coefficient is positive but only cubic in temperature. Thus there should be a (perhaps very low) temperature at which the Casimir force should dominate. Data for many substances, although still positive at the achieved degree of cooling, do extrapolate to negative values of α at finite temperatures. This is not the case, however, for all substances [122]. Even excluding the possibility of error in these difficult experiments, this is not necessarily inconsistent with our theory for the following reasons. As the temperature is lowered, it takes a long time (proportional to T^{-3}) for the tunneling transitions to occur and appear thermally activated. For these same reasons, like the amorphous heat capacity, the direct interaction effect is time dependent at low temperatures. It may therefore take an excessively long time to actually observe the effects, discussed in this section, at very low temperatures, thus making it difficult to see a sign change in α for lattices with relatively large anharmonicity. Incidentally, this analysis predicts that the response of the length of an amorphous sample to a temperature change at subplateau temperatures must be time dependent (such time dependence, acompanied by heat release, has been observed in polycrystalline NbTi [131]). Since the interaction effect is quadratic in concentration, one expects qualitatively that the relative rate of the expansion's time dependence should be twice that of the specific heat.

VII. CONCLUSIONS

This chapter elucidates the origin of the thermal phenomena observed in the amorphous materials at temperatures $\sim T_D/3$ and below, down to the so far reached millikelvins. The nature of these phenomena can be boiled down to the existence of excitations other than elastic strains of a stable lattice. The peculiarity of these excitations is exhibited most conspicuously in the following phenomena: the specific heat obeys a nearly linear dependence on the temperature at the lowest T, greatly exceeding the Debye contribution. At the same temperatures, the heat conductivity is nearly quadratic in T and is universal if scaled in terms of the elastic constants. At higher temperatures ($\sim T_D/30$), the density of these mysterious excitations grows considerably, leading to enhanced phonon scattering and thus a plateau in the temperature dependence of the heat conductance. This increase in the density of states is also directly observed as the so-called boson peak in the heat capacity data, as well as inelastic scattering experiments.

We have argued that the origin of these excitations is a necessary consequence of the nonequilibrium nature of the structural glass transition. This transition, not strictly being a phase transition at all in a regular equilibrium sense, occurs if the barriers for molecular motions in a supercooled liquid become so high as to prohibit any macroscopic shape changes in the material on the scales of hours and longer [1]. The origin of these high barriers lies in a cooperative character of the molecular motions, which involve around 200 molecules at the glass transition temperature. Unlike regular crystals, where the correlation between the molecular motions is rather long range, thus leading to the emergence of translational symmetry below solidification, the motions within the cooperative regions in a supercooled liquid, or entropic droplets, are only weakly correlated with their surrounding. In the language of the energy landscape paradigm, a crystal is a (possibly nonunique) ground state of the sample (thus the long-range correlation!), whereas a glass is caught in a high-energy state, not being able to reach the true ground state for kinetic reasons. The respective dense energy spectrum at these energies exhibits itself in the existence of alternative mutually accessible conformational states of regions, or domains, of about 200 molecules in size. It was argued that quantum transitions between these alternative states are the additional excitations observed in glasses at low temperatures. The knowledge of the spectral and spatial density of these excitations allowed us to estimate from first principles the magnitude of the observed linear specific heat. The relevant energy scale here is the glass transition temperature T_g itself.

Stability requirements for the existence of these alternative conformational states at T_g allowed us also to estimate the strength of their coupling to the regular lattice vibrations, which is determined by T_g, the material mass density, and the speed of sound. This enabled us to understand the universality of the phonon scattering at the low temperatures.

The novelty of this picture is that we have established rather generally a *multiparticle* character of the tunneling events. This is counterintuitive because, naively, the larger the number of particles involved in a tunneling event, the larger the tunneling mass is, and the harder the tunneling becomes. This is indeed the case for systems like disordered crystals or crystals with substitutional impurities, where the tunneling mass is that of an atom, and the barrier heights are determined by the energy of stretching a chemical bond by a molecular distance; this virtually excludes the possibility of tunneling. The existence of structural rearrangements in a macroscopically rigid system is a sign of the system being in a high-energy state in which the available phase space is potentially macroscopically large. However, a decrease in this density of states for glass transitions occurring at a slower pace of quenching would result in the necessity to engage a larger number of atoms in these structural rearrangements. Transitions between the internal states of a domain involve only a very minor length change of each individual bond and atomic displacements not exceeding the Lindemann length, which is on the order of one-tenth the atomic length scale. It is not particularly beneficial to picture the tunneling events as individual atomic motions but rather as the motion of an interface between the alternative states of the domain. This domain wall is a quasiparticle of a sort, which has a low mass indeed: per molecule in the domain, it is only about *one-hundredth* the atomic mass. This contributes to the ease of the tunneling events that are thermally relevant at cryogenic temperatures: these events are subject to only very *mild* potential variations and are possible, again, because the lattice is frozen-in in a high-energy state.

The spatially extended character of the domain wall excitations along with their strongly anharmonic nature explains also higher temperature phenomena, such as the boson peak and the plateau in the heat conductivity. By using our knowledge of the surface tension and the mass density of the domain wall, we were able to calculate the energy spectrum of vibrational excitations of the active domain walls, or ripplons. This spectrum is in good agreement with the observed frequency of the boson peak. The ripplonic excitations accompany the transitions between the domain's internal states and thus are strongly coupled to the phonons. This has enabled us to understand the experimentally observed rapid drop in the phonon mean free path at the plateau temperatures. In addition, we have investigated the effects of phonon coupling on the spectrum of the ripplons. These spectral shifts scale with T_g and seem to be the cause of the nonuniversal position of the plateau.

We have carrried out an analysis of the multilevel structure of the tunneling centers that goes beyond a semiclassical picture of the formation of those centers at the glass transition, which was primarily employed in this chapter. These effects exhibit themselves in a deviation of the heat capacity and conductivity from the nearly linear and quadratic laws, respectively, that are predicted by the semiclassical theory.

A van der Waals attraction between the domain walls undergoing tunneling motions was argued to contribute to the puzzling negative expansivity, observed in a number of low T glasses.

Finally, we note that the conclusions of this chapter strictly apply only to glasses made by quenching a supercooled liquid. One may ask, nevertheless, to what extent the present results are pertinent to *other* types of disordered solids, such as "amorphous" films made chemically or by vapor depositions, or, say, disordered crystals. Indeed, phenomena reminiscent of real glasses, such as an excess density of states, are observed in many types of disordered materials, although they do not appear to be as universal as in true glasses (e.g., see Ref. [124]). In this regard, we note that most of the phenomena discussed in this chapter should indeed take place in other types of aperiodic structures. What makes quenched glasses special is the *intrinsic* character of their additional degrees of freedom that stems from the nonequilibrium nature of the glass transition. Since the characteristics of this transition (while not being a transition in a strict thermodynamic sense!) are nearly universal from substance to substance, many low (and not so low) temperature properties of all those substances can be understood within a unified approach.

Acknowledgments

We thank J. Schmalian, A. Leggett, A. C. Anderson, and R. J. Silbey for helpful discussions. This work was supported by NSF Grant CHE 0317017.

APPENDIX A: RAYLEIGH SCATTERING OF THE PHONONS DUE TO THE ELASTIC COMPONENT OF RIPPLON–PHONON INTERACTION

In this appendix, we present an argument on the strength of the phonon scattering due to the direct coupling with the ripplons via lattice distortions, but not due to the inelastic momentum-absorbing transition in which the internal state of the domain changes. We thus consider phonon scattering processes that do obey selection rules and couple to the lattice strain only in the second and higher order. This scattering is of the Rayleigh type (and higher order) and occurs off the domain walls as localized modes. Importantly, we will use only derived quantities and no adjustable parameters in this estimate. We show here that, indeed, this absorption mechanism is not significant compared to the resonant scattering by the inelastic transitions between the internal states of a thermally active domain.

First, it proves handy to rederive the ripplon spectrum from Eq. (34) in the less general case $\rho_g = 0$ (but nonzero pressure!). As argued in Section IV, the droplet wall is at equilibrium pressure

$$p = \frac{3}{2}\frac{\sigma}{R} = \frac{3}{2}\frac{\sigma_0 a^{1/2}}{R^{3/2}}$$

If the surface is distorted locally by Ω, this results in an extra force on this portion of the wall due to a changed curvature [109]. The second Newton's law (as applied per unit area) then yields

$$\frac{9}{8}\frac{\sigma}{R^2}\left[2+\frac{1}{\sin\theta}\frac{\partial}{\partial\theta}\left(\sin\theta\frac{\partial\Omega}{\partial\theta}\right)+\frac{1}{\sin^2\theta}\frac{\partial^2\Omega}{\partial\phi^2}\right]=\rho_W\frac{\partial^2\Omega}{\partial^2 t} \tag{A.1}$$

where θ and ϕ are the usual polar and azimuth angular coordinates on the surface and we took into account the r dependence of pressure. Equation (A.1) can be solved by a linear combination of the eigenfunctions of angular momentum in 3D:

$$\chi \equiv \sum_{lm} \Omega_{lm}(t) Y_{lm}(\theta,\phi) \tag{A.2}$$

$Y_{lm}(\theta,\phi)$ are the spherical Laplace functions ($m=-l,\ldots,1$). Substituting a harmonic of lth order in Eq. (A.1) yields the equation for ω_l derived in text as Eq. (34). We will absorb the $\frac{9}{8}$ factor into the definition of σ in the rest of the appendix.

A (fake) potential energy, yielding the equation of motion (A.1), is (cf. the discussion of surface waves on a spherical liquid droplet in Landau adn Lifshitz [116]):

$$f_{\text{surf}} = \sigma\int d\phi\int d(\cos\theta)\left\{(R+\Omega)^2+\frac{1}{2}\left[\left(\frac{\partial\Omega}{\partial\theta}\right)^2+\frac{1}{\sin^2\theta}\left(\frac{\partial\Omega}{\partial\phi}\right)^2\right]\right\} \tag{A.3}$$

Although varying Eq. (A.3) w.r.t. Ω does produce Eq. (A.1), note that it differs (by a factor of $\frac{9}{8}$!) from the original surface energy $\sigma 4\pi r^2$. The resulting error is sufficiently small for our purposes; however, this subtlety may be worth thinking about as this could reveal an extra friction mechanism due to the wetting phenomenon and surface tension renormalization mentioned in our discussion of the random first-order transition in Section II.

While the domain wall positions are not strictly tied to the atomic locations, they *are* tied to the lattice as a continuum and follow the lattice distortions. Let us employ our usual "scalar" phonons descibed by Hamiltonian

$$H_{\text{ph}} = \int d^3\mathbf{r}\left[\frac{\pi^2}{2\rho}+\frac{\rho c_s^2(\nabla\psi)^2}{2}\right] \tag{A.4}$$

where $[\psi(\mathbf{r}_1), \pi(\mathbf{r}_2)] = i\hbar\delta(\mathbf{r}_1 - \mathbf{r}_2)$. The surface energy due to the presence of both Ω and ψ is

$$H_{\text{surf}} = \sigma \int d\phi \int d(\cos\theta) \left\{ (R + [\psi - \psi(r_i)] + \Omega)^2 + \frac{1}{2}\left[\left(\frac{\partial(\psi+\Omega)}{\partial\theta} \right)^2 + \frac{1}{\sin^2\theta} \left(\frac{\partial(\psi+\Omega)}{\partial\phi} \right)^2 \right] \right\} \tag{A.5}$$

where ψ is taken on the sphere of radius R with the center located at $\mathbf{r}_i$. The potential energy in Eq. (A.5) thus provides an explicit form of phonon–ripplon interaction due to the liquid free energy functional solutions being embedded in the real space.

If we expand the value of the displacement field ϕ in terms of spherical harmonics according to $\psi_{lm} \equiv \int d\phi\, d(\cos\theta)\psi(r = R)Y^*_{lm}(\phi, \theta)$, it is then possible to write down equations of motion for the (l, m) components of both ripplon and phonon displacements:

$$\frac{\partial^2 \Omega_{lm}}{\partial t^2} + \omega_l^2(\Omega_{lm} + \psi_{lm}) = 0 \tag{A.6}$$

The equation of motion for the phonon field can be obtained, for example, from $\ddot{\psi} = i[H_{\text{ph}} + H_{\text{surf}}, \pi/\rho]$ to yield

$$\ddot{\psi} - c_s^2 \Delta\psi = -\frac{\sigma}{\rho} \int_0^{2\pi} d\phi' \int_{-1}^{1} d(\cos\theta') \int dr' \delta(r' - R) \left\{ 2(R + [\psi(\mathbf{r}') - \psi(\mathbf{r}_i)]) - \frac{1}{\sin\theta'} \left(\sin\theta' \frac{\partial\psi}{\partial\theta'} \right) - \frac{1}{\sin^2\theta'} \left(\frac{\partial^2\psi}{\partial\phi'^2} \right) + \sum_{lm} \Omega_{lm}[2 + l(l+1)] Y_{lm}(\theta', \phi') \right\} \delta(\mathbf{r} - \mathbf{r}') \tag{A.7}$$

The terms with ψ on the r.h.s. serve only to modify the local elastic constants and therefore give rise to the regular Rayleigh scattering, so we will ignore them from now on.

Equations (A.6) and (A.7) can be used to write down equations of motion for the retarded Green's functions, which are preferable due to their convenient analytical properties (see Zubarev [132] for our conventions). We are interested in the system's response to "plucking" the lattice at site $\mathbf{r} = \mathbf{0}$ at time zero; hence the choice of the Green's function corresponding to an operator X: $-i\theta(t - t')\langle[X(t), \psi(\mathbf{r} = \mathbf{0}, t' = 0)]\rangle$. Equations (A.6) and (A.7), if rewritten for the corresponding Green's functions, will remain the same except there will be an additional term $-(1/\rho)\,\delta(t)\delta^3(\mathbf{r})$, corresponding to the "plucking" event, on the right-hand side of Eq. (A.7) (note also a change in units). The obtained equations can be rewritten in the Fourier space:

$$-\omega^2\tilde{\Omega}^i_{lm} + \omega_l^2[\tilde{\Omega}^i_{lm} + \tilde{\psi}^i_{lm}] = 0 \tag{A.8}$$

and

$$-\omega^2\tilde{\psi}_{\mathbf{k}} + c_s^2k^2\tilde{\psi}_{\mathbf{k}} = -\sum_i \frac{\sigma}{\rho}\sum_{lm}\tilde{\Omega}^i_{lm}[2+l(l+1)]\frac{e^{-i\mathbf{k}\mathbf{r}_i}}{2\pi^2}Y_{lm}(-\mathbf{k}/k)i^l j_l(kR) - \frac{1}{(2\pi)^4\rho} \tag{A.9}$$

where $\tilde{\psi}^i_{lm} \equiv \int d^3\mathbf{k}\,\tilde{\psi}_{\mathbf{k}}e^{i\mathbf{k}\mathbf{r}_i}(4\pi)i^l j_l(kR)Y^*_{lm}(\mathbf{k}/k)$ and we used the expansion of a plane wave in terms of the spherical harmonics: $e^{i\mathbf{k}\mathbf{r}} = 4\pi\sum_{l=0}^{\infty}\sum_{m=-l}^{l} i^l j_l(kr)Y^*_{lm}(\mathbf{k}/k)Y_{lm}(\mathbf{r}/r)$. Here $j_l(x) \equiv \sqrt{\pi/2x}\,J_{l+1/2}(x)$ is the spherical Bessel function, which scales as x^l for small x; hence we see that the ripplons' coupling with the phonons is quadratic or higher order in $\mathbf{k}$ as the second harmonic is the lowest order term allowed. Modes $l = 0$ and $l = 1$ have the meaning of the droplet's growth and translation, respectively, as was discussed in Section IVC. These modes are not covered by this section's formalism. Even though the theory as a whole could be thought of as a multipole expansion of a molecular cluster interacting with the rest of the lattice, the modes of different orders end up being described by different theories.

The system of Eqs. (A.8) and (A.9) can now be used to determine the sound dissipation due to the interaction with the ripplons. Since the system is infinite and has a continuous spectrum, all excitations will have finite lifetimes, which in principle can be obtained self-consistently by using, for example, the Feenberg's perturbative expansion [118, 121] (in the end one arrives at Green's functions that are well behaved at infinity, as implied in the greatly simplified derivation). We do not have to do this self-consistent self-energy determination as long as we are interested in the lowest order estimate, as justified in the end by the smallness of the obtained value of the perturbation. Substituting Eq. (A.8) into Eq. (A.9) yields

$$\begin{aligned}-\omega^2\tilde{\psi}_{\mathbf{k}} + c_s^2k^2\tilde{\psi}_{\mathbf{k}} &= (4\pi)^2\frac{\sigma}{\rho}\sum_i\sum_{lm}\frac{[2+l(l+1)]\omega_l^2}{\omega_l^2-\omega^2} \\ &\times \int\frac{d^3\mathbf{k}_1}{(2\pi)^3}e^{i(\mathbf{k}_1-\mathbf{k})\mathbf{r}_i}(-1)^l j_l(k_1R)j_l(kR)Y_{lm}(-\mathbf{k}/k)Y^*_{lm}(\mathbf{k}_1/k_1)\tilde{\psi}_{\mathbf{k}_1}\end{aligned} \tag{A.10}$$

Since the spatial locations $\mathbf{r}_i$ of active droplets are not correlated,[8] we can replace the summation over the droplets by a continuous integral, assuming at the same time that the ripplon frequency corresponding to ω_l varies from droplet to droplet within a (normalized) distribution $\mathcal{P}_l(\omega)$ centered around ω_l and having a characteristic width $\delta\omega_l$, whose value will be discussed shortly. There is no reason to believe that the frequency and location of the tunneling centers are correlated; therefore one obtains

$$-\omega^2\tilde{\psi}_{\mathbf{k}} + c_s^2 k^2 \tilde{\psi}_{\mathbf{k}} = n\frac{\sigma}{\rho}\sum_l \int d\omega' \times \mathcal{P}_l(\omega')\frac{4\pi[2+l(l+1)](2l+1)\omega'^2}{\omega'^2 - (\omega + i\epsilon)^2} j_l^2(kR)\tilde{\psi}_{\mathbf{k}} \tag{A.11}$$

where n is the concentration of the active domain walls to be estimated shortly and we have displaced ω by ϵ into the upper half-plane because we are looking for the *retarded* Green's function. Also, in order to derive Eq. (A.11), we have used the summation theorem for the spherical functions

$$P_l(\mathbf{n}\mathbf{n}') = \frac{4\pi}{2l+1}\sum_{m=-1}^{l} Y_{lm}^*(\mathbf{n}')Y_{lm}(\mathbf{n})$$

as well as $P_l(-1) = (-1)^l$, where P_l is the Legendre polynomial. If we ignore the real part on the right-hand side of Eq. (A.11), responsible only for the dispersion, the poles of the resultant phonon Green's function are found by solving $\omega^2 - c_s^2 k^2 + i\,2\omega\tau_\omega^{-1} = 0$, where τ_ω^{-1} clearly has the meaning of the inverse lifetime of a phonon of frequency ω and is given by

$$\tau_\omega^{-1} = n\frac{\sigma}{\rho}\sum_{l=2}^{9}\pi^2[2+l(l+1)](2l+1)j_l^2(kR)\mathcal{P}_l(\omega) \tag{A.12}$$

where we have ignored the contribution of the peaks centered around $(-\omega_l)$. We remind the reader that $l_{\max} \simeq 9$ is dictated by the finite size of a droplet.

One can find the value of $\delta\omega_l$ from an argument identical to the one used in Xia and Wolynes [45] to obtain the width of the distribution of the barriers for the droplet growth free energy profile. At the glass transition, a liquid breaks up into dynamically cooperative regions, so that a translation of one atom involves moving about 200 atoms around it, which involves overcoming a large (on average) barrier. This barrier's height is determined, together with the domain surface tension coefficient, by the configurational entropy density, which in its turn reflects the number of metastable states available to a particular volume of liquid at this temperature. Even though a good description of freezing

[8] This is not strictly true—they, of course, cannot be on top of each other.

is achieved by assuming that this number of available states does not strongly depend on where exactly on the free energy surface a particular molecular cluster is [1], it should vary from domain to domain. The size of the variation can be estimated from the known magnitude of the entropy fluctuations at constant energy, so that the ratio of the variance to the mean is related to the jump in the heat capacity at T_g and subsequently turns out to be $1/2\sqrt{D}$ [45], where D is the liquid's fragility, entering the Vogel–Fulcher law for relaxation times in a supercooled liquid $\tau_{\text{relaxation}} \propto e^{DT_K/(T-T_K)}$. We conclude then that the lower bound on the fluctuations of the ripplon frequency ω_l is given by $\delta\omega_l \simeq \omega_l/2\sqrt{D}$.

Lastly, in order to use Eq. (A.12) to compute the phonon absorption due to this particular mechanism, we need to estimate the density of the active domain walls. It will suffice for our purposes here to consider as active the defects that contribute to the specific heat, that is, roughly, $n \simeq (1/\xi^3)T/T_g$. A more accurate estimate would be similar to the one we made when calculating the bump in the heat capacity in Section IVC.

We are now ready to give a numerical estimate of the expression in Eq. (A.12). We will compute here the contribution of the $l = 2$ term in the plateau region. It is convenient to represent kR from Eq. (A.12) as $kR \sim \omega/0.4\,(a/\xi)\omega_D$. For the reference, $(a/\xi)T_D \sim 0.2T_D$ is at the high-temperature end of the plateau, whereas its middle is about an order of magnitude lower depending on the substance (see κ vs. T/T_D plot in Fig. 1). We can now use our usual expressions connecting $\sigma, T_g, \omega_D, c_s, \rho, a$, and so on to obtain a numerical estimate of Eq. (A.12) at the plateau frequencies $\omega_{\text{plateau}} \sim 10^{-1.5}\omega_D$. Even if one favorably assumes that $\omega_2 \sim \omega_{\text{plateau}}$ (it is somewhat larger according to Section IVD), one still gets $l_{\text{mfp}}/\lambda > \sim 10^4$ at the plateau frequency, whereas the *resonant* absorption by the TLS would give $l_{\text{mfp}}/\lambda \sim 10^2$. The amplitude of this type of absorption is small due to the weakness of direct coupling to the ripplons for the processes not accompanied by a change in the domain's internal state.

APPENDIX B: FREQUENCY CUTOFF IN THE INTERACTION BETWEEN THE TUNNELING CENTERS AND THE LINEAR STRAIN

As argued in Section IIIB, the coupling of the tunneling transition to a phonon can be found from an additional energy cost of moving the molecules within the domain in the presence of a strain and is given by an integral over the droplet's volume (we consider only longitudinal strain for simplicity):

$$g = \rho c_s^2 \int_V d^3\mathbf{r}(\nabla\vec{\phi})(\nabla\mathbf{d}) \tag{B.1}$$

where ρc_s^2 is basically the elastic modulus, and $\vec{\phi}$ and $\mathbf{d}$ are elastic and inelastic components of the atomic displacements, respectively. If the phonon's wavelength is much larger than ξ, the elastic component is constant throughout the integration region and the integral reduces to one over the droplet's surface and thus the g estimate obtained in text. Otherwise, one obtains

$$g = \rho c_s^2 \left\{ \int_S d\mathbf{S}\, \mathbf{d}\, (\nabla \vec{\phi}) - \int_V d^3\mathbf{r}\, (\mathbf{d}\nabla)(\nabla \vec{\phi}) \right\} \tag{B.2}$$

The volume integral will give a higher order term in k, so for now, we focus on the surface integral. The displacement due to the phonon is conveniently expanded in terms of the spherical waves: $e^{i\mathbf{kr}} = 4\pi \sum_{l=0}^{\infty} \sum_{m=-l}^{l} i^l j_l(kr) Y_{lm}^*(\mathbf{k}/k) Y_{lm}(\mathbf{r}/r)$. Since it is the first derivative with respect to $\mathbf{r}$ that we are interested in, we only need the $l = 1$ term from this expansion. The angular part contributes only to the overall constant, but it is the spherical function $j_1(kr)$ that sets the cutoff value of the wavevector, above which the phonons do not produce significant linear uniform stress on the domain. In Fig. 24, we plot the derivative $\partial j_1(x)/\partial x$ (or, rather, we plot the square of it, which enters into all the final expressions).

We see that it is not unreasonable to assume that only the phonons with $kR < \sim 6$ will exert an appreciable linear strain on the domain. $kR = 6$ translates into $\omega_c \sim 2.5(a/\xi)\omega_D$.

While we are at it, we estimate the interaction of the domain with the higher order strain, at least due to the term (B.1), in the frequency region of interest. The next order term in the k expansion in the surface integral from Eq. (B.2) has the same structure but is scaled down from the linear term by a factor of kR. At the plateau frequencies $\sim\omega_D/30$, $kR < 0.5$ as immediately follows from the previous paragraph. While this is not a large number, it is not very small either. Therefore this interaction term is of potential importance.

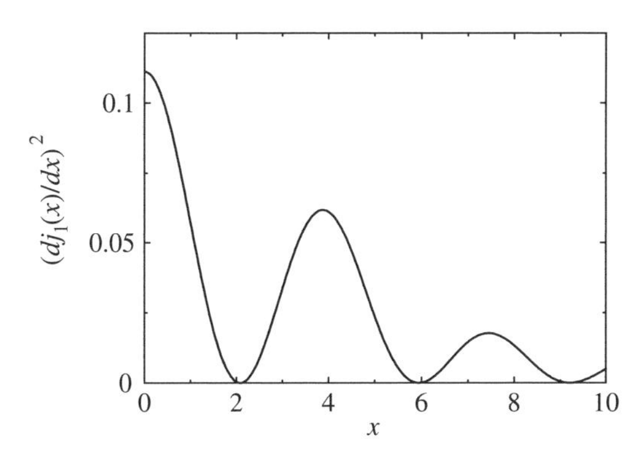

Figure 24. Shown is the derivative of the first-order spherical Bessel function determining the effective decrease in the elastic field gradient produced by a phonon of wavelength k ($x = kR$).

The volume integral in Eq. (B.2) produces a quadratic term, which is roughly equal to $(\vec{\nabla}\phi) \int_V d^3\mathbf{r}\,(\mathbf{d}\,\mathbf{k})$. We then proceed in a completely identical fashion to our earlier estimate of g. Assuming the diplacements within the droplet are random, one gets for the integral $\frac{1}{4}\sqrt{N^*}\,a^3\,d_L k$, where the factor of $\frac{1}{4}$ comes about because the displacement is assumed to decrease from d_L in the center of the droplet to zero at the edge [4]. This term becomes comparable to the linear one at frequencies $\omega \simeq \omega_D\sqrt{(a/\xi)}\,4/(6\pi^2)^{1/2} \simeq 0.4\,\omega_D$—well beyond the high T end of the plateau.

There are other sources of nonlinearity in the system, such as the intrinsic anharmonicity of the molecular interactions present also in the corresponding crystals. While these issues are of potential importance to other problems, such as the Grüneisen parameter, expression (B.1) only considers the lowest order harmonic interactions and thus does not account for this nonlinear effect. We must note that if this nonlinearity is significant, it could contribute to the nonuniversality of the plateau, in addition to the variation in T_g/ω_D ratio. It would thus be helpful to conduct an experiment comparing the thermal expansion of different glasses and see whether there is any correlation with the plateau's location.

NOTE

During the production of the chapter, a current review of the RFOT theory has appeared in print [V. Lubchenko and P. G. Wolynes, *Annu. Rev. Phys. Chem.* **58**, 235 (2007)]. In addition, microscopic descriptions of the onset of activationless reconfigurations [J. D. Stevenson, J. Schmalian, and P. G Wolynes, *Nat. Phys.* **2**, 268 (2006)] and prefactors for viscosity and ionic conductivity of deeply supercooled melts [V. Lubchenko, *J. Chem. Phys.* **126**, 174503 (2007)] are now available.

References

1. X. Xia and P. G. Wolynes, *Proc. Natl. Acad. Sci.* **97**, 2990 (2000).
2. T. R. Kirkpatrick, D. Thirumalai, and P. G. Wolynes, *Phys. Rev. A* **40**, 1045 (1989).
3. V. Lubchenko, *Quantum Theory of Glasses*, Ph.D. thesis, University of Illinois, 2002.
4. V. Lubchenko and P. G. Wolynes, *Phys. Rev. Lett.* **87**, 195901 (2001).
5. V. Lubchenko and P. G. Wolynes, *Proc. Natl. Acad. Sci.* **100**, 1515 (2003).
6. A. C. Anderson, private communication, 1999.
7. R. C. Zeller and R. O. Pohl, *Phys. Rev. B* **4**, 2029 (1971).
8. P. W. Anderson, B. I. Halperin, and C. M. Varma, *Philos. Mag.* **25**, 1 (1972).
9. W. A. Phillips, *J. Low Temp. Phys.* **7**, 351 (1972).
10. M. P. Zaitlin and A. C. Anderson, *Phys. Rev. B* **12**, 4475 (1975).
11. W. A. Phillips (ed.), *Amorphous Solids: Low-Temperature Properties*, Springer-Verlag, Berlin, 1981.
12. B. Golding and J. E. Graebner, *Phys. Rev. Lett.* **37**, 853 (1976).

13. S. Hunklinger, W. Arnold, S. Stein, R. Nava, and K. Dransfeld, *Phys. Lett.* **42A**, 253 (1976).
14. L. Guttman, and S. M. Rahman, *Phys. Rev. B* **33**, 1506 (1986).
15. K. K. Mon and N. W. Ashcroft, *Solid State Commun.* **27**, 609 (1978).
16. A. K. Raychaudhuri and R. O. Pohl, *Phys. Rev. B* **25**, 1310 (1981).
17. C. L. Reynolds, Jr., *J. Non-Cryst. Solids* **30**, 371 (1979).
18. C. L. Reynolds, Jr., *J. Non-Cryst. Solids* **37**, 125 (1980).
19. J. J. Freeman, and A. C. Anderson, *Phys. Rev. B* **34**, 5684 (1986).
20. J. E. Graebner, B. Golding, and L. C. Allen, *Phys. Rev. B* **34**, 5696 (1986).
21. A. Einstein, *Ann. Phys.* **35**, 679 (1911).
22. A. J. Leggett, *Physica B* **169**, 322 (1991).
23. C. C. Yu and A. J. Leggett, *Comments Cond. Mat. Phys.* **14**, 231 (1988).
24. S. N. Coppersmith, *Phys. Rev. Lett.* **67**, 2315 (1991).
25. M. Meissner and K. Spitzmann, *Phys. Rev. Lett.* **46**, 265 (1981).
26. A. L. Burin and Y. Kagan, *Sov. Phys. JETP* **82**, 159 (1996).
27. P. Neu, D. R. Reichman, and R. J. Silbey, *Phys. Rev. B* **56**, 5250 (1997).
28. H. M. Caruzzo, *On the Collective Model of Glasses at Low Temperature*, Ph.D. thesis, University of Illinois, 1994.
29. V. Lubchenko and P. G. Wolynes, unpublished, 2000.
30. M. Foret, E. Courtens, R. Vacher, and J.-B. Suck, *Phys. Rev. Lett.* **77**, 3831 (1996).
31. P. Benassi, M. Krisch, C. Masciovecchio, G. Monaco, G. Ruocco, F. Sette, and R. Verbeni, *Phys. Rev. Lett.* **77**, 3835 (1996).
32. O. Pilla, A. Cunsol, A. Fontana, C. Masciovecchio, G. Monaco, M. Montagna, G. Ruocco, T. Scopigno, and F. Sette, *Phys. Rev. Lett.* **85**, 2136 (2000).
33. T. S. Grigera, V. Martín-Mayor, G. Parisi, and P. Verocchio, eprint cond-mat/0110129, 2001.
34. U. Buchenau, V. L. Gurevich, D. A. Parshin, M. A. Ramos, and H. R. Schober, *Phys. Rev. B* **46**, 2798 (1992).
35. V. G. Karpov, M. I. Klinger, and F. N. Ignat'ev, *Sov. Phys. JETP* **57**, 439 (1983).
36. J. Horbach, W. Kob, and K. Binder, *J. Phys. Chem.* **103**, 4104 (1999).
37. A. C. Anderson, in *Amorphous Solids: Low-Temperature Properties*, W. A. Phillips (ed.), Springer-Verlag, Berlin, 1981.
38. Y. P. Joshi, *Phys. Status Solidi* B **95**, 317 (1979).
39. J. P. Stoessel and P. G. Wolynes, *J. Chem. Phys.* **80**, 4502 (1984).
40. Y. Singh, J. P. Stoessel, and P. G. Wolynes, *Phys. Rev. Lett.* **54**, 1059 (1985).
41. T. R. Kirkpatrick and P. G. Wolynes, *Phys. Rev. A* **35**, 3072 (1987).
42. T. R. Kirkpatrick and P. G. Wolynes, *Phys. Rev. B* **36**, 8552 (1987).
43. T. R. Kirkpatrick and D. Thirumalai, *Phys. Rev. Lett.* **58**, 2091 (1987).
44. T. R. Kirkpatrick and D. Thirumalai, *Phys. Rev. B* **36**, 5388 (1987).
45. X. Xia and P. G. Wolynes, *Phys. Rev. Lett.* **86**, 5526 (2001).
46. X. Xia and P. G. Wolynes, *J. Phys. Chem.* **105**, 6570 (2001).
47. V. Lubchenko and P. G. Wolynes, *J. Chem. Phys.* **119**, 9088 (2003).
48. V. Lubchenko and P. G. Wolynes, *J. Chem. Phys.* **121**, 2852 (2004).
49. C. A. Angell, K. L. Ngai, G. B. McKenna, P. F. Millan, and S. W. Martin, *Appl. Phys.* **88**, 3113 (2000).
50. B. Böhmer, K. L. Ngai, C. A. Angell, and D. J. Plazek, *J. Chem. Phys.* **99**, 4201 (1993).

51. C. A. Angell, *J. Non-Cryst. Solids* **73**, 1 (1985).
52. R. W. Hall and P. G. Wolynes, *Phys. Rev. Lett.* **90**, 085505 (2003).
53. V. N. Novikov and A. P. Sokolov, *Nature* **431**, 961 (2004).
54. H. G. Pfaender, *Schott Guide to Glass*, Chapman & Hall, New York, 1996.
55. F. Simon, *Trans. Faraday Soc.* **33**, 65 (1937).
56. W. Kauzmann, *Chem. Rev.* **43**, 219 (1948).
57. D. J. Gross, I. Kanter, and H. Sompolinsky, *Phys. Rev. Lett.* **55**, 304 (1985).
58. D. J. Gross and M. Mézard, *Nucl. Phys. B* **240**, 431 (1984).
59. M. P. Eastwood and P. G. Wolynes, *Europhys. Lett.* **60**, 587 (2002).
60. F. H. Stillinger, *J. Chem. Phys.* **88**, 7818 (1988).
61. R. Richert and C. A. Angell, *J. Chem. Phys.* **108**, 9016 (1998).
62. Adam, G. and J. H. Gibbs, *J. Chem. Phys.* **43**, 139 (1965).
63. F. Mezei, *Liquids, Freezing and the Glass Transition*, North–Holland, Amsterdam, 1991, p. 629.
64. M. Mézard and G. Parisi, *Phys. Rev. Lett.* **82**, 747 (1999).
65. J. Villain, *J. Physique* **46**, 1843 (1985).
66. D. P. Belanger, in *Spin Glasses and Random Fields*, A. P. Young (ed.), World Scientific, Singapore, 1998, p. 251.
67. T. Nattermann, in *Spin Glasses and Random Fields*, A. P. Young (ed.), World Scientific, Singapore, 1998, p. 277.
68. J.-P. Bouchaud and G. Biroli, *J. Chem. Phys.* **121**, 7347 (2004).
69. B. Derrida, *Phys. Rev. B* **24**, 2613 (1981).
70. F. A. Lindemann, *Phys. Z.* **11**, 609 (1910).
71. U. Bengtzelius, W. Gotze, and A. Sjolander, *J. Phys. C: Solid State Physics* **17**, 5915 (1984).
72. P. G. Wolynes, *Acc. Chem. Res.* **25**, 513 (1992).
73. J. Stevenson and P. G. Wolynes, *J. Phys. Chem. B* **109**, 15093 (2005).
74. E. V. Russel and N. E. Israeloff, *Nature* **408**, 695 (2000).
75. H. Silescu, *J. Non-Cryst. Solids* **243**, 81 (1999),
76. U. Tracht, M. Wilhelm, A. Heuer, H. Feng, K. Schmidt-Rohr, and H. W. Spiess, *Phys. Rev. Lett.* **81**, 2727 (1998).
77. L. D. Landau and E. M. Lifshitz, *Statistical Mechanics*, Pergamon Press, New York, 1980.
78. D. J. Plazek and J. H. Magill, *J. Chem. Phys.* **49**, 3678 (1968).
79. S. Perry, private communication, 2004.
80. C. T. Moynihan, A. J. Easteal, M. A. Debolt, and J. Tucker, *J. Am. Ceram. Soc.* **59**, 12 (1976).
81. O. S. Narayanaswamy, *J. Am. Ceram. Soc.* **54**, 491 (1971).
82. A. Q. Tool, *J. Am. Ceram. Soc.* **29**, 240 (1946).
83. J. P. Hansen and I. R. McDonald, *Theory of Simple Liquids*, Academic Press, New York, 1976.
84. A. Heuer and R. J. Silbey, *Phys. Rev. Lett.* **70**, 3911 (1993).
85. T. Vegge, J. P. Sethna, S.-A. Cheong, K. W. Jacobsen, C. R. Myers, and D. C. Ralph, *Phys. Rev. Lett.* **86**, 1546 (2001).
86. M. Mézard, G. Parisi, and M. A. Virasoro, *J. Physique Lett.* **46**, L217 (1985).
87. J.-P. Bouchaud and M. Mézard, *J. Phys. A* **30**, 7997 (1997).
88. J. Schmalian and P. G. Wolynes, unpublished, 2000.

89. K. Trachenko, M. T. Dove, M. J. Harris, and V. Heine, *J. Phys. Condens. Matter* **12**, 8041 (2000).
90. L. D. Landau and E. M. Lifshitz, *Theory of Elasticity*, Pergamon Press, New York, 1986.
91. A. Heuer and R. J. Silbey, *Phys. Rev. B* **48**, 9411 (1993).
92. V. Lubchenko, R. J. Silbey, and P. G. Wolynes, *Mol. Phys.* **104**, 1325 (2006).
93. J. Jäckie, *Z. Phys.* **257**, 212 (1972).
94. S. Hunklinger and A. K. Raychaudhuri, in *Progress in Low Temperature Physics*, D. F. Brewer (ed.), Elsevier, New York, 1986, vol. 9.
95. R. O. Pohl, in *Amorphous Solids: Low-Temperature Properties*, W. A. Phillips (ed.), Springer-Verlag, Berlin, 1981.
96. J. L. Black and B. I. Halperin, *Phys. Rev. B* **16**, 2879 (1977).
97. T. Geszti, *J. Phys. C***9**, 481 (1982).
98. A. Leggett, private communication, 1999.
99. A. Nittke, S. Sahling, and P. Esquinazi, Heat release in solids, in *Tunneling Systems in Amorphous and Crystalline Solids*, P. Esquinazi (ed.), Springer-Verlag, Heidelberg, 1998.
100. S. Sahling, S. Abens, and T. Eggert, *J. Low Temp. Phys.* **127**, 215 (2002).
101. W. M. Goubau and R. A. Tait, *Phys. Rev. Lett.* **34**, 1220 (1975).
102. P. Strehlow and M. Meissner, *Physica B* **263–264**, 273 (1999).
103. R. P. Feynman, *Phys. Rev.* **94**, 262 (1954).
104. J. P. Wittmer, A. Tanguy, J.-L. Barrat, and L. Lewis, 2001, eprint cond-mat/0104509.
105. A.-M. Boiron, P. Tamarat, B. Lounis, R. Brown, and M. Orrit, *Chem. Phys.* **247**, 119 (1999).
106. R. P. Feynman, *Phys. Rev.* **91**, 1291 (1953).
107. A. J. Heeger, S. Kivelson, J. R. Schrieffer, and W. P. Su, *Rev. Mod. Phys.* **60**, 781 (1988).
108. P. G. Wolynes, *Phys. Rev. Lett.* **47**, 968 (1981).
109. P. M. Morse and H. Feshbach, *Methods of Theoretical Physics*, Volume 2, McGraw-Hill, New York, 1953, p. 1469.
110. A. Wischnewski, U. Buchenau, A. J. Dianoux, W. A. Kamitakahara, and J. L. Zarestky, *Phys. Rev. B* **57**, 2663 (1998).
111. C. Kittel, *Introduction to Solid State Physics*, John Wiley & Sons, Hoboken, NJ, 1956.
112. T. L. Smith, Ph.D. thesis, University of Illinois, 1974.
113. L. Gamaitoni, P. Hanggi, P. Jung, and F. Marchesoni, *Rev. Mod. Phys.* **70**, 223 (1998).
114. J. Jäckie, L. Piché, W. Arnold, and S. Hunklinger, *J. Non-Cryst. Solids* **20**, 365 (1976).
115. R. Maynard, in *Phonon Scattering in Solids*, L. J. Challis, V. W. Rampton, and A. F. G. Wyatt (eds.), Plenum Press, New York, 1975.
116. L. D. Landau and E. M. Lifshitz, *Fluid Mechanics*, Pergamon Press, New York, 1987.
117. G. Belessa, *Phys. Rev. Lett.* **40**, 1456 (1978).
118. E. Feenberg, *Phys. Rev.* **74**, 206 (1948).
119. A. J. Leggett, S. Chakravarty, A. T. Dorsey, M. P. A. Fisher, A. Garg, and W. Zwerger, *Rev. Mod. Phys.* **59**, 1 (1987).
120. Y. M. Galperin, V. G. Karpov, and V. I. Kozub, *Adv. Phys.* **38**, 669 (1991).
121. R. Abou-Chacra, P. W. Anderson, and D. J. Thouless, *J. Phys. C* **6**, 1734 (1973).
122. D. A. Ackerman, A. C. Anderson, E. J. Cotts, J. N. Dobbs, W. M. MacDonald, and F. J. Walker, *Phys. Rev. B* **29**, 966 (1984).

123. S. Wei, C. Li, and M. Y. Chou, *Phys. Rev. B* **50**, 14587 (1994).
124. R. O. Pohl, X. Liu, and E. Thompson, *Rev. Mod. Phys.* **74**, 991 (2002).
125. J. C. Lasjaunas, R. Maynard, and M. Vandorpe, *J. Physique C6* **39**, 973 (1978).
126. E. Thompson, G. Lawes, J. M. Parpia, and R. O. Pohl, *Phys. Rev. Lett.* **84**, 4601 (2000).
127. Y. M. Galperin, V. L. Gurevich, and D. A. Parshin, *Phys. Rev. B* **32**, 6873 (1985).
128. M. Papoular, *J. Phys. C* **5**, 1943 (1972).
129. W. A. Phillips, *J. Low Temp. Phys.* **11**, 757 (1973).
130. J. T. Krause and C. R. Kurkjian, *J. Am. Ceram. Soc.* **51**, 226 (1968).
131. U. Escher, S. Abens, A. Gladun, C. Koeckert, S. Sahling, and M. Schneider, *Physica B* **284**, 1159 (2000).
132. D. N. Zubarev, *Sov. Phys. Uspekhi* **3**, 320 (1960).

DIAMONDOID MOLECULES

G. ALI MANSOORI

Departments of BioEngineering and Chemical Engineering, University of Illinois at Chicago, Chicago, Illinois 60607-7052, USA

CONTENTS

I. MOLECULAR STRUCTURE OF DIAMONDOIDS

Diamondoid molecules are cage-like, ultrastable, saturated hydrocarbons. These molecules are ringed compounds, which have a diamond-like structure consisting of a number of six-member carbon rings fused together (see Fig. 1). More explicitly, they consist of repeating units of ten carbon atoms forming a tetracyclic cage system [1–3]. They are called "diamondoid" because their

Advances in Chemical Physics, Volume 136, edited by Stuart A. Rice

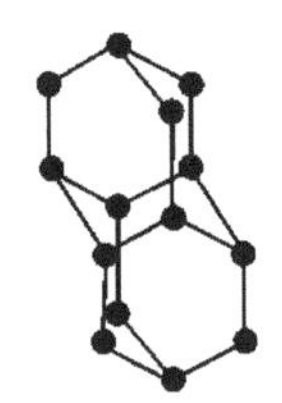
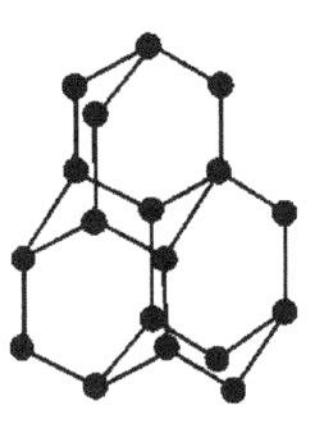

Figure 1. Molecular structures of adamantane, diamantane, and triamantane, the smaller diamondoids, with chemical formulas $C_{10}H_{16}$, $C_{14}H_{20}$, and $C_{18}H_{24}$, respectively.

carbon–carbon framework constitutes the fundamental repeating unit in the diamond lattice structure. This structure was first determined in 1913 by Bragg and Bragg [4] using X-ray diffraction analysis.

The first and simplest member of the diamondoid group, adamantane, is a tricyclic saturated hydrocarbon (tricyclo[3.3.1.1]decane). Adamantane is followed by its polymantane homologs (adamantologs): diamantane, tria-, tetra-, penta-, and hexamantane. Figure 1 illustrates the smaller diamondoid molecules, with the general chemical formula $C_{4n+6}H_{4n+12}$: adamantane ($C_{10}H_{16}$), diamantane ($C_{14}H_{20}$), and triamantane ($C_{18}H_{24}$). Each of these lower adamantologs has only one isomer.

Depending on the spatial arrangement of the adamantane units, higher polymantanes ($n \geq 4$) can have several isomers and nonisomeric equivalents. There are three possible tetramantanes, all of which are isomeric. They are depicted in Fig. 2 as iso-, anti-, and skew-tetramantane. Anti- and skew-tetramantanes each possess two quaternary carbon atoms, whereas iso-tetramantane has three.

With regard to the remaining members of the diamondoid group, there are seven possible pentamantanes, six of which are isomeric ($C_{26}H_{32}$) and obeying the molecular formula of the homologous series and one is nonisomeric ($C_{25}H_{30}$). For hexamantane, there are 24 possible structures: 17 are regular cata-condensed isomers with the chemical formula $C_{30}H_{36}$, six are irregular cata-condensed isomers with the chemical formula $C_{29}H_{34}$, and one is

Figure 2. There are three possible tetramantanes, all of which are isomeric, respectively, from left to right as anti-, iso-, and skew-tetramantane. Anti- and skew-tetramantanes possess two quaternary carbon atoms, whereas iso-tetramantane has three quaternary carbon atoms. The number of diamondoid isomers increases appreciably after tetramantane.

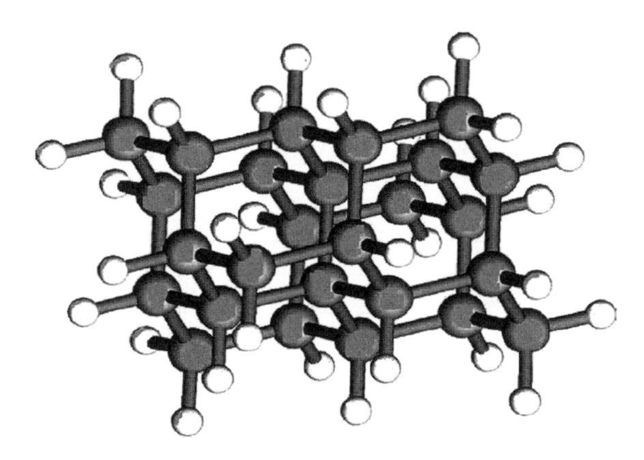

Figure 3. Molecular structure of (*peri-condensed*) cyclohexamantane ($C_{26}H_{30}$). Darker spheres represent carbon atoms while lighter spheres are hydrogen atoms.

peri-condensed with the chemical formula $C_{26}H_{30}$, as shown in Fig. 3. The top and side views of a peri-condensed cyclohexamantane cage hydrocarbon are illustrated in Table I. The table also lists some physical properties of diamondoids mostly compiled by ChevronTexaco.

Diamondoids, when in the solid state, melt at much higher temperatures than other hydrocarbon molecules with the same number of carbon atoms in their structures. Since they also possess low strain energy, they are more stable and stiff, resembling diamond in a broad sense. They contain dense, three-dimensional networks of covalent bonds, formed chiefly from first and second row atoms with a valence of three or more. Many of the diamondoids possess structures rich in tetrahedrally coordinated carbon. They are materials with superior strength-to-weight ratio.

It has been found that adamantane crystallizes in a face-centered cubic lattice, which is extremely unusual for an organic compound. The molecule therefore should be completely free from both angle strain (since all carbon atoms are perfectly tetrahedral) and torsional strain (since all C—C bonds are perfectly staggered), making it a very stable compound and an excellent candidate for various applications, as will be discussed later.

At the initial growth stage, crystals of adamantane show only cubic and octahedral faces. The effects of this unusual structure on physical properties are interesting [5].

Many of the diamondoids can be brought to macroscopic crystalline forms with some special properties. For example, in its crystalline lattice, the pyramidal-shaped [1(2,3)4]pentamantane (see Table I) has a large void in comparison to similar crystals. Although it has a diamond-like macroscopic structure, it possesses the weak, noncovalent, intermolecular van der Waals

TABLE I
Some Physical Properties of Diamondoids[a, b]

Diamondoid Chemical Formula	Molecular Structure	MW	MP (°C)	aBP (°C)	ρ (g/cc)	Crystal Structures
Adamantane $C_{10}H_{16}$		136.240	269.	135.5 @ 10 *mm Hg*	1.07	Cubic, fcc
Diamantane $C_{14}H_{20}$		188.314	236.5	272	1.21	Cubic, Pa_3
Triamantane $C_{18}H_{24}$		240.390	221.5	330	1.24	Orthorhombic, Fddd
Tetramantanes $C_{22}H_{28}$	[1(2)3]	292.466	NA	NA	NA	NA
	[121]		174	NA	1.27	Monoclinic, $P2_1/n$
	[123]		NA	NA	1.32	Triclinic, P1
Pentamantanes $C_{26}H_{32}$	[1212]	344.543	NA	NA	1.26	Orthorhombic, $P2_12_12_1$
	[12(3)4]		NA	NA	NA	Monoclinic, $P2_1/n$
	[1234]		NA	NA	1.30	NA
	[1(2,3)4]		NA	NA	1.33	Orthorhombic, Pnma
	[1213]		NA	NA	1.36	Triclinic, P-1
	[12(1)3]		NA	NA	NA	NA
Cyclohexamantane (*peri-condensed*) $C_{26}H_{30}$	*Top* [12312] *Side*	342.528	>314	NA	1.38	Orthorhombic, Pnma
Heptamantanes $C_{30}H_{34}$	[121321]	394.602	NA	NA	1.35	Monoclinic, C2/m (#12)
	[123124]	394.602	NA	NA	NA	NA

[a]aBP = apparent boiling point; MP = melting point; MW = molecular weight; ρ = normal density.

[b]Properties mostly compiled by Chevron Texaco (www.moleculardiamond.com).

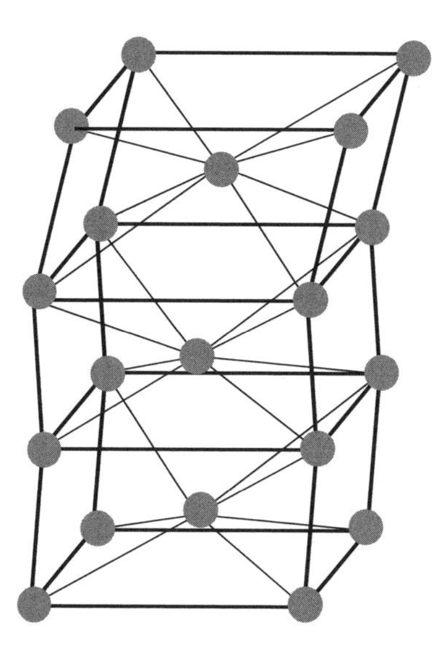

Figure 4. The quasi-cubic units of crystalline network for 1,3,5,7- tetrahydroxyadamantane. Molecules are shown as spheres and hydrogen bonds as solid linking lines. This crystalline structure is similar to that of CsCl. Taken from Ref. [8] with permission.

attractive forces involved in forming a crystalline lattice [6, 7]. Another example is the crystalline structure of 1,3,5,7-tetracarboxyadamantine, which is formed via carboxyl hydrogen bonds of each molecule with four tetrahedral nearest neighbors. The similar structure in 1,3,5,7-tetraiodoadamantane crystal would be formed by I–I interactions. In 1,3,5,7-tetrahydroxyadamantane, the hydrogen bonds of hydroxyl groups produce a crystalline structure similar to inorganic compounds, like cesium chloride (CsCl) lattice [8] (see Fig. 4).

The presence of chirality is another important feature in many derivatives of diamondoids. It should be pointed out that tetramantane[123] is the smallest of the lower diamondoids to possess chirality [6] (see Table I for chemical structures).

The vast number of structural isomers and stereoisomers is another property of diamondoids. For instance, octamantane possesses hundreds of isomers in five molecular weight classes. The octamantane class with formula $C_{34}H_{38}$ and molecular weight 446 has 18 chiral and achiral isomeric structures. Furthermore, there is unique and great geometric diversity with these isomers. For example, rod-shaped diamondoids (with the shortest one being 1.0 nm long) and disk-shaped and screw-shaped diamondoids (with different helical pitches and diameters) have been recognized [6], as shown in Table I.

II. CHEMICAL AND PHYSICAL PROPERTIES OF DIAMONDOIDS

Diamondoids show unique properties due to their exceptional atomic arrangements. Adamantane consists of cyclohexane rings in "chair" conformation. The name adamantine is derived from the Greek word for diamond since its chemical structure is like the three-dimensional diamond subunit, as shown in Fig. 5.

Later, the name diamondoids was chosen for all the higher cage hydrocarbon compounds of this series because they have the same structure as the diamond lattice: highly symmetrical and strain-free so that their carbon atom structure can be superimposed on a diamond lattice, as shown in Fig. 5 for adamantane, diamantane, and triamantane. These compounds are also known as adamantologs and polymantanes.

These compounds are chemically and thermally stable and strain-free. These characteristics cause high melting points (m.p.) in comparison to other hydrocarbons. For instance, the m.p. of adamantane is estimated to be ~269 °C, yet it sublimes easily, even at atmospheric pressure and room temperature. The melting point of diamantane is about 236.5 °C and the melting point of triamantane is estimated to be 221.5 °C. The available melting point data for diamondoids are reported in Table I.

Limited amounts of other chemical and physical property data have been reported in the literature for diamondoids [5, 9–30]. What is available is mostly for low molecular weight diamondoids. In what follows we report and analyze a selection of the available property data for diamondoids.

Adamantane {(CAS No: 281-23-2) 1-tricyclo[3.3.1.$1^{3,7}$]decane} is a cage hydrocarbon with a white or almost white crystalline solid nature, like solid wax, at normal conditions. Its odor resembles that of camphor. It is a stable and nonbiodegradable compound that is combustible due to its hydrocarbon nature. It has not been found to be hazardous or toxic to living entities [14, 15]. It should be pointed out that adamantane can exist in gas, liquid, and two solid crystalline states.

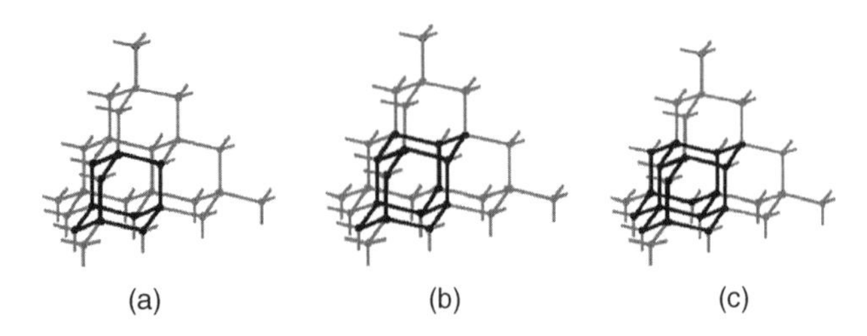

Figure 5. The relation between lattice diamond structure and (a) adamantane, (b) diamantane, and (c) triamantane structures.

TABLE II
Vapor Pressure Equations of Adamantane and Diamantane for Liquid–Vapor and Solid–Vapor Phase Transitions

Diamondoid	Phase Transition Kind	ln P (kPa) =	Temperature Range (K)	Reference
Adamantane	Liquid-Vapor	$-4670/T + 14.75$	$T > 543$ K	[13]
	Solid–Vapor	$-6570/T + 18.18$	483–543	[13]
		$-6324.7/T + 17.827$	366–443	[16]
		$-9335.6/T + 65.206 - 15.349 \log T$	313–443	[16]
		$-7300/T + 31.583 - 4.376 \log T$	333–499	[17]
Diamantane	Liquid–Vapor	$-5680/T + 14.858$	516–716	[13]
	Solid–Vapor	$-7330/T + 18.00$	498–516	[13]
		$-7632.5/T + 18.333$	353–493	[17]
		$-18981.3/T + 190.735 - 55.4418 \log T$	332–423	[17]

Diamantane {(CAS No: 2292-79-7) pentacyclo[7.3.1.1$^{4.12}$.0$^{2.7}$.0$^{6.11}$]tetradecane}, also known as decahydro-3,5,1,7-[1.2.3.4]-butanetetraylnaphtalene, can exist in gas, liquid, and three different solid crystalline states. Higher diamondoids possess two or more solid crystalline states [5].

The solid and liquid vapor pressures for adamantane and diamantane have been determined between ambient temperature and their estimated critical points using various measurement techniques by a number of investigators [13, 16, 17]. In Table II, equations representing the natural logarithm of adamantane and diamantane vapor pressures, ln P, as a function of absolute temperature as measured by various investigators are reported along with the temperature ranges of their validity. Figures 6 and 7 are based on the vapor pressure data reported in Table II. In these figures the smoothed vapor–liquid–solid phase transition lines (vapor pressures) for adamantane and diamantane are reported. There is a limited amount of available data for vapor pressures of binary mixtures of adamantane and diamantane with other hydrocarbons [18].

In Table III we also report the average of the available data for molar enthalpies, molar entropies, and molar heat capacities of adamantane and diamantane as measured and reported by various investigators [10, 19–30].

In Table IV we report the available enthalpies of formation, sublimation, and combustion of methyl-adamantanes, dimethyl-adamantane, trimethyl-adamantane, and tetramethyl-adamantane and compare them with the same properties of adamantane as measured by various investigators [29, 30]. Also reported in Table IV are the same properties for 1-methyl-diamantane, 3-methyl-diamantane, and 4-methyl-diamantane as compared with the diamantane data.

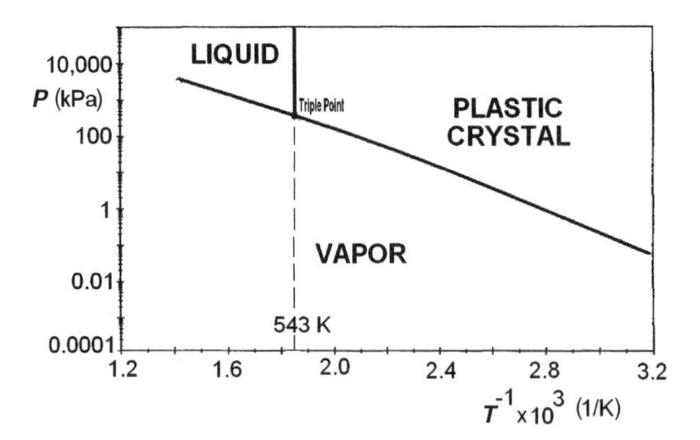

Figure 6. Vapor–liquid–solid (plastic crystal) phase diagram of adamantane. The phase transition from plastic crystal to rigid crystal phase occurs at 208.6 K ($1/T = 0.004794\,K^{-1}$). This diagram is based on the data of Table II.

A number of other thermodynamic properties of adamantane and diamantane in different phases are reported by Kabo et al. [5]. They include (1) standard molar thermodynamic functions for adamantane in the ideal gas state as calculated by statistical thermodynamics methods; and (2) temperature dependence of the heat capacities of adamantane in the condensed state between 340 and 600 K as measured by a scanning calorimeter and reported here in Fig. 8. According to this figure, liquid adamantane converts to a solid plastic with simple cubic crystal structure upon freezing. After further cooling it moves into another solid state, an fcc crystalline phase.

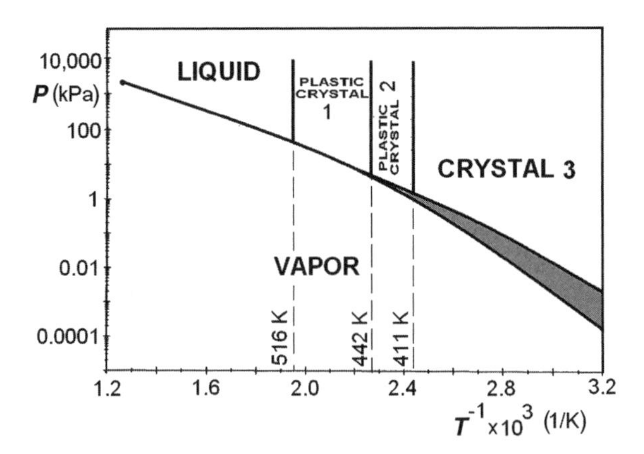

Figure 7. Vapor–liquid–solids (plastic crystal 2, plastic crystal 2, crystal 3) phase diagram of diamantane. This diagram is based on the data of Table II. The shaded area between vapor and plastic crystal 2 and crystal 3 phase transitions is indicative of the error range of the available data.

TABLE III
Thermodynamic Properties of Adamantane and Diamantane[a]

Property	Diamondoid		Value	Units	T (K)	References
ΔH^{of}_{gas}	Adamantane		−133.6	kJ/gmol		[10,19–21]
	Diamantane		−145.9	kJ/gmol		[10]
ΔH^{of}_{solid}	Adamantane		−191.1	kJ/gmol		[10,19–22]
	Diamantane		−241.9	kJ/gmol		[10]
S^{o}_{solid}	Adamantane (crystalline phase II)		195.83	J/gmol•K		[23,24]
	Diamantane (crystalline phase III)		200.16	J/gmol•K		[25]
C_P^{solid} (solid phase I)	Adamantane		189.74	J/gmol•K	298.15	[23]
	Diamantane		220.2	J/gmol•K	295.56	[26]
			223.22		298.15	[25]
$\Delta H^{o}_{sublimation}$	Adamantane		59.9	kJ/gmol		[10, 16, 21, 27, 28]
	Diamantane		96	kJ/gmol	305.-333	[29]
$\Delta H_{phase\ transition}$	Adamantane	(solid I–solid II)	3.376	kJ/gmol	208.62	[23, 24]
	Diamantane	(solid I–solid III)	4.445	kJ/gmol	407.22	
		(solid I–solid II)	8.960	kJ/gmol	440.43	[26]
		(solid I–liquid)	8.646	kJ/gmol	517.92	
$\Delta S_{phase\ transition}$	Adamantane	(solid I–solid II)	16.18	J/mol•K	208.62	
	Diamantane	(solid I–solid III)	10.92	J/mol•K	407.22	[26]
		(solid I–solid II)	20.34	J/mol•K	440.43	
		(solid I–liquid)	16.69	J/mol•K	517.92	
$\Delta H^{o}_{combustion}$	Adamantane	(solid phase)	−6030.04	kJ/mol		[10, 16, 20, 21]
	Diamantane	(solid phase)	−8125.58	kJ/gmol		[29]

[a]Values in table are the average of values reported by various researchers.
Note: Solid adamantane possesses two crystalline phases and diamantane exists in three crystalline phases.

For higher diamondoids very limited data are available in the literature. For triamantane, for example, the enthalpy and entropy of transition from crystalline phase II to crystalline phase I are reported [31] as follows:

$$\Delta H_{\text{phase II–I}} \quad 1.06\,\text{kJ/mol (at 293.65 K)}$$
$$\Delta S_{\text{phase II–I}} \quad 3.77\,\text{kJ/mol (at 293.65 K)}$$

TABLE IV
Thermodynamic Properties of Methyl Derivatives of Adamantane and Diamantane.[a] In this table the average of the values reported by various investigators are reported.

Diamondoid	ΔH^{of}_{gas} (kJ/gmol)	ΔH^{of}_{solid} (kJ/gmol)	$\Delta H^{o}_{sublimation}$ (kJ/gmol)	$\Delta H^{o}_{combustion}$ (kJ/gmol) (solid phase)	References
Adamantane	−133.6	−191.1	59.9	−6030.04	See Table II
1-Methyl-adamantane	−171.6	−240.1	67.7	−6661.1	[10, 30]
1,3-Dimethyl-adamantane	−219.0	−287.3	67.8	−7294.0	[30]
1,3,5-Trimethyl-adamantane	−255.0	−333.0	77.8	−7927.4	[30]
1,3,5,7-Tetramethyl-adamantane	−283.3	− 370.7	82.4	− 8568.7	[10, 30]
2-Methyl-adamantane	−151.7	− 220.8	67.7	− 6680.4	[10, 30]
Diamantane	−145.9	−241.9	96.	− 8125.6	See Table II
1-Methyl-diamantane	−166.7	−247.4	80.6	− 8799.4	[10]
3-Methyl-diamantane	−157.3	−260.4	103.1	−8786.36	[10]
4-Methyl-diamantane	−182.1	−261.5	79.4	−8786.2	[10]

[a]Values in table are the average of values reported by various researchers.

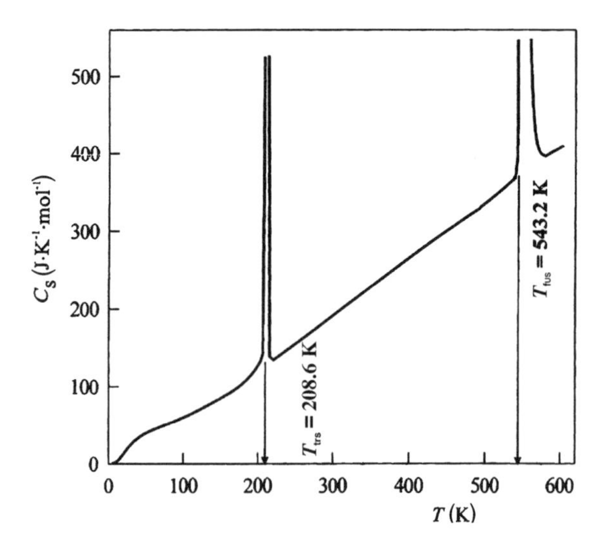

Figure 8. The temperature dependence of the heat capacity in the condensed state for adamantane [5] as measured by a scanning calorimeter. T_{trs} stands for temperature of transition from rigid crystal (fcc) to plastic crystal (cubic) state of adamantane and T_{fus} stands for fusion temperature.

Little data is reported on tetra-, penta-, and hexamantane and other higher diamondoids. What is available is compiled by ChevronTexaco scientists as reported in Table I. This is possibly due to the fact that of these compounds only anti-tetramantane has been successfully synthesized in the laboratory in small quantities [32, 33].

A. Solubilities

A limited amount of solubility data for diamondoids in liquid solvents is available and it is given in Table V. In this table the solubility limits of adamantane and diamantane in various liquid solvents at normal conditions are reported. In producing this data a known amount of diamondoid was titrated with various liquids at 25 °C. Continuous stirring was used until all the diamondoid was dissolved, which defined the solubility limit [13]. According to Table V, adamantane solubility in tetrahydrofuran (THF) is higher than in other organic solvents. Overall, cyclohexane is a better solvent for diamondoids among the liquids tested due to the similarities in the molecular structure of cyclohexane to diamondoids. It should also be pointed out that since diamondoid molecules are substantially hydrophobic, their solubility in organic solvents is a function of their hydrophobicity [13]. Carbon atoms on diamondoid surfaces possess

TABLE V
Solubilities of Diamondoids in Liquid Solvents at 25 °C

Solvent	Adamantane (wt%)	Diamantane (wt%)
Carbon tetrachloride	7.0	5.0
Pentane	11.6	4.0
Hexane	10.8	3.9
Heptane	10.4	3.7
Octane	10.0	3.9
Decane	8.9	3.5
Undecane	7.9	3.2
Tridecane	7.3	2.7
Tetradecane	7.5	2.3
Pentadecane	7.1	2.2
Cyclohexane	11.1	6.3
Benzene	10.9	4.3
Toluene	9.9	4.5
m-Xylene	9.8	4.5
p-Xylene	9.6	4.5
o-Xylene	9.6	4.1
THF	12.0	4.0
Diesel oil	7.5	2.7
1,3, Dimethyl-adamantane	6.0	2.0

Source: Taken from Ref. [13].

primary carbons in the methylated analogs, as well as secondary and tertiary carbons. This makes it possible to use a wide and selective array of derivatization strategies on diamondoids [34]. Diamondoids' derivatization could enhance their solubility and as a result their potential applications in separation schemes, chromatography, and pharmaceuticals.

The solubilities of adamantane and diamantane in supercritical (dense) methane, ethane, and carbon dioxide gases have been measured by a number of investigators [35–37] at a few temperatures with various pressures and solvent densities. These measurements are reported in Figs. 9–12.

Experimental data [36] on the effect of temperature and pressure on the supercritical solubility of adamantane in dense (supercritical) carbon dioxide gas is reported in Fig.9.

In Fig. 10 the experimental data [35, 37] showing the effect of temperature and supercritical solvent density on the solubility of adamantane in dense (supercritical) carbon dioxide at various temperatures are reported.

The data of Smith [35] is reported graphically in Fig. 11 and shows the effect of pressure on the solubility of adamantane in various supercritical solvents (carbon dioxide, methane, and ethane) at 333 K.

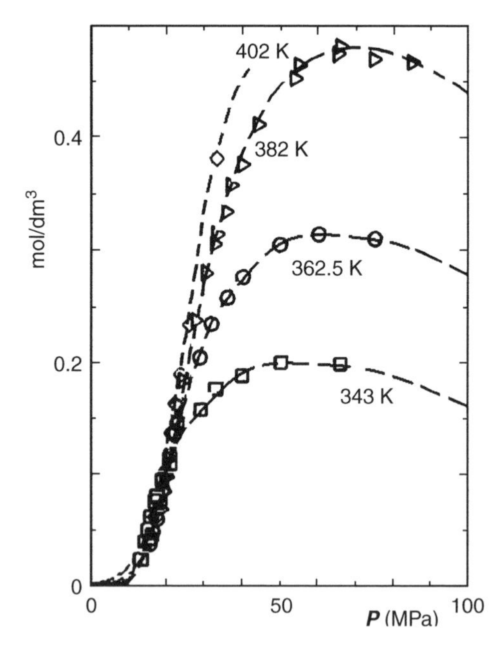

Figure 9. Effect of temperature and pressure on solubility (in units of mol/dm^3) of adamantane in dense (supercritical) carbon dioxide gas. Data from Ref. [36].

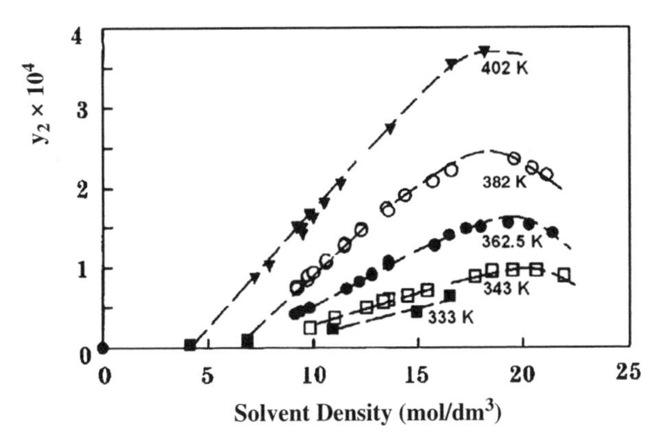

Figure 10. Effect of temperature and supercritical solvent density on solubility of adamantane (in units of mole fraction) in dense (supercritical) carbon dioxide. Data of isotherm at 333 K is from Ref. [35]. Data of isotherms at 343 K, 362.5 K, 382 K, and 402 K are from Ref. [37].

A graphical representation of diamantane solubility data [36] in various supercritical solvents (carbon dioxide and ethane at 333 K and methane at 353 K) is shown in Fig. 12.

Trends of solubility enhancement for each diamondoid follow regular behavior like other heavy hydrocarbon solutes in supercritical solvents with respect to variations in pressure and density [38, 39]. Supercritical solubilities of

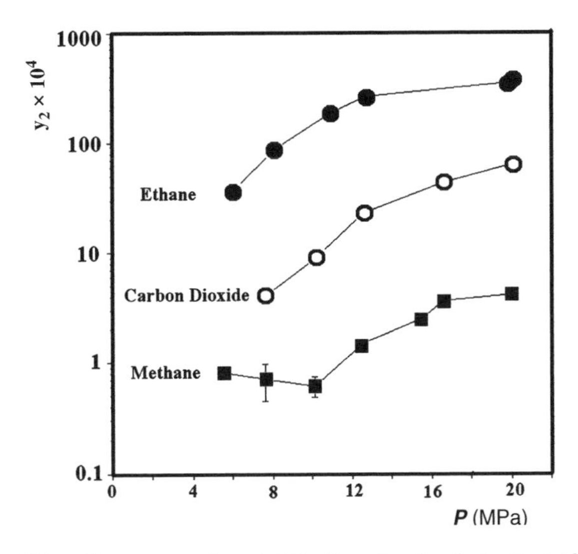

Figure 11. Effect of pressure on the solubility (in units of mole fraction) of adamantane in dense (supercritical) carbon dioxide, methane, and ethane gases at 333 K. Data from Ref. [35].

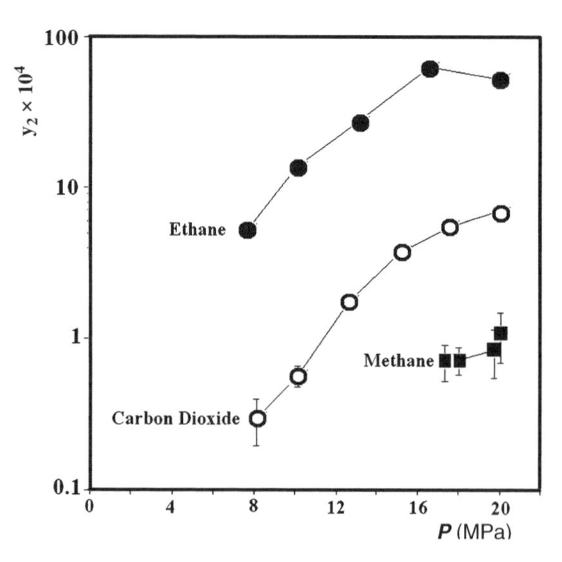

Figure 12. Effect of pressure on solubility (in units of mole fraction) of diamantane in dense (supercritical) gases at 333 K (for carbon dioxide and ethane) and at 353 K (for methane). Data from Ref. [35].

these lower diamondoids have been successfully correlated through cubic equations of state [35].

The supercritical fluid and liquid solubilities reported in Figs. 9–12 suggest that diamondoids will preferentially partition themselves into the high-pressure, high-temperature, and rather low-boiling fraction of any mixture including crude oil.

III. SYNTHESIS OF DIAMONDOIDS

Originally, diamondoids were considered hypothetical molecules since they could neither be isolated from a natural repository nor made through rational organic synthesis [3] until 1933 when adamantane was discovered and isolated from petroleum [40]. The natural source of diamondoids was the only effective source until 1941 when adamantane was synthesized for the first time, although the yield was very low [32, 41, 42]. In 1957 Schleyer introduced a Lewis acid–catalyzed rearrangement of hydrocarbons to produce adamantane [43, 44]. A Lewis acid is any acid that can accept a pair of electrons and form a coordinate covalent bond [45]. According to Schleyer, *endo*-trimethylenenorbornane, which could be produced readily, rearranged to adamantane when refluxed overnight with aluminum bromide or aluminum chloride, as shown in Fig. 13. Later, in 1965, Cupas, Schleyer, and Track [46] used the same Lewis acid–catalyzed rearrangement synthesis approach to produce diamantane.

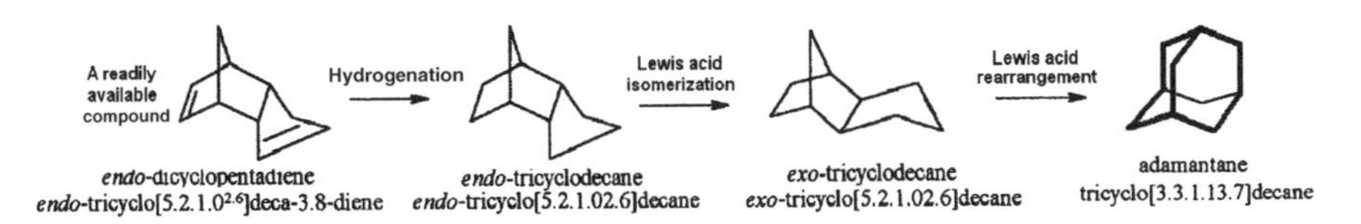

Figure 13. Various stages in synthesis of adamantane based on the work reported in Refs. [43 and 44].

Schleyer's Lewis acid–catalyzed rearrangement method, which is based on diamondoid thermodynamic stability during carbocation rearrangements, has had little or no success in synthesizing diamondoids beyond triamantane. In recent years, outstanding successes have been achieved in the synthesis of adamantane and other lower molecular weight diamondoids [42–49]. Some new methods have been developed and the yield has been increased to 60%.

While attempts to synthesize lower diamondoids (adamantane, diamantane, triamantane) have been successful through Lewis acid-catalyzed rearrangement, it is no longer a method of choice for synthesis of higher polymantanes [32]. The fourth member of diamondoid homologs, tetramantane ($C_{22}H_{28}$), has three isomers. The fifth member, pentamantane ($C_{26}H_{32}$), has six isomers, while the sixth member, hexamantane ($C_{30}H_{36}$), has as many as seventeen possible isomers. Producing these heavier diamondoids via the above-mentioned rearrangement synthesis method has not been convenient because of their close structural properties and their lower thermal stability.

The synthesis of higher diamondoids, by its nature, is a challenging and complex process. For example, after numerous efforts the anti-isomer of tetramantane was synthesized with only 10% yield [32]. The usage of zeolites as the catalyst in the synthesis of adamantane has been investigated and different types of zeolites have been tested for achieving better catalyst activity and selectivity in adamantane formation reactions [42]. Recently, Shibuya et al. [50] have reported two convenient methods for the synthesis of enantiomeric adamantane derivatives. High molecular weight diamondoids have not been synthesized as of this date. Present day efforts are geared toward more economical separation of high molecular weight diamondoids from petroleum fluids and more economical synthesis of lower molecular weight diamondoids [51].

IV. NATURAL OCCURRENCE OF DIAMONDOIDS

Adamantane and other diamondoids are constituents of petroleum, gas condensate (also called NGL or natural gas liquid), and natural gas reservoirs [52–56]. Adamantane was originally discovered [40] and isolated from petroleum fractions of the Hodonin oil fields in Czechoslovakia in 1933.

Naturally occurring adamantane is generally accompanied by small amounts of alkylated adamantane: 2-methyl-; 1-ethyl-; and probably 1-methyl-; 1,3-dimethyl- adamantane; and others [3]. Diamantane, triamantane, and their alkyl-substituted compounds are also present in certain petroleum crude oils. Their concentrations in crude oils are generally lower than that of adamantane and its alkyl-substituted compounds.

Tetramantane, pentamantanes, and hexamantanes are found in some deep natural gas deposits [57, 58] but are not readily synthesized in the laboratory.

The question of how diamondoids (polymantanes) come to be present in petroleum fluids is an interesting one. Diamond in nature is formed from abiogenic carbon and is abiogenic. However, the structurally related diamondoids in oil are biogenic. Diamondoids in petroleum are believed to be formed from enzymatically created lipids with subsequent structural rearrangement during the process of source rock maturation and oil generation. Because of this, the diamondoid content of petroleum is applied to distinguish source rock facies [59–62].

Due to the particular structure of diamondoids, they could be useful in making new biomarkers with more stability than the existing ones. New findings indicate that diamondoids are the appropriate alternatives for analyzing reservoirs that could not be assessed with conventional techniques. They appear to be resistant to biodegradation. Following biodegradation the remaining oil is enriched with diamondoids. Then the level of biodegradation is estimated by determination of the ratio of diamondoids to their derivatives, particularly when the main part of hydrocarbons has been degraded [56, 59–62].

It is believed [57, 63] that the diamondoids found in petroleum result from carbonium ion rearrangements of suitable organic precursors (such as multi-ringed terpene hydrocarbons) on clay mineral from the same source. In view of the Lewis acid-catalyzed isomerization (rearrangement) of hydrocarbons as discussed earlier, it is speculated that diamondoids may have been formed via homologation of the lower adamantologs at high pressure and temperature in the natural underground oil and gas reservoirs. The lower adamantologs are believed to have been formed originally by the catalytic rearrangement of tricycloalkanes (Fig. 13) during or after oil generation [64].

Despite the above-mentioned hypothesis aimed at finding the adamantane precursors, the adamantane-forming substance in natural petroleum fluids remains unknown. In rare cases, tetra-, penta-, and hexamantanes are also found in petroleum crude oils. Tetramantane, pentamantane, and hexamantane were discovered for the first time in 1995 in a gas condensate (NGL) produced from a very deep ($\sim$6800 m below the surface) petroleum reservoir located in the U.S. Gulf Coast [64]. A group of investigators [65] recently reported isolation of crystals of the lower-order diamondoid cyclohexamantane ($C_{26}H_{30}$), Fig. 3, from distilled Gulf of Mexico petroleum using reverse-phase HPLC (high-performance liquid chromatography). They determined the structure of $C_{26}H_{30}$ using X-ray diffraction, mass spectroscopy, and ^{1}H,^{13}C-NMR spectroscopy. They also used the experimental Raman spectra of crystalline diamond, adamantane, and nanophase diamond to indirectly identify frequencies in the experimental Raman spectra of $C_{26}H_{30}$.

Recent identification and isolation of crystals of many new medium- and higher-order diamondoids from petroleum have been reported [6]. These investigators reported the isolation of multiple families of higher diamondoid molecules containing 4–11 diamond-crystal cages from petroleum and they supplied X-ray structures for representatives from three families. They reported separation of higher diamondoids from selected petroleum feedstock by distillation. They also reported the removal of nondiamondoids by pyrolysis at 400–450 °C. Aromatic and polar compounds were removed from pyrolysis products by argentic silica gel liquid chromatography. Higher diamondoids were then isolated by a combination of reverse-phase HPLC on octadecyl silane columns and highly shape-selective Hypercarb® high-performance liquid chromatography (HPLC) columns. Individual higher diamondoids were recrystallized to high purity [6].

A group of investigators recently suggested that the density-functional theory (DFT), which calculates IR and Raman spectra, is a useful tool for direct characterization of the structures of diamondoids with increasing complexity [66]. They applied DFT to calculate Raman spectra whose frequencies and relative intensities were shown to be in excellent agreement with the experimental Raman spectra for $C_{26}H_{30}$, thus providing direct vibrational proof of its existence.

Characteristics of diamondoids present in petroleum fluids are considered a proof of the biogenic origin of petroleum. Since diamondoids are resistant to biodegradation, use of diamondoid compounds for the characterization of heavily biodegraded oils has been suggested and demonstrated. For example, it is found that the kinds of diamondoids present in crude oil are indicators of the level of petroleum biodegradation, which are not affected by reservoir maturity [67, 68]. The presence of diamondoids is considered to be closely related to the geological maturity of an oil field [69–72]. Some investigators have used the relative abundance of diamondoids to fingerprint, identify, and estimate the degree of oil cracking in an underground oil reservoir [73].

The relatively high melting points of diamondoids have caused their precipitation in oil wells, transport pipelines, and processing equipment during production, transportation, and refining of diamondoid-containing petroleum crude oil and natural gas [74–76]. This may cause fouling of pipelines and other oil processing facilities. Diamondoid deposition and possible fouling problems are usually associated with deep natural underground petroleum reservoirs that are rather hot and at high pressure. Other hydrocarbons with molecular weights in the same range as diamondoids are generally less stable and crack at high temperatures [74].

The practice of petroleum production may lead to an environment that favors the reduction of diamondoids' solubility in petroleum, their separation from petroleum fluid phase, and their precipitation. For instance, it has been observed that phase segregation of diamondoids from the "dry petroleum" (meaning petroleum fluids low in light hydrocarbons) streams takes place upon reduction of pressure and/or temperature of the stream. It has also been observed that in "wet petroleum" streams diamondoids partition themselves among the existing phases (vapor, liquid, solid) [74–76]. As a result, diamondoids may nucleate out of solution upon drastic changes of pressure and/or temperature. Instabilities of this sort in crude oils may potentially initiate a sudden precipitation of other heavy organic compounds (including asphaltenes, paraffins, and resins) on such nuclei [74–76]. Therefore, knowledge about the solubility behavior of diamondoids in organic solvents and dense gases, as reported previously, becomes important. It should be pointed out that among heavy organics in crude oil, asphaltenes, paraffins, and diamondoids are propounded in nanotechnology subjects. By using the proper solvents, asphaltene molecules can form nanometer-sized micelles and micelle self-assembly (coacervate) particles [77].

Adamantane and diamantane are usually the dominant diamondoids found in petroleum and natural gas pipeline deposits [74, 75]. This is because diamondoids are soluble in light hydrocarbons at high pressures and temperatures. Upon expansion of the petroleum fluid coming out of the underground reservoir and a drop in its temperature and pressure, diamondoids could deposit.

Deposition of adamantane from petroleum streams is associated with phase transitions resulting from changes in temperature, pressure, and/or composition of reservoir fluid. Generally, these phase transitions result in a solid phase from a gas or a liquid petroleum fluid. Deposition problems are particularly cumbersome when the fluid stream is dry (i.e., low LPG content in the stream). Phase segregation of solids takes place when the fluid is cooled and/or depressurized. In a wet reservoir fluid (i.e., high LPG content in the stream) the diamondoids partition into the LPG-rich phase and the gas phase. Deposition of diamondoids from a wet reservoir fluid is not as problematic as in the case of dry streams [74, 75].

A. Separation, Detection, and Measurement

Petroleum crude oil, gas condensate, and natural gas are generally complex mixtures of various hydrocarbons and nonhydrocarbons with diverse molecular weights. In order to analyze the contents of a petroleum fluid it is a general practice to separate it first into five basic fractions: namely, *volatiles*, *saturates*, *aromatics*, *resins*, and *asphaltenes* [74, 77]. Volatiles consist of the low-boiling

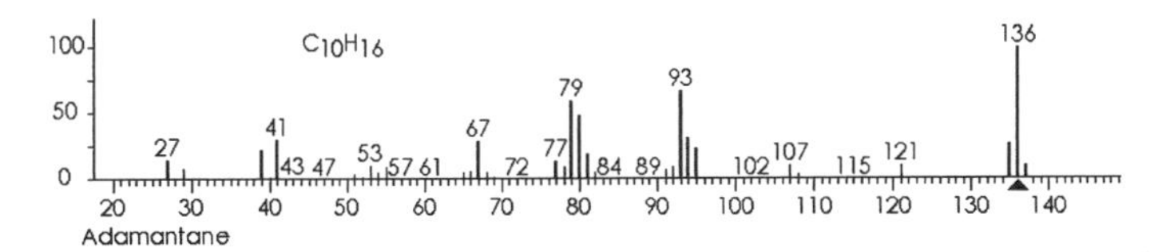

Figure 14. Standard molecular fragmentation spectrum of adamantane (136 m/z).

fraction of crude oil separated at room temperature and under vacuum (~27 °C and 10 mm Hg) from crude oil. The contents of the volatiles fraction can be further analyzed using gas chromatography. The remaining four fractions are separated from the vacuum residue with the use of a liquid chromatography column (i.e., SARA separation [78]). Initially the asphaltenes fraction of the sample is removed by the ASTM D3279-90 separation method [79]. Then the saturates fraction is extracted with *n*-hexane solvent by passing the sample through a liquid chromatography column that is packed with silica gel and alumina powder. Diamondoids are saturated hydrocarbons. Therefore analysis to determine their presence in a crude oil must be performed on the saturates fraction. Mass spectrometry analyses must then be performed on the low-boiling part of the saturates fraction to determine whether diamondoids are present in this petroleum fluid. To achieve this, the saturates fraction is analyzed through gas chromatography/mass spectrometry (GC/MS) [11, 74, 75, 80].

After the GC/MS analyzer produces the standard molecular fragmentation spectra of the diamondoids (like Fig. 14), the diamondoids' molecular weights are determined by GC/MS. Some representative chromatograms of petroleum fluids containing diamondoids obtained by various investigators [11, 74, 75, 81] from the GC/MS analyses are shown in Figs. 15–18. Adamantane, if present in the sample, will elute between nC_{10} and nC_{11}; diamantane will elute between nC_{15} and nC_{16}; triamantane will elute between nC_{19} and nC_{20}, and so on [75]. For example, the unknown peak with retention time of 5.70 eluted between nC_{15} and nC_{16} in Fig. 15 is indicative of the probable existence of diamantane in the sample. A full scan total ion chromatogram (TIC) of the sample is turned around from 100 m/z to 1000 m/z using molecular fragmentation spectra (like Fig. 14 for adamantane). With the availability of standard molecular fragmentation spectra of diamondoids (like Fig. 14 for adamantane) [81, 82], diamondoids are identified in the sample.

There exist a number of other methods for the separation of diamondoids from petroleum fluids or natural gas streams: (1) a gradient thermal diffusion process [54] is proposed for separation of diamondoids; (2) a number of extraction and absorption methods [53,83] have been recommended for removing diamondoid compounds from natural gas streams; and (3) separation of certain diamondoids from petroleum fluids has been achieved using zeolites [56, 84] and a number of other solid adsorbents.

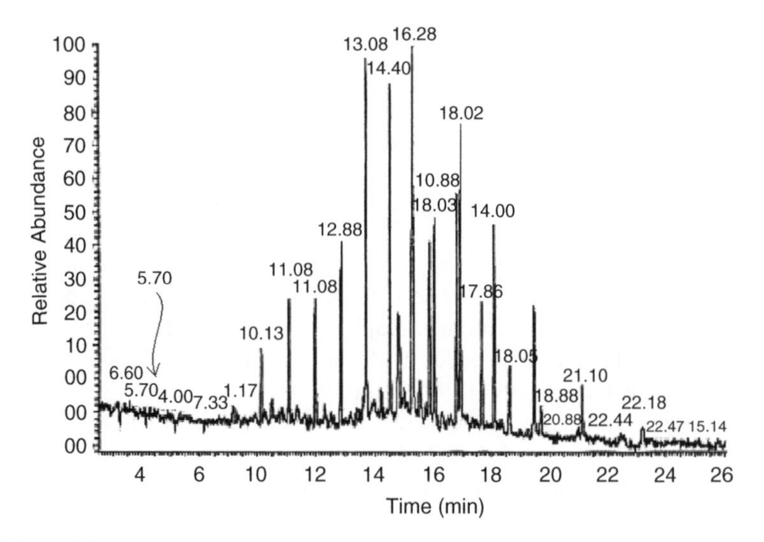

Figure 15. Gas chromatogram of a gas condensate (NGL = natural gas liquid) sample [74]. The peak with retention time of 5.70 eluted between nC_{15} and nC_{16} is indicative of the probable existence of diamantane in the sample.

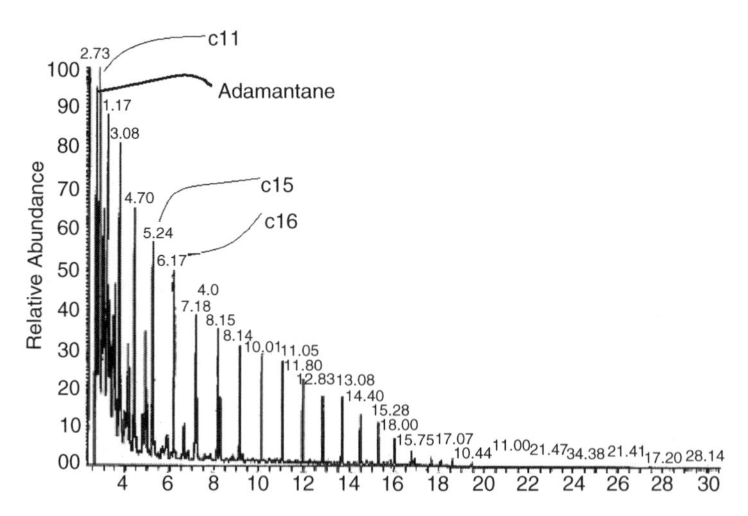

Figure 16. Gas chromatogram of a crude oil sample showing the possible existence of adamantane and diamantane in the sample.

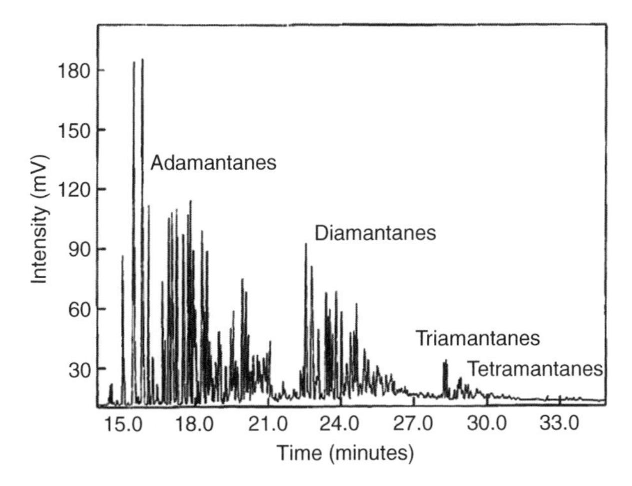

Figure 17. Gas chromatogram of a diamondoid-rich gas condensate (NGL) sample showing clusters of peaks representing adamantanes, diamantanes, triamantanes, and tetramantanes. Taken from Ref. [11] with permission.

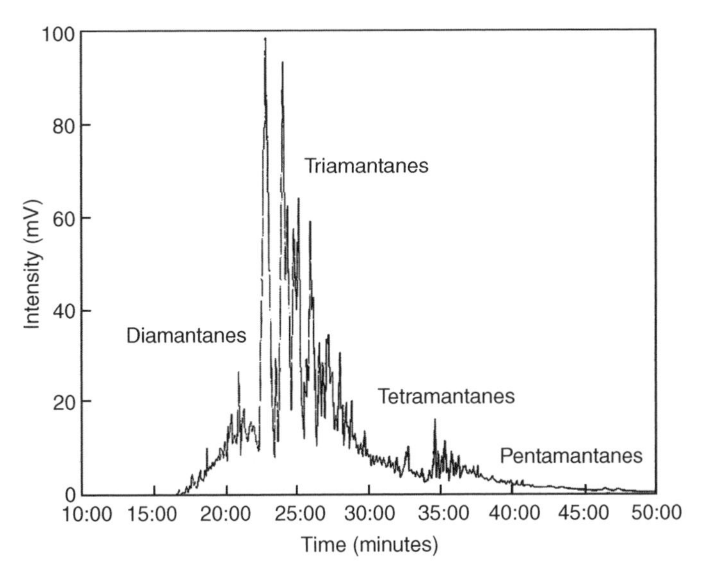

Figure 18. Gas chromatogram from the full-scan GC/MS analysis of a high-temperature distillation fraction (343 °C) containing diamondoids. Taken from Ref. [11] with permission.

V. DIAMONDOIDS AS TEMPLATES AND MOLECULAR BUILDING BLOCKS

Each successively higher diamondoid family shows increasing structural complexity and varieties of molecular geometries. Sui generis properties of diamondoids have provoked an extensive range of inquiries in different fields of science and technology. They have been used as templates for crystallization of zeolite catalysts [85], in the synthesis of high-temperature polymers [86], in drug delivery and drug targeting, in nanotechnology, in DNA-directed assembly, in DNA–protein nanostructures, and in host–guest chemistry.

A. In Polymers

Diamondoids, especially adamantane, its derivatives, and diamantane, can be used for improvement of thermal stability and other physicochemical properties of polymers and preparation of thermosetting resins, which are stable at high temperatures. For example, it is demonstrated that acetylene groups on highly hindered diamondoid cage compounds (like 1,3-diethynyladamantane) can be polymerized thermally to give thermoset polymers. The resulting thermoset polymer was shown to be stable to 475 °C in air and exhibited less than 5% weight loss in air after 100 h at 301 °C. This unusually high thermal stability for an aliphatic hydrocarbon polymer obviously results from the presence of adamantane units in the polymer backbone, which due to their "diamond-like" structure retards degradation reactions resulting from either nucleophilic or electrophilic attack, or from elimination reactions. It is also demonstrated that synthesis of acetylenic derivatives of diamantane and their thermal polymerization give materials with thermooxidative stabilities significantly greater than those of the corresponding adamantane polymers. The presence of diamantyl groups enhances the thermal stability of the resulting thermoset resins by nearly 50 °C over the corresponding adamantane-based polymers [87]. Another example of high-temperature polymers in this category is polymerization of diethynyldiamantane [88]. Adamantyl-substituted poly(*m*-phenylene) is synthesized starting with 1,3-dichloro-5-(1- adamantyl)benzene monomers (Fig. 19) and it is also shown to have a high degree of polymerization and stability, decomposing at high temperatures of around 350 °C [89].

Diamantane-based polymers are synthesized to take advantage of their stiffness, chemical and thermal stability, high glass transition temperature, improved solubility in organic solvents, and retention of their physical properties at high temperatures. All these special properties result from their diamantane-based molecular structure [90]. Polyamides are high-temperature polymers with a broad range of applications in different scientific and industrial fields. However, their process is very difficult because of poor solubility and lack of adequate thermal stability retention [90]. Incorporation of 1,6- or

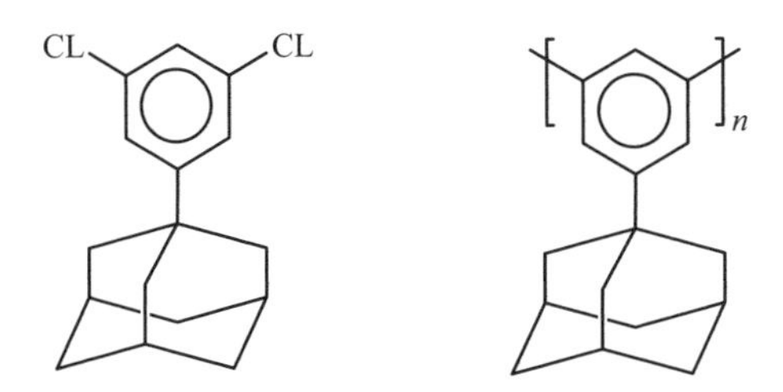

Figure 19. (*Left*) 1,3-Dichloro-5-(1-adamantyl)benzene monomer and (*Right*) adamantyl-substituted poly(*m*-phenylene), which is shown to have a high degree of polymerization and stability, decomposing at high temperatures of around 350 °C. Taken from Ref. [89] with permission.

4,9-diamantylene groups into the polyamides results in improvement in their thermal stability and satisfactory retention of their storage modulus (~ stress/strain) above 350 °C [90]. These characteristics, as well as improved viscosity and solubility in organic solvents, are specific to polyamides, which are derived from 4,9-bis(4-aminophenyl)diamantane and 4,9-bis[4-(4-aminophenoxy)phenyl]diamantane (Fig. 20).

Star polymers are a class of polymers with interesting rheological and physical properties. The tetra-functionalized adamantane cores (adamantyls) have been employed as initiators in the atom transfer radical polymerization (ATRP) method applied to styrene and various acrylate monomers (see Fig. 21).

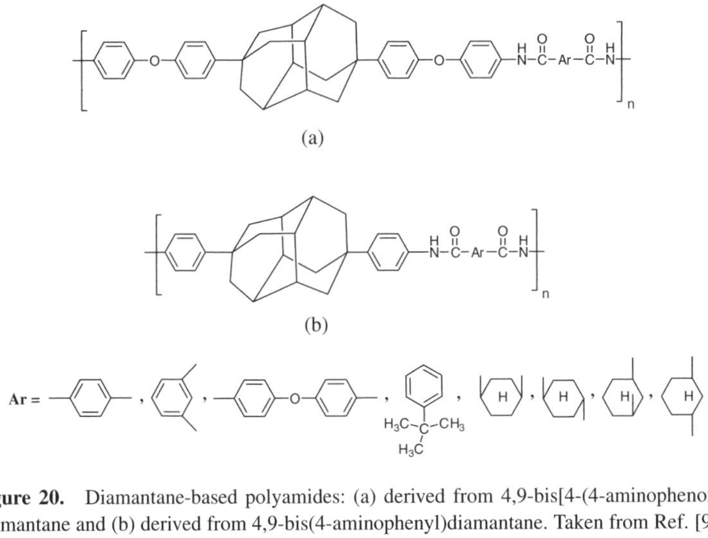

(a)

(b)

Figure 20. Diamantane-based polyamides: (a) derived from 4,9-bis[4-(4-aminophenoxy)phenyl]diamantane and (b) derived from 4,9-bis(4-aminophenyl)diamantane. Taken from Ref. [90] with permission.

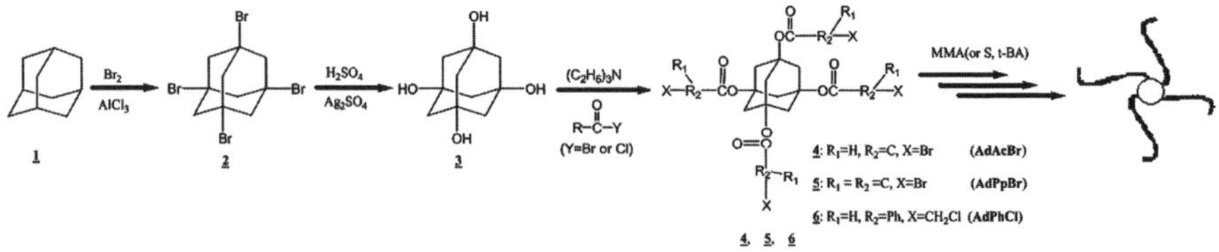

Figure 21. Atom transfer radical polymerization (ATRP) synthetic route to tetrafunctional initiators of a star polymer with adamantyl (adamantane core). Taken from Ref. [91] with permission.

As a result of this process, star-like polymers have been produced with a wide range of molecular weights [91].

In another study, the introduction of an adamantyl group to the poly(etherimide) structure caused polymer glass transition temperature, T_g, and solubility enhancements in some solvents like chloroform and other aprotic solvents [92].

The introduction of bulky side chains that contain adamantyl groups to poly(*p*-phenylenevinylene) (PPV), a semiconducting conjugated polymer, decreases the number of interchain interactions. This action will reduce the aggregation quenching and polymer photoluminescence properties would be improved [93].

Substitution of the bulky adamantyl group on the C(10) position of the biliverdin pigment structure leads to the distortion of helical conformation and hence the pigment color would shift from blue to red [94].

A three-dimensional (3D) fourfold interpenetrating coordination diamondoid polymer framework with adjustable porous structure was synthesized recently. It was demonstrated that this polymer framework is capable of trapping gaseous molecules and may be useful for gas storage [95].

B. In Nanotechnology

Nanotechnology is the branch of engineering that deals with the manipulation of individual atoms, molecules, and systems smaller than 100 nanometers. Two different methods are envisioned for nanotechnology to build nanostructured systems, components, and materials. One method is the "top–down" approach and the other method is called the "bottom–up" approach. In the top–down approach the idea is to miniaturize the macroscopic structures, components, and systems toward a nanoscale of the same. In the bottom–up approach the atoms and molecules constituting the building blocks are the starting point to build the desired nanostructure [96–98].

Various illustrations are available in the literature depicting the comparison of top–down and bottom–up approaches [96, 97]. In the top–down method a macrosized material is reduced in size to reach nanoscale dimensions. The photolithography used in the semiconductor industry is an example of the top–down approach. Bottom–up nanotechnology is the engineered manipulation of atoms and molecules in a user-defined and repeatable manner to build objects with certain desired properties. To achieve this goal, a number of molecules are identified as the molecular building blocks (MBBs) of nanotechnology, among which diamondoids are the most important ones owing to their unique properties [6, 99–104]. Diamondoids can be divided into two major clusters based on their size: lower diamondoids (1–2 nm in diameter) and higher diamondoids (>2 nm in diameter).

The building blocks of all materials in any phase are atoms and molecules. Their arrangements and how they interact with one another define many properties of the material. The nanotechnology MBBs, because of their sizes of a few nanometers, impart to the nanostructures created from them new and possibly preferred properties and characteristics heretofore unavailable in conventional materials and devices. These nanosize building blocks are intermediate in size, lying between atoms and microscopic and macroscopic systems. These building blocks contain a limited and countable number of atoms. They constitute the basis of our entry into new realms of bottom–up nanotechnology [97, 98].

The controlled and directed organization of MBBs and their subsequent assembly into nanostructures is one fundamental theme of bottom–up nanotechnology. Such an organization can be in the form of association, aggregation, arrangement, or synthesis of MBBs through noncovalent van der Waals forces, hydrogen bonding, attractive intermolecular polar interactions, or electrostatic interactions, or hydrophobic effects [97].

The ultimate goal of assemblies of nanoscale MBBs is to create nanostructures with improved properties and functionality heretofore unavailable to conventional materials and devices. As a result, one should be able to alter and engineer materials with desired properties. For example, ceramics and metals produced through controlled consolidation of their MBBs are shown to possess properties substantially improved and different from materials with coarse microstructures. Such different and improved properties include greater hardness and higher yield strength in the case of metals and better ductility in the case of ceramic materials [102].

Considering that nanoparticles have much higher specific surface areas, in their assembled forms there are large areas of interfaces. One needs to know in detail not only the structures of these interfaces, but also their local chemistries and the effects of segregation and interaction between MBBs and their surroundings. The knowledge of ways to control nanostructure sizes, size distributions, compositions, and assemblies are important aspects of bottom–up nanotechnology [97].

In general, nanotechnology MBBs are distinguished for their unique properties. They include, for example, graphite, fullerene molecules made of various numbers of carbon atoms (C60, C70, C76, C240, etc.), carbon nanotubes, nanowires, nanocrystals, amino acids, and diamondoids [97]. All these molecular building blocks are candidates for various applications in nanotechnology.

One of the properties used to distinguish MBBs from one another is the number of their available linking groups. MBBs with three linking groups, like graphite, could only produce planar or tubular structures. MBBs with four linking groups may form three-dimensional diamond lattices. MBBs with five linking groups may create three-dimensional solids and hexagonal planes. The ultimate possibility is presently MBBs with six or more linking groups. Adamantane with six linking groups (see Fig. 22) and higher diamondoids are of the latter category, which can be used to construct many complex three-dimensional structures [102]. Such MBBs can have numerous applications in nanotechnology and they are of major interest in designing shape-targeted nanostructures including synthesis of supramolecules with manipulated architectures [105–109].

In addition to possessing six or more linking groups, diamondoids have high strength, toughness, and stiffness when compared to other known MBBs. They are tetrahedrally symmetric stiff hydrocarbons. The strain-free structures of diamondoids give them high molecular rigidity, which is quite important for an MBB. High density, low surface energy, and oxidation stability are some other preferred properties of diamondoids as MBBs. Diamondoids have noticeable electronic properties [110]. In fact, they are H-terminated diamond and the only semiconductors that show a negative electron affinity [6].

Since diamondoids possess the capability for derivatization, they can be used to achieve suitable molecular geometries needed for MBBs of nanotechnology. Functionalization by different groups can produce appropriate reactants for desired reactions, microelectronics, and optics, by employing polymers, films, and crystal engineering.

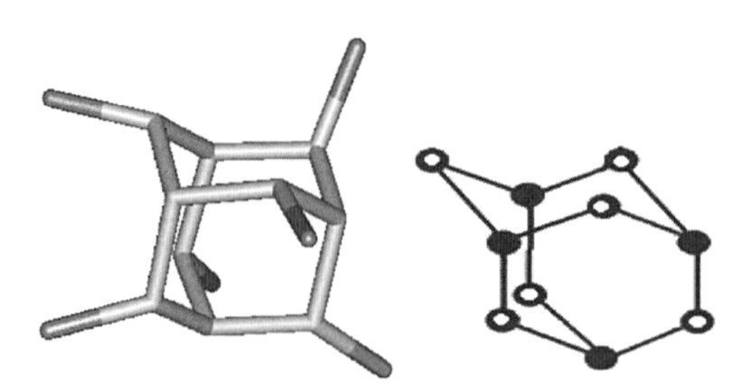

Figure 22. Demonstration of the six linking groups of adamantane.

Over 20,000 variants of diamondoids have been identified and synthesized and even more are possible [102], providing a rich and well-studied set of MBBs.

Adamantane can be used in molecular studies and preparation of fluorescent molecular probes [111]. Because of its incomparable geometric structure, the adamantane core (adamantyl) can impede interactions of fluorophore groups and self-quenching would diminish due to steric hindrance. Hence, mutual quenching would be diminished, allowing the introduction of several fluorescent groups to the same molecular probe in order to amplify their signals. Figure 23 shows the general scheme of an adamantane molecule with three fluorophore groups (F1) and a targeting group for attachment of biomolecules. Such a molecular probe can be very useful in DNA probing and especially in fluorescent in-situ hybridization (FISH) diagnostics [112].

Due to their demanding synthesis, diamondoids are helpful models to study structure–activity relationships in carbocations and radicals, to develop empirical computational methods for hydrocarbons, and to investigate orientational disorders in molecular crystals as well [5,32].

Atomic force microscope (AFM) is a powerful nanotechnology tool for molecular imaging and manipulations. One major factor limiting resolution in AFM to observe individual biomolecules such as DNA is the low sharpness of the AFM tip that scans the sample. Nanoscale 1,3,5,7-tetrasubstituted adamantane is found to serve as the molecular tip for AFM and may also find application in chemically well-defined objects for calibration of commercial AFM tips [113].

One of the branches of nanotechnology is called "crystal engineering." Crystal engineering is a new concept through which the power of noncovalent

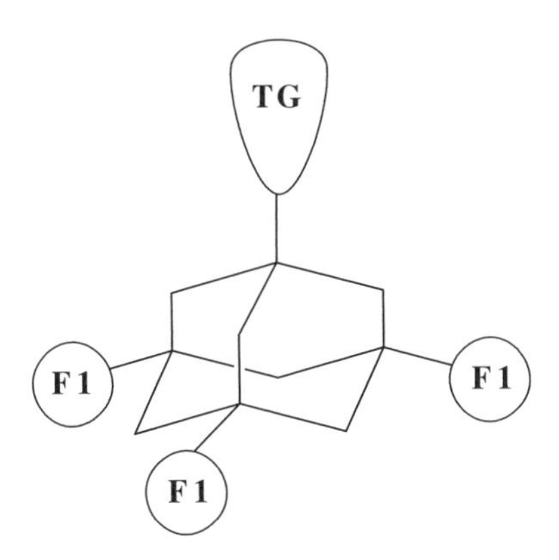

Figure 23. Schematic drawing that shows adamantane as a molecular probe with three fluorophore groups (F1) and a targeting part (TG) for specific molecular recognition. Taken from Ref. [112] with permission.

intermolecular forces is used in the solid state to design new nanomaterials with desired functions.

The approach in crystal engineering is to learn from known crystalline structures of, for example, minerals in order to design compounds with desired properties. Crystal engineering is considered to be a key new technology with applications in pharmaceuticals, catalysis, and materials science. The structures of adamantane and other diamondoids have received considerable attention in crystal engineering due to their molecular stiffness, derivatization capabilities, and their six or more linking groups [114–117].

A concept named "molecular manufacturing," which was originally proposed by K. Eric Drexler [99] in 1992, has attracted the attention of some investigators [100, 118–121]. Molecular manufacturing is defined as "the production of complex structures via non-biological mechanosynthesis (and subsequent assembly operations)" [99]. A chemical synthesis controlled by mechanical systems operating on the atomic scale and performing direct positional selection of reaction sites by atomic-precision manipulation systems is known as mechanosynthesis.

Due to their strong stiff structures containing dense, three-dimensional networks of covalent bonds, diamondoids are one of the favorite sets of molecules considered for molecular manufacturing as originally proposed, described, and analyzed in "nanosystems." Diamondoid materials, if they could be synthesized as proposed, could be quite strong but light. For example, diamondoids are being considered to build stronger, but lighter, rockets and other space components and a variety of other earth-bound articles for which the combination of weight and strength is a consideration [99, 100, 121–123].

Some of the applications of molecular manufacturing based on diamondoids are the design of an artificial red blood cell called respirocyte, nanomotors, nanogears, molecular machines, and nanorobots [102, 123–126]. The other potential application of molecular manufacturing of diamondoids is in the design of molecular capsules and cages for various applications including drug delivery.

For the concept of molecular manufacturing to become successful, a systematic study of the fundamental theory of the molecular processes involved and the possible technological and product capabilities are needed [127].

C. In Drug Delivery and Drug Targeting

The unique structure of adamantane is reflected in its highly unusual physical and chemical properties, which can have many applications including drug design and drug delivery. The carbon skeleton of adamantane comprises a cage structure, which may be used for the encapsulation of other compounds, like drugs. Although adamantane has been the subject of many research projects in the field of pharmacophore-based drug design, its application to drug delivery and drug targeting systems is a new matter of considerable importance [128].

Diamondoids could also be used as encapsulated, cage-shaped molecules in designing drug delivery systems. If the drug doesn't accumulate at its exact sight of action, it would not be able to produce the intended therapeutic effects even by using it at a high concentration. The nanometer size of diamondoid molecules allows them to enter living cells while carrying the drug into the cells. Commonly, particles with of size less than 100 nm can enter cells, whereas diamondoids are even smaller than 10 nm. Furthermore, due to their high stability and solubility in blood plasma, they are expected to play an important part in future drug delivery systems [128].

Furthermore, polymantanes have the potential to be utilized in the rational design of multifunctional drug systems and drug carriers. In host–guest chemistry, there is plenty of room for working with diamondoids.

In pharmacology, two adamantane derivatives, Amantadine (1-adamantaneamine hydrochloride) and Rimantadine (α-methyl-1-adamantane methylamine hydrochloride) (see Fig. 24), have been well known because of their antiviral activity [129]. The main application of these drugs is prophylaxis (treatment to prevent the onset of a particular disease) and treatment of influenza-A viral infections. They are also used in the treatment of parkinsonism and inhibition of hepatitis-C virus. Memantine (1-amino-3,5-dimethyladamantane) (see Fig. 24) has been reported effective in slowing the progression of Alzheimer's disease [130].

The site of action for Memantine is the central nervous system (CNS) and it has CNS affinity. Amantadine and Rimantadine can penetrate to the CNS and cause some adverse effects.

These observations lead one to conclude that the adamantane nucleus, which is present in all of the aforesaid drugs, might be responsible for their penetration to the BBB and their accumulation in the CNS. In other words, the adamantane nucleus is likely to possess a so-called intimate CNS tropism (because of its hydrophobicity) and cause these drugs to show such a CNS affinity [131].

Furthermore, because the half-life of Amantadine and Rimantadine in the bloodstream is long (12–18 hours for Amantadine and 24–36 hours for

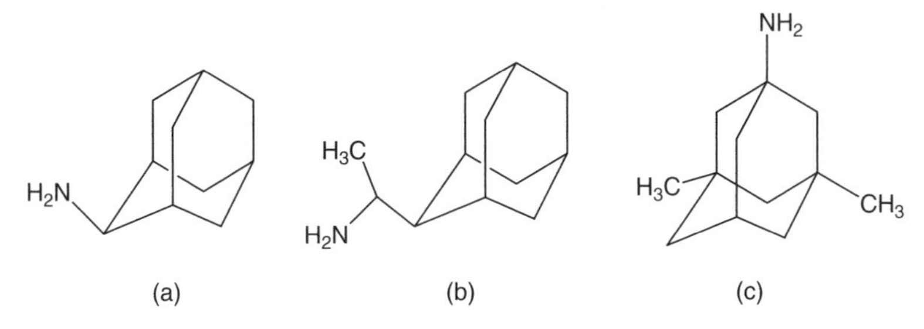

Figure 24. Chemical structures of (a) Amantadine, (b) Rimantadine, and (c) Memantine.

Rimantadine in young adults), utilization of adamantane derivative carriers can probably prolong drug presence time in blood circulation.

Extensive investigations have been performed related to the synthesis of new adamantane derivatives with better therapeutic actions and less adverse effects. For example, it has been proved that adamantylamino-pyrimidines and -pyridines are strong stimulants of tumor necrosis factor-α (TNF-α) [132]. TNF is a substance that can improve the body's natural response to cancer by killing cancer cells. Another example is 1,6-diaminodiamantane [87], which possesses an antitumor and antibacterial activity. Also, many derivatives of aminoadamantanes have antiviral activity like 3-(2-adamantyl) pyrolidines with two pharmacophoric amine groups, which have antiviral activity against influenza-A virus [133].

Some derivatives of adamantane with antagonist or agonist effects have also been synthesized. For instance, monocationic and dicationic adamantane derivatives block the α-amino-3-hydroxy-5-methylisoxazole-4-propionic acid (AMPA) receptors, *N*-methyl-D-aspartate (NMDA) receptors [134–136] and 5-hydroxytryptamine (5-HT3) receptors [137].

The monocationic and dicationic adamantane derivatives have been used to investigate the topography of the channel binding sites of AMPA and NMDA receptors [135].

A dicationic adamantane derivative has been exploited as a selective and specific marker of native AMPA receptor assembly to determine the distribution of AMPA receptor subtypes among populations of rat brain cells [134, 136]. Other examples include antagonism of 5-HT3 and agonism of 5-HT4 receptors by aza(nor)adamantanes [138], P2X7 receptors antagonism by adamantane amides [139], antagonism of voltage gated calcium channels and probably activation of γ-aminobutyric acid (GABA) receptors by an adamantane amine derivative that results in its anticonvulsive and antinociceptive actions [140], and inhibition of glucosylceramidase enzyme and glycolipid biosynthesis by a deoxynojirimycin-bearing adamantane derivative leading to strong anti-inflammatory and immunosuppressive activities [141].

Attaching some short peptidic sequences to adamantane makes it possible to design novel antagonists. The bradykinin antagonist, which is used as an anticancer agent, is an example. The adamantane-based peptidic bradykinin analog was utilized in structure–activity relationship (SAR) studies on the bradykinin receptors and showed a potent activity in inhibition of bradykinin-induced cytokine release and stimulation of histamine release [142].

In an attempt to design the β-turn-peptide-mimics, aspartic acid (an amino acid also known as aspartate) and lysine (an amino acid especially found in gelatin and casein) were attached to each amine group of 1,3-diaminoadamantane in the form of amide bonds. The term β–turn refers to a peptide chain that forms a tight loop such that the carbonyl oxygen of one residue is hydrogen bonded with the amide nitrogen of a residue located three positions down the chain. The β-turn-peptide-mimics display some degree of fibrinogen-GPIIB/IIIA antagonism [143].

Adamantane derivatives can be employed as carriers for drug delivery and targeting systems. Due to their high lipophilicity, attachment of such groups to drugs with low hydrophobicity could lead to a substantial increase of drug solubility in lipidic membranes and thus increases of its uptake.

Furthermore, the large number of linking group possibilities of adamantane and other diamondoids (six or more) makes it possible to introduce several functional groups (consisting of drug, targeting part, linkers, etc.) to them without undesirable interactions. In fact, adamantane derivatives can act as a central core for such drug systems.

As an example, short peptidic sequences can be bound to adamantane and provide a binding site for connection of macromolecular drugs (like proteins, nucleic acids, lipids, and polysaccharides) as well as small molecules. Hence, short amino acid sequences can have linker roles, which are capable of drug release at the target site. There are some successful examples of adamantyl (adamantane core) application for delivery of drugs to the brain [131]. For this purpose, 1-adamantyl was attached to several AZT (azidothymidine) drugs via an ester spacer and these prodrugs could pass the blood–brain barrier (BBB) easily to reach the brain. The drug's concentration after using such lipophilized prodrugs was measured in the brain tissue and showed an increase of 7–18 fold in comparison with AZT drugs without adamantane vector. The ester spacer link is resistant to the plasma esterases, but it is cleaved after passing through the BBB by brain tissue esterases. Overall, it is important to note that adamantane is now considered a successful brain-directing drug carrier.

Another example of adamantane utilization for poorly absorbed-drug delivery to the brain is the conjugation of [D-Ala2]Leu-enkephalin derivatives with 1-adamantyl [144]. The antinociceptive effect of Leu-enkephalin disappears when it is administered peripherally since proteolytic enzymes would decompose it. As a result, it cannot penetrate into the CNS (central nervous system). It is feasible to conjugate the [D-Ala2]Leu-enkephalin with a 1-adamantane vector via an ester, amide, or a carbamate linkage in order to enhance the drug lipophilicity and thus facilitate its delivery across the blood–brain barrier (BBB) to the brain [144]. The adamantane-conjugated [D-Ala2]Leu-enkephalin prodrugs (Fig. 25) are highly lipophilic and show a significant antinociceptive effect because of their ability to cross the BBB [132]. These results suggest that adamantyl is a promising brain-directing drug vector providing a high lipophilicity, low toxicity, and high BBB permeability for sensitive and poorly absorbed drugs [144, 145].

Adamantane has also been used for lipidic nucleic acid synthesis as a hydrophobic group. Two major problems in gene delivery are the low uptake of

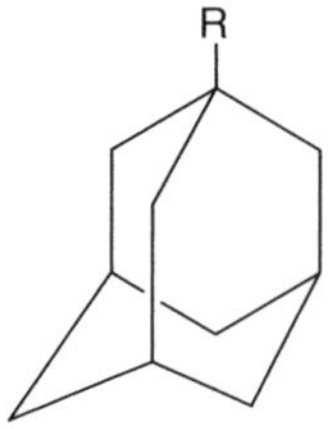

R

(I) R = NH_2-Tyr-(D-Ala)-Gly-Phe-Leu-CO-O-
(II) R= NH_2-Tyr-(D-Ala)-Gly-Phe-Leu-CO-NH-

Figure 25. The adamantane-conjugated [D-Ala2]Leu-enkephalin prodrugs. Taken from Ref. [144] with permission.

nucleic acids by cells and their instability in blood medium. An increase in lipophilicity using hydrophobic groups would probably lead to improvement of uptake and an increase in intracellular concentration of nucleic acids. In this case, an amide linker is used to attach the adamantane derivatives to a nucleic acid sequence. Such a nucleic acid derivatization has no significant effect on hybridization with the target RNA. Lipidic nucleic acids possessing adamantane derivative groups can also be exploited for gene delivery [146].

Recently, synthesis of a polyamine adamantane derivative has been reported, which has a special affinity for binding to the major grooves in double-stranded DNA [147]. It should be pointed out that most of the polyamines have affinity for binding to double-stranded RNA, thus making RNA stabilized. DNA selectivity is one of the outstanding features of the said ligand. This positive nitrogen-bearing ligand has a tendency to establish hydrophobic interactions with deeper DNA grooves due to its size and steric properties. Such an exclusive behavior occurs because the ligand fits better in the DNA major grooves. This bulky ligand size is the same as zinc-finger protein, which also binds to DNA major grooves.

Higher affinity of adamantane-bearing ligand to DNA, instead of RNA, probably arises from the presence of adamantane and leads to DNA stabilization. This fact may be exploited for using such ligands as stabilizing carriers in gene delivery. Adamantane causes lipophilicity to increase as well as DNA stabilization. Furthermore, ligand/groove size-based targeting might also be possible with less specificity by changing the bulk and conformation of ligand.

Polymers conjugated with 1-adamantyl moieties as lipophilic pendent groups can be utilized to design nanoparticulate drug delivery systems. Polymer (1) in Fig. 26, which is synthesized by homopolymerization of ethyladamantyl malolactonate, can be employed as highly hydrophobic blocks to construct

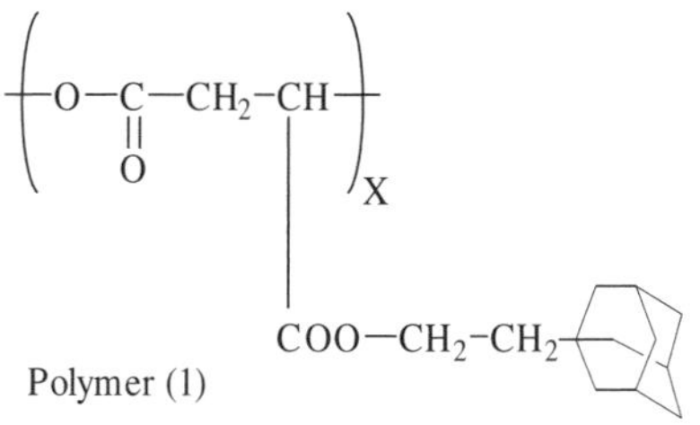

$$\left[O-\underset{\underset{O}{\|}}{C}-CH_2-\underset{|}{CH} \right]_X$$
$COO-CH_2-CH_2-$(adamantyl)

Polymer (1)

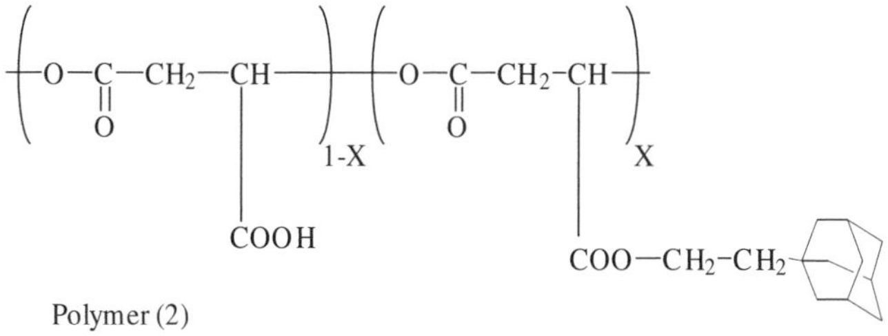

$$\left[O-\underset{\underset{O}{\|}}{C}-CH_2-\underset{|}{CH} \right]_{1-X} \left[O-\underset{\underset{O}{\|}}{C}-CH_2-\underset{|}{CH} \right]_X$$
$COOH$; $COO-CH_2-CH_2-$(adamantyl)

Polymer (2)

Figure 26 Polymer (1) {poly(ethyladamantyl β-malate)} is hydrophobic and polymer (2) {poly(β-malic acid-*co*-ethyladamantyl β-malate} is hydrophilic. Both of these polymers are used as carriers for different drugs [148].

polymeric drug carries. In contrast, polymer (2) (Fig. 26), which is synthesized by copolymerization of polymer (1) with benzyl malolactonate, is water soluble and its lateral carboxylic acid functions can be used to bind biologically active molecules in order to achieve targeting as well. These represent examples of the possibilities of producing adamantane-based pH-dependent hydrogels and intelligent polymeric systems [148].

D. In DNA-Directed Assembly and DNA–Protein Nanostructures

Due to the ability of adamantane and its derivatives to attach to DNA, it is possible to construct well-defined nanostructures consisting of DNA fragments as linkers between adamantane cores. This could be a powerful tool to design DNA-directed nanostructured self-assemblies [149].

A unique feature of such DNA-directed self-assemblies is their site-selective immobilization, which makes it possible to construct well-defined nanostructures. On the other hand, the possibility of the introduction of a vast number of substitutes (like peptidic sequences, nucleoproteins, of hydrophobic hydrocarbon chains) to an adamantane core (adamantyl) makes such a process capable of designing steric colloidal and supramolecular conformations by setting hydrophobic/hydrophilic and other interactions. In addition, the rigidity of the adamantane structure can provide strength and rigidity to such self-assemblies [150].

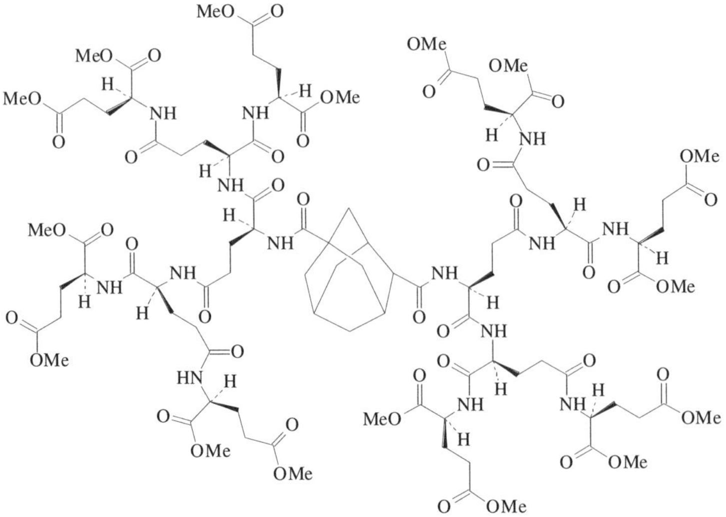

Figure 27 Adamantane nucleus with amino acid substituents creates a peptidic matrix [151]. The represented structure is Glu4-Glu2-Glu-[ADM]-Glu-Glu2-Glu4.

Bifunctional adamantyl, as a hydrophobic central core, can be used to construct peptidic scaffolding [151], as shown in Fig. 27. This is the reason why adamantane is considered one of the best MBBs. This may be considered an effective and practical strategy to substitute different amino acids or DNA segments on the adamantane core (Fig. 28). In other words, one may exploit nucleic acid (DNA or RNA) sequences as linkers and DNA hybridization (DNA probe) to attach to these modules with an adamantane core. Thus a DNA–adamantane–amino acid nanostructure may be produced.

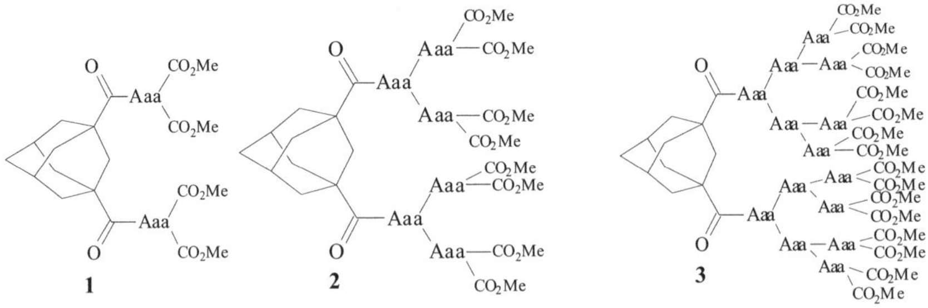

a :Aaa = Asp ; b : Aaa = Glu

Figure 28 A dendrimer-based approach for the design of globular protein mimic using glutamic (Glu) and aspartic (Asp) acids as the building blocks and adamantyl as the core [151].

Knowledge about protein folding and conformation in biological systems can be used to mimic the design of a desired nanostructure conformation from a particular MBB and to predict the ultimate conformation of the nanostructure [152]. Such biomimetic nano-assembly is generally performed step by step. This will allow observation of the effect of each new MBB on the nanostructure. As a result, it is possible to control accurate formation of the desired nanostructure. Biomimetic controlled and directed assembly can be utilized to investigate molecular interactions, molecular modeling, and study of relationships between the composition of MBBs and the final conformation of the nanostructures. Immobilization of molecules on a surface could facilitate such studies [153].

Nucleic acid attachments to an adamantane core (adamantyl) can be achieved in several ways. At least two nucleic acid sequences, as linkage groups, are necessary for each adamantyl to form a nanostructured self-assembly. Various geometrical structures may be formed by changing the position of the two nucleic acid sequences with respect to each other on two of the six adamantyl linking groups or by the addition of more nucleic acid sequences on the other linking groups. A number of alterations can be imposed on the nucleic acid sequences utilizing the new techniques developed in solid-phase genetic engineering for immobilized DNA alteration [154]. For instance, ligated DNA (two linear DNA fragments enzymatically joined through covalent bonds) may be employed to join the adamantyl nanomodules similar to what has been done on immobilized DNA in the case of gene assembly [154], provided some essential requirements could be met for the retention of enzyme activities. Instead of ligated DNA, one may use hybridized DNA as well. It may be possible to modify the amino acid parts, adamantane cores, and DNA sequences of the resulting nanostructure. For example, by using some unnatural (synthetic) amino acids [155, 156] with appropriate folding characteristics, the ability of conformation fine-tuning could be improved. Hence, the assembling and composing of adamantyls as central cores, DNA sequences as linkers, and amino acid substituents (on the adamantane) as conformation controllers may lead to the design of DNA–adamantane–amino acid nanostructures with desired and predictable properties.

By all accounts, the hypothesis of formation of DNA+adamantane+amino acid nanoarchitectures is currently immature and amenable to many technical modifications. Advancement in this subject requires a challenging combination of state-of-the-art approaches of organic chemistry, biochemistry, proteomics, and surface science.

A dendrimer-based approach for the design of globular protein mimics using glutamic (Glu) and aspartic (Asp) acids as building blocks has been developed [151]. The preassembled Glu/Asp dendrones were attached to a 1,3-bifunctional adamantyl based on a convergent dendrimer synthesis strategy (see Fig. 28).

Three successive generations of dendrimers composed of an adamantane central core and two, six, and fourteen chiral centers (all L-type amino acids) and thus four, eight, and sixteen peripheral carbomethoxy groups, respectively, were synthesized. The adamantane core was selected to render the dendrimer structures spherical with the capability of different ligands being incorporated into their peripheral reactive arms. The Glu dendritic scaffoldings 2b and 3b (Fig. 28) showed a noticeable feature from the solubility standpoint. The resulting dendrimers dissolved slowly in warm water to form a clear solution in spite of the fact that Glu dendrones, on their own, are quite insoluble in water. This solubility enhancement probably results from the double layer structures of dendritic scaffoldings 2b and 3b in which a hydrophilic outer shell encircles the hydrophobic adamantane core. In other words, this property can be attributed to the role of the adamantane core due to its ability to bring about a wide range of changes in physicochemical properties and turn a completely hydrophobic molecule into a hydrophilic one.

E. In Host–Guest Chemistry

The main aim in host–guest chemistry is to construct molecular receptors by a self-assembly process so that such receptors could, to some extent, gain molecular recognition capability. The goal of such molecular recognition capability is to either mimic or block a biological effect caused by molecular interactions [157].

Calixarenes, which are macrocyclic compounds, are one of the best building blocks to design molecular hosts in supramolecular chemistry [158]. Synthesis of calix[4]arenes, which have been adamantylated, has been reported [105, 109]. In calix[4]arenes, adamantane or its ester/carboxylic acid derivatives were introduced as substituents (Fig. 29). The purpose of this synthesis was to learn how to employ the flexible chemistry of adamantane in order to construct different kinds of molecular hosts. The X-ray structure analysis of *p*-(1-adamantyl)thiacalix[4]arene [109] demonstrated that it contained four $CHCl_3$ molecules, one of which was located inside the host molecule cavity, and the host molecule assumed the cone-like conformational shape (Fig. 30).

Some other types of macrocycle compounds have been synthesized using adamantane and its derivatives. Recently, a new class of cyclobisamides has been synthesized using adamantane derivatives, which shows the general profiles of amino acid (serine or cystine)–ether composites. They were shown to be efficient ion transporters (especially for Na^+ ions) in the model membranes [159]. Another interesting family of compounds to which adamantane derivatives have been introduced in order to obtain cyclic frameworks is "crown ethers" [160]. The outstanding feature of these adamantane-bearing crown ethers (which are also called "diamond crowns") is that α-amino acids can be incorporated into the adamantano-crown backbone [160]. This family of

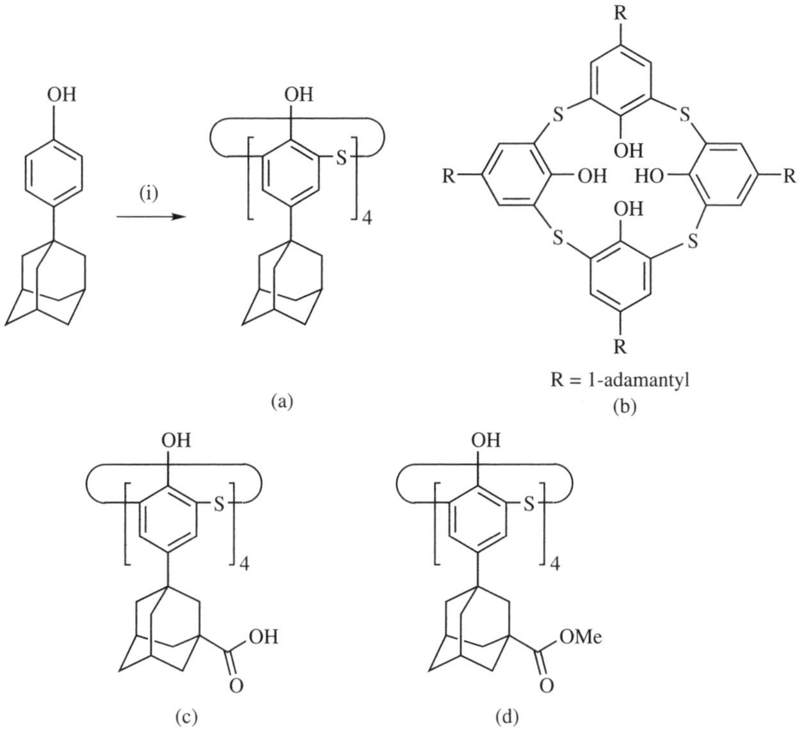

Figure 29 (a) Synthesis route of the molecule (b): (i) S8, NaOH, tetraethyleneglycol dimethyl ether, heat, (28%). (b) Adamantane upper rim derivative based on the thiacalix[4]arene platform. (c,d) The carboxylic acid and ester derivative of adamantane can also be used as substituents. Taken from Ref. [109] with permission.

compounds provides valuable models for studying selective host–guest chemistry, ion transport, and ion complexation [160].

Adamantane has also been used as a cage-like alicyclic (both aliphatic and cyclic) bridge to construct a new class of tyrosine-based cyclodepsipeptides (tyrosinophanes) [161]. Macrocyclic peptides composed of an even number of D and L amino acids can self-assemble to form a tube through which ions and molecules can be transported across the lipid bilayers. Although they rarely exist in nature, they are synthesized to be employed in the host–guest studies (Fig. 31a) and to act as ion transporters in the model membranes (Fig. 31b,c) [161]. The adamantane-bridged, leucine-containing macrocycle in Fig. 31b shows a modest ability to transport Na^+/K^+ ions across the model membranes [161]. The adamantane-constrained macrocycle in Fig. 31c is also suitable for attachment of different functional groups to design artificial proteins [161]. The

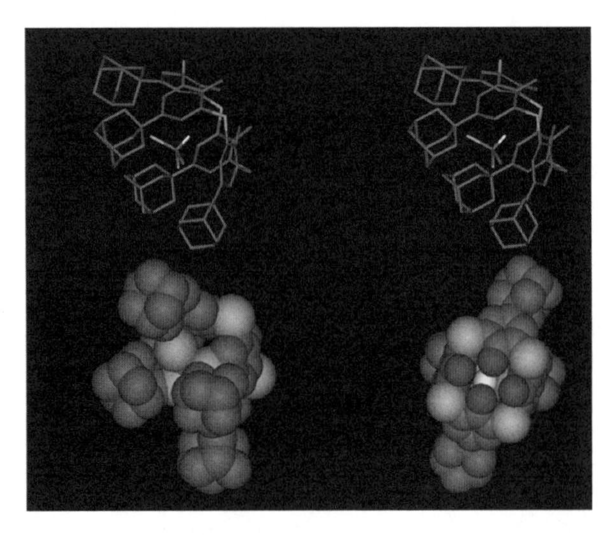

Figure 30 Lateral stereo views of adamantane derivative thiacalix[4]arene (*top*) presented in Fig. 29. A $CHCl_3$ molecule has been entrapped inside the inclusion compound. The bottom view (*left bottom*) and top view (*right bottom*) are also shown. H atoms have been removed from the inclusion compound for more clarity [128].

adamantane-containing cyclic peptides are efficient metal ion transporters and utilization of adamantane in such compounds improves their lipophilicity and thus membrane permeability [162]. A new class of norbornene-constrained cyclic peptides has been synthesized using adamantane as a second bridging ligand (Fig. 32). The macrocycle in Fig. 32a is a specific ion transporter for monovalent cations, while the cyclic peptide in Fig. 32b is able to transport both mono- and divalent cations across the model membranes [162].

Peptidic macrocycles are especially useful models for discovering protein folding mechanisms and designing novel peptide-made nanotubes as well as other biologically important molecules. These large cyclic peptides tend to fold in such a way that they can adopt a secondary structure like β-turns, β-sheets, and helical motifs. A new series of double-helical cyclic peptides have been synthesized, among them the adamantane-constrained cystine-based cyclic trimers {cyclo(Adm-Cyst)3}. They have attracted a great deal of attention due to their figure eight-like helical topologies and special way of hydrogen binding and symmetries [163, 164] (Fig. 33). The cyclo(Adm-Cyst)3 molecule was able to transport K^+ ions through the model membranes and it was a valuable model to study the mechanism of secondary structure formation in proteins [165].

Cyclodextrins (CDs) are inclusion compounds formed by enzymatic decomposition of starch to the cyclic oligosaccharides containing six to eight

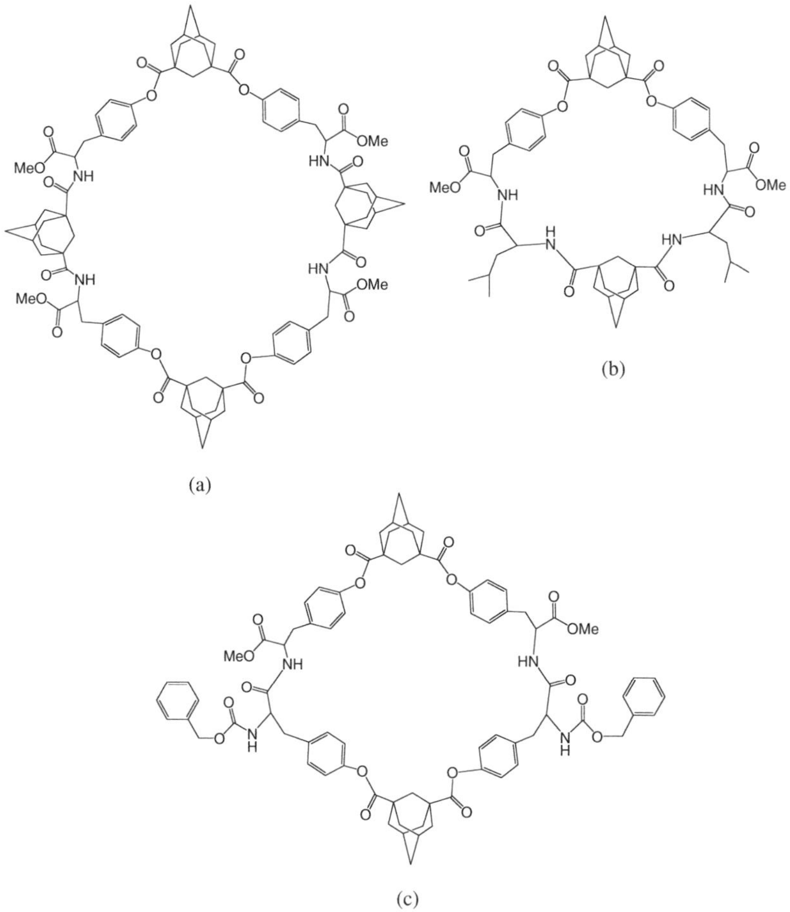

(a)

(b)

(c)

Figure 31 Adamantane-bridged tyrosine-based cyclodepsipeptides are suitable models for host–guest studies and they are also able to act as ion transporters. Taken from Ref. [161] with permission.

glucose units. Depending on the number of glucose units, there are three types of natural CDs, namely, α, β, and γ consisting of six, seven, and eight glucoses, respectively. The interior lining of the parent CDs' cavities is somehow hydrophobic. Adamantane is one of the best guests entrapped within the CDs' cavities [166–168]. Its noticeable association constant with CDs ($\sim 10^4$ to 10^5 M^{-1}) denotes a high affinity to interact with a hydrophobic pocket of CDs, which is a valuable linking system to join different molecules together (see Fig. 34). Interestingly, this system adsorbs and immobilizes

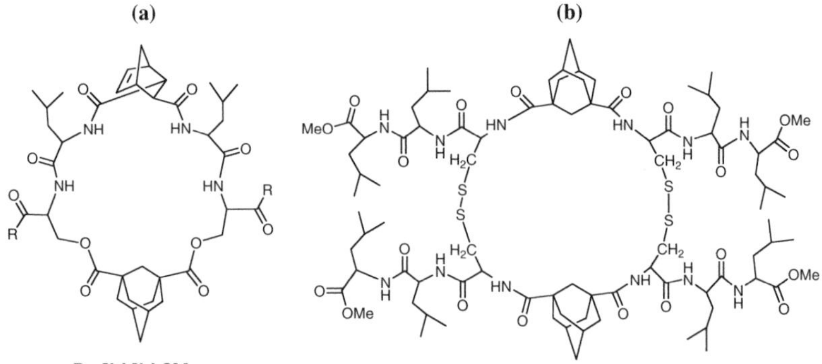

(a) **(b)**

R = Val-Val-OMe

Figure 32 Adamantane-containing norbornene ( )-constrained cyclic peptides possess the ability to transport ions across the model membranes in both specific and nonspecific ways [162].

molecules on a solid support and has been exploited to immobilize an adamantane-bearing polymer onto the surface of a β-CD-incorporated silica support [169]. In this case, adamantane acts as a linker to attach a dextran–adamantane–COOH polymer to a solid support through a physical entrapment mechanism and thus contributes to formation of a stationary phase for

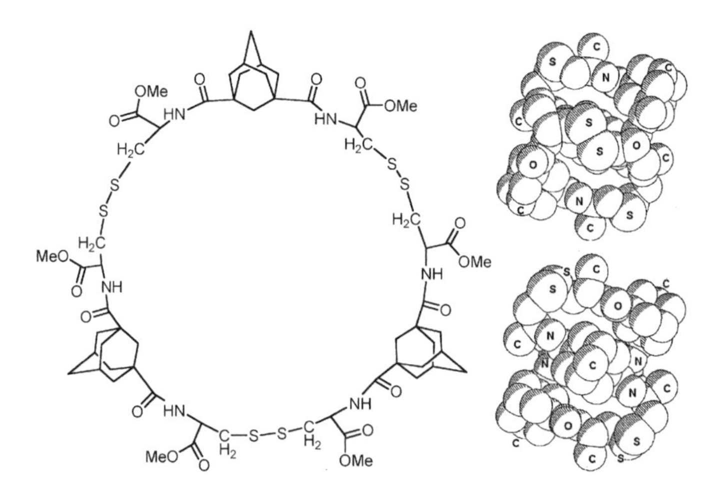

Figure 33 The cyclo(Adm-Cyst)3 adopts a figure-eight-like helical structure. The chiral amino acid, cystine, configuration determines the helix disposition (right-handed or left-handed helix). Adamantane plays an important role as a ring size controlling agent. Taken from Ref. [163] with permission.

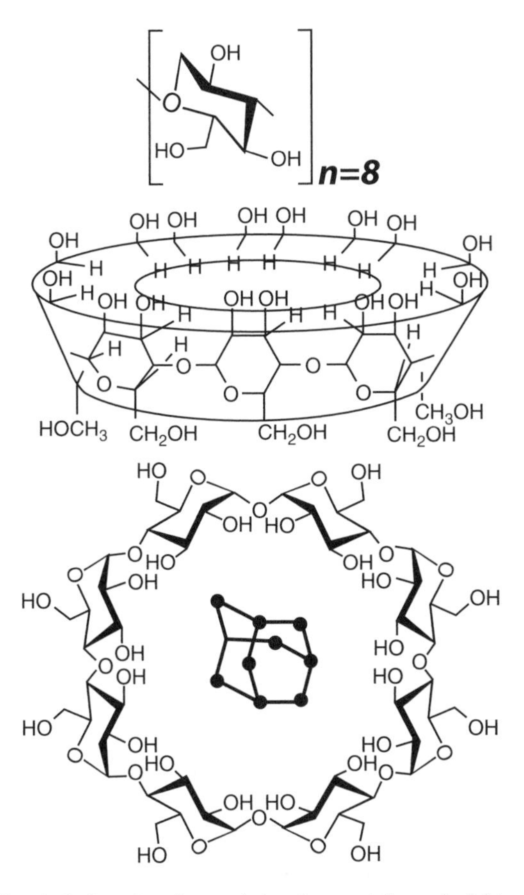

Figure 34 Chemical formula of γ-cyclodextrin consisting of eight glucose molecules with adamantane as the guest entrapped within its hydrophobic cavity. Structures of α- and β-cyclodextrins will be similar but made up of six and seven ($n = 6$, 7) glucoses, respectively.

chromatographic purposes [169]. The aforementioned stationary phase could readily be prepared under mild conditions and is stable in aqueous media. It revealed some cation-exchange properties, suggesting its application to the chromatography of proteins [169].

Covalent attachment of adamantane molecules is a key strategy to string them together and construct molecular rods. The McMurry coupling reaction was employed to obtain polyadamantane molecular rods (see Fig. 35) [170]. As another example, synthesis of tetrameric 1,3-adamantane and its butyl derivative has been reported [171] (see Fig. 36).

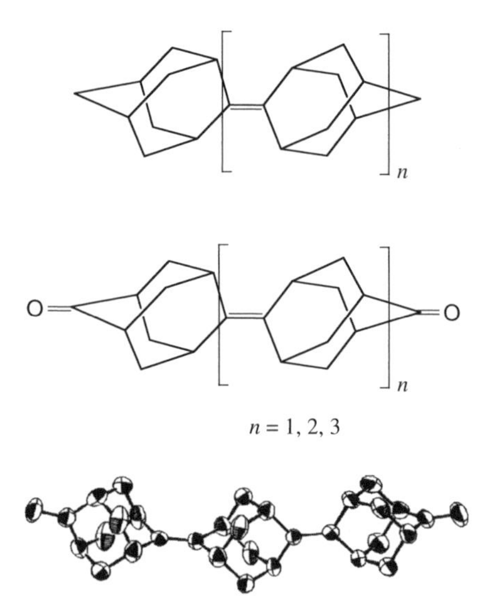

Figure 35 Polyadamantane molecular rods. Taken from Ref. [170] with permission.

VI. DISCUSSION

Diamondoids are organic compounds (hydrocarbons) with unique structures and properties. This family of compounds (with over 20,000 variants) is one of the best candidates for templates and molecular building blocks for synthesis of high-temperature polymers and in nanotechnology, drug delivery, drug targeting, DNA-directed assembly, DNA–protein nanostructure formation, and host–guest chemistry.

Some of their derivatives have been used as antiviral drugs. Due to their flexible chemistry, they can be exploited to design drug delivery systems and in molecular nanotechnology. In such systems, they can act as a central lipophilic core and different parts like targeting segments, linkers, spacers, or therapeutic agents can be attached to the said central nucleus. Their central core can be functionalized by peptidic and nucleic acid sequences and also by numerous important biomolecules.

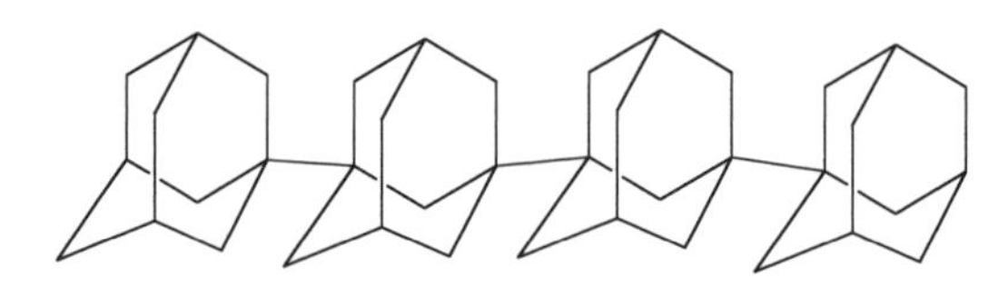

Figure 36 Synthetic design of a molecular rod made of adamantanes: the tetrameric 1,3-adamantane. Taken from Ref. [171] with permission.

Furthermore, some adamantane derivatives possess special affinity to bind to DNA, thereby stabilizing it. This is an essential feature for a gene vector. Some polymers have been synthesized using adamantane derivatives, the application of which is under investigation for drug delivery.

Adamantane can be used to construct peptidic scaffolding and synthesis of artificial proteins. It has been introduced into different types of synthetic peptidic macrocycles, which are useful tools in peptide chemistry and stereochemistry studies and have many other applications as well. Introduction of amino acid-functionalized adamantane to the DNA nanostructures might lead to construction of DNA–adamantane–amino acid nanostructures with desirable stiffness and integrity. Diamondoids can be employed to construct molecular rods, cages, and containers and also for utilization in different methods of self-assembly. In fact, through the development of self-assembly approaches and utilization of diamondoids in these processes, it would be possible to design and construct novel nanostructures for effective and specific carriers for each drug.

The phase transition boundaries (phase envelope) of adamantane need to be investigated and constructed. Predictable and diverse geometries are important features for molecular self-assembly and pharmacophore-based drug design. Incorporation of higher diamondoids in solid-state systems and polymers should provide high-temperature stability, a property already found in polymers synthesized from lower diamondoids.

Diamondoids offer the possibility of producing a variety of nanostructural shapes including molecular-scale components of machinery such as rotors, propellers, ratches, gears, and toothed cogs. We expect them to have the potential for even more possibilities for applications in molding and cavity formation characteristics due to their organic nature and their sublimation properties. The diverse geometries and possibility of six or more attachment sites (linking groups) in diamondoids provide an extraordinary potential for the production of shape derivatives.

VII. GLOSSARY

5-HT3 receptor. 5-Hydroxytryptamine receptor; a receptor for serotonin (a neurotransmitter), which activates a variety of second messenger signaling systems and through them indirectly regulates the function of ion channels.

Adamantologs. Diamondoids.

Adamantyl. The tetra-functionalized adamantane core.

AMPAR. (α-Amino-3-hydroxy-5-methyl-4-isoxazole propionic acid) receptor; a non-NMDA-type ionotropic transmembrane receptor for glutamate that mediates fast synaptic transmission in the central nervous system.

Antagonist. A compound that is able to bind to a drug or endogenous chemical's receptors and block them. Hence, it competes with the drug/

endogenous chemical to occupy the receptors and inhibits the pharmacological effects of a drug/endogenous chemical to appear.

Antinociceptive. Analgesic; painlessness.

Aprotic solvent. A solvent that has no OH groups and therefore cannot donate a hydrogen bond.

Association constant. In host–guest chemistry, between host (H) and guest (G) molecules it is defined as $K = [\text{H-G}] \cdot [\text{H}]^{-1} \cdot [\text{G}]^{-1}$.

BBB. Blood–brain barrier.

Carbocation. Carbonium ion; an ion with a positively charged carbon atom.

CNS. Central nervous system.

Crystal engineering. Utilization of noncovalent intermolecular forces in the solid state to design new nanomaterials with desired functions.

Cyclodextrin (CD). An inclusion cyclic oligosaccharide compound containing six to eight glucose units.

Cytokines. Nonspecific water-soluble glycoproteins with a short half-life produced and secreted abruptly by white blood cells in response to an external stimulus, and which act as chemical messengers between cells.

DFT. Density-functional theory.

Enantiomeric. Optical antipode; enantiomers are a pair of optical isomers.

Esterases. Enzymes that catalyze the hydrolysis of an ester bond.

Facies. A rock or stratified body distinguished from others by its appearance or composition.

FCC. Face–centered cubic; the FCC structure is a close-packed structure.

Fibrinogen-GPIIB/IIIA. A complex responsible for aggregation of platelets in the bloodstream.

FISH. Fluorescent in-situ hybridization; a method utilizing fluorescently labeled DNA probes to detect or confirm gene or chromosome abnormalities that are generally beyond the resolution of routine cytogenetics.

GABA receptors. Receptors for γ-aminobutyric acid. GABA is an amino acid that acts as an inhibitory neurotransmitter in the CNS.

Histamine. A chemical present in cells throughout the body and released during an allergic reaction.

LPG. Gas condensate; liquid petroleum gas.

MBB. Molecular building block.

McMurry coupling reaction. The mechanism of the McMurry coupling reaction consists of two defined steps: the reductive dimerization of the aldehyde or ketone, and then subsequent deoxygenation of the 1,2-diolate intermediate, yielding the alkene.

Mechanosynthesis. Chemical synthesis controlled by mechanical systems operating on the atomic scale and performing direct positional selection of reaction sites by atomic-precision manipulation systems.

NMDAR. An ionotropic receptor for glutamate. It plays a critical role in synaptic plasticity mechanisms and thus is necessary for several types of learning and memory.

NMDAR. *N*-methyl-D-aspartate receptor—an ionotropic receptor for glutamate. It plays a critical role in synaptic plasticity mechanisms and thus is necessary for several types of learning and memory.

NMR. Nuclear magnetic resonance.

Norbornene. A bicyclic olefin.

Pharmacophore. A part within a drug molecule which is believed to play a major role in interaction of that drug with its target. There may be more than one pharmacophoric site within the chemical structure of a drug molecule.

Polymantanes. Diamondoids.

Prodrug. Inactive precursor of a drug which is converted into its active form in the body by normal metabolic processes.

Proteolytic enzymes. An enzyme that catalyzes the breakdown of proteins into their building blocks, the amino acids.

RNA. Ribonucleic acid—a molecule present in the cell of all living beings and essential for the synthesis of proteins.

TNF. Tumor necrosis factor. TNFs are among the important cytokines playing a key role in activation and induction of some immune system cells and cellular immunity processes responsible for proinflammatory and inflammatory response reactions as well.

Synaptic transmission. Transmission through the junction across which a nerve impulse passes from an axon terminal to a neuron, muscle cell, or gland cell.

Tropism. Response/reaction of an organism to an external stimulus shown as movement.

Zinc-finger protein. A DNA-binding protein that contains a zinc atom.

Acknowledgements

This work was supported by the U.S. Army Research Office. The content of the information does not necessarily reflect the position or the policy of the Federal Government, and no official endorsement should be inferred. The author would like to thank L. Assoufid, A. Fazeli, M. Genio Goluch, T. F. George, N. Gruszauskas, M. A. Moradi, H. Ramezani, G. R. Vakili-Nezhaad and G. Zhang for their collaborations, comments, and reading of the manuscript for this chapter.

References

1. R. C. Fort, *Adamantane, The Chemistry of Diamond Molecules* (Studies in Organic Chemistry; Vol. 5), Marcel Dekker, New York, 1976.
2. G. A. Olah (ed.), *Cage Hydrocarbons*, John Wiley & Sons, Hoboken, NJ, 1990.
3. A. P. Marchand, *Science* **299** (3 Jan), 52–53 (2003).
4. W. H. Bragg and W. L. Bragg, *Nature* **91**, 554–556 (1913).
5. G. J. Kabo, A. V. Blokhin, M. B. Charapennikau, A. G. Kabo, and V. M. Sevruk, *Thermochimica Acta* **345**, 125–133 (2000).
6. J. E. Dahl, S. G. Liu, and R. M. Carlson, *Science* **299**, 96–99 (2003).
7. G. A. Mansoori, L. Assoufid, T. F. George, and G. Zhang, Measurement, simulation and prediction of intermolecular interactions and structural characterization of organic nanostructures; in *Proceedings of the Conference on Nanodevices & Systems, Nanotech 2003*, San Francisco, CA, 2003.
8. G. R. Desiraju, *J. Mol. Structure* **374**, 191–198 (1996).
9. R. C. Fort and P. v. R. Schleyer, *Chem. Rev.* **64**, 277–300 (1964).
10. T. Clark, T. Mc O. Knox, M. A. McKervey, H. Mackle, and J. J. Rooney, *J. Am. Chem. Soc.* **101**(9), 2404–2420 (1979).
11. W. S. Wingert, *Fuel.* **71** (Jan.), 37–43 (1992).
12. *Aldrich—Catalog Handbook of Fine Chemicals*, Milwaukee, Aldrich Chemical Co., Inc., **1992**, p. 30.
13. J. Reiser, E. McGregor, J. Jones, R. Enick, and G. Holder, *Fluid Phase Equil.* **117**, 160–167 (1996).
14. K. W. Hedberg, *Part II. Determination of Some Molecular Structures by the Method of Electron Diffraction. A. Adamantane*, PhD Dissertation, Chemistry, Cal Tech, 1948.
15. R. C. Fort, *Adamantane, The Chemistry of Diamond Molecules*, Marcel Dekker, New York, **1976**.
16. R. H. Boyd, S. N. Sanwal, S. Shary-Tehrany, and D. McNally, *J. Phys. Chem.* **75**, 1264–1271 (1971).
17. A. S. Cullick, J. L. Magouirk, and H. J. Ng, Paper #P1994.01, in *Proceedings of the 73rd Annual GPA Convention*, Gas Processors Association, Tulsa, OK, 1994.
18. A. van Miltenburg, W. Poot, and T. W. de Loos, *J. Chem. Eng. Data* **45**(5), 977–979 (2000); W. Poot, K. M. Kruger, and T. W. de Loos, *J. Chem. Thermod.* **35**(4), 591–604 (2003); W. Poot, and T. W. de Loos, *Fluid Phase Equil.* **221**(1-2), 165–174 (2004).
19. R. H. Boyd, S. N. Sanwal, S. Shary-Tehrany, and D. McNally, *J. Phys. Chem.* **75**, 1264–1271 (1971).
20. R. S. Buttler, A. S. Carson, P. G. Laye, and W. V. Steele, *J. Phys. Chem.* **3**, 277–280 (1971).
21. M. Manson, N. Rapport, and E. F. Westrum, Jr., *J. Am. Chem. Soc.* **92**, 7296–7299 (1970).
22. E. E. Baroody and G. A. Carpenter, *Rpt. Naval Ordnance Systems Command Task No. 331-003/067-1/UR2402-001* for Naval Ordnance Station, Indian Head, MD, 1972, pp. 1–9.
23. E. F. Westrum, Jr., *J. Phys. Chem. Solids* **18**, 83–85 (1961).
24. S.-S. Chang and E. F. Westrum, Jr., *J. Phys. Chem.* **64**, 1547–1551 (1960).
25. E. F. Westrum, Jr., M. A. McKervey, J. T. S. Andrews, R. C. Fort, Jr., and T. Clark, *J. Chem. Thermod.* **10**(10), 959–965 (1978).
26. G. M. Spinella, J. T. S. Andrews, and W. E. Bacon, *J. Chem. Thermodyn.* **10**, 1023–1032 (1978).

27. R. Jochems, H. Dekker, C. Mosselman, and G. Somesen, *J. Chem. Thermodyn.* **14**, 395–398 (1982).
28. W. K. Bratton, I. Szilard, and C. A. Cupas, *J. Org. Chem.* **32**, 2019–2021 (1967).
29. T. Clark, T. M. O. Knox, M. A. McKervey, H. Mackle, and J. J. Rooney, *J. Am. Chem. Soc.* **97**, 3835–3836 (1975).
30. W. V. Steele and I. Watt, *J. Chem. Thermodyn.* **9**, 843 (1977).
31. P. O'Brien and T. E. Jenkins, *Phys. Status Solidi A* **67**, K161–162 (1981).
32. M. A. McKervey, *Tetrahedron* **36**, 971–992 (1980).
33. P. v. R. Schleyer, *Angew. Chem.* (*Int. engl. Ed.*) **8**(7), 529–529 (2003) (published online).
34. S.-G. Liu, J. E. Dahl, and R. M. K. Carlson, *Nanotech* **2004** *Conference Technical Program Abstract*, NSTI, Cambridge, MA. See also: P. A. Cahill, *Tetrahedron Lett.* **31**, 5417–5420 (1990).
35. V. S. Smith and A. S. Teja, *J. Chem. Eng. Data* **41**, 923–925 (1996).
36. T. Kraska, K. O. Leonhard, D. Tuma, and G. M. Schneider, *J. Supercrit. Fluids* **23**, 209–224 (2002).
37. I. Swaid, D. Nickel, and G. M. Schneider, *Fluid Phase Equil.* **21**, 95–112 (1985).
38. S. J. Park, T. Y. Kwak, and G. A. Mansoori, *Int. J. Thermophys.* **8**, 449–471 (1987).
39. R. Hartono, G. A. Mansoori, and A. Suwono, *Chem. Eng. Commum.* **173**, 23–42 (1999).
40. S. Landa, V. Machacek, M. Mzourek, and M. Landa, *Chim. Ind.* (Publ. No. 506), 1933.
41. V. Prelog and R. Seiwerth, *Ber. Dtsch. Chem. Ges.* **74**, 1644–1648 (1941).
42. M. Navratilova and K. Sporka, *Appl. Catalysis A, General* **203**, 127–132 (2000).
43. P. v. R. Schleyer, *J. Am. Chem. Soc.* **79**, 3292 (1957).
44. P. v. R. Schleyer and M. M. Donaldson, *J. Am. Chem. Soc.* **82**, 4645–4651 (1960).
45. W. B. Jensen, *The Lewis Acid–Base Concepts, An Overview*, John Wiley & Sons, Hoboken, NJ, 1980.
46. C. Cupas, P. v. R. Schleyer, and D. J. Trecker, *J. Am. Chem. Soc.* **87**, 917–918 (1965).
47. A. Schwartz, *Synthetic and Mechanistic Studies in Adamantane Chemistry*, Educational Research Institute, British Columbia, Canada, 1973.
48. R. B Morland, *Heteroadamantanes, The Improved Synthesis and Some Reactions of 2-Hetero-adamantanes*, PhD Dissertation, Chemistry, Kent State University, 1976.
49. Z. Kafka and L. Vodicka, *Collec. Czechoslovak Chem. Commun.* **55**(8), 2043–2045 (1990). See also: H. Hopf, *Classics in Hydrocarbon Chemistry, Syntheses, Concepts, Perspective*, Wiley-VCH Verlag GmbH & Co, Berlin, 2000.
50. M. Shibuya, T. Taniguchi, M. Takahashi, and K. Ogasawara, *Tetrahedron Lett.* **43**, 4145–4147 (2002).
51. J. E. Dahl, R. M. Carlson, and L. Shenggao, *Sci. J.* **299**, 23–25 (2003).
52. W. J. King, SPE Paper #17761, *SPE Gas Technical Symposium Proceedings*, Society of Petroleum Engineers International, Richardson, TX, 1988, pp. 469–490.
53. R. A. Alexander, C. E. Knight, and D. D. Whitehurst, U.S. Patent No. 4982049, 1991.
54. R. A. Alexander, and C. E. Knight, and D. D. Whitehurst, U.S. Patent No. 4952747, 1990.
55. R. A. Alexander and C. E. Knight, U.S. Patent No. 4952748, 1990.
56. R. A. Alexander, C. E. Knight, and D. D. Whitehurst, U.S. Patent No. 4952749, 1990.
57. J. E. Dahl, J. M. Moldowan, K. E. Peters, G. E. Claypool, M. A. Rooney, G. E. Michael, M. R. Mello, and M. L. Kohnen, *Nature* **399** (6 May), 54–57 (1999).

58. M. Schoell and R. M. Carlson, *Nature* **399**, 15–16 (1999).
59. K. E. Peters, C. M. Walters, and J. M. Moldowan, *The Biomarker Guide*, 2nd ed., parts 1 & 2, Cambridge University Press, Cambridge, UK, 2005. See also: L. K. Schulz, A. Wilhelms, E. Rein, and A. S. Steen, *Org. Geochem.* **32**(3), 365–375 (2001).
60. J. J. Brocks and R. E. Summons, Sedimentary hydrocarbons, biomarkers for early life, *Treatise Geochem.* **8**, 63–115 (2003).
61. J. A. William, M. Bjoroy, D. J. Dolcater, and J. A. Winters, *Org. Geochem.* **10**, 451–461 (1986).
62. W. S. Wignert, *Fuel* **71**, 37–43 (1992).
63. A. Petrov, O. A. Arefjev, and Z. V. Yakubson, *Adv. Org. Geochem.* (*Ed. Techniq, Paris*), 517–522 (1974).
64. R. Lin and Z. A. Wilk, *Fuel* **74**(10), 1512–1521 (1995).
65. J. E. P. Dahl, J. M. Moldowan, T. M. Peakman, J. C. Cardy, E. Lobkovsky, M. M. Olmstead, P. W. May, T. J. Davis, J. W. Steeds, K. E. Peters, A. Pepper, A. Ekuan, and R. M. K. Carlson, *Angew. Chem., Int. Ed.* **42**, 2040–2044 (2003).
66. S. L. Richardson, T. Baruah, M. J. Mehl, and M. R. Pederson, *Chem. Phys. Lett* **403**, 83–88 (2005). See also: Y. F. Chang, Y. L. Zhao, M. Zhao M, et al., *Acta Chim. Sinica* **62**(19), 1867–1870 (2004).
67. K. Grice, R. Alexander, and R. I. Kagi, *Org. Geochem.* **31**, 67–73 (2000).
68. L. K. Schulz, A Wilhelms, E. Rein, and A. S. Steen, *Org. Geochem.* **32**, 365–375 (2001).
69. J. H. Chen, J. M. Fu, G. Y. Sheng, D. H. Liu, and J. J. Zheng, *Org. Geochem.* **25**, 170–190 (1996).
70. Z.-N. Gao, Y.-Y. Chen, and F. Niu, *Geochem. J.* **35**, 155–168 (2001).
71. A. Shimoyama and H. Yabuta, *Geochem. J.* **36**, 173–189 (2002).
72. J. H. Chen, J. M. Fu, G. Y. Sheng, D. H. Liu, and J. J. Zheng, *Org. Geochem.* **25**, 170–190 (1996).
73. S. A. Stout and G. A. Douglas, *Eniro.NVIRO Forensics* **5**(4), 225–235 (2004).
74. D. Vazquez, J. Escobedo, and G. A. Mansoori, Characterization of crude oils from southern mexican oilfields, in *Proceedings of the EXITEP 98, International Petroleum Techniques*, Exhibition, Placio de Los Deportes, 15–18 November 1998, Mexico City, Mexico.
75. D. Vazquez and G. A. Mansoori, *J. Petroleum Sci. Eng.* **26**, 49–55 (2000).
76. G. A. Mansoori, *J. Petroleum Sci. Eng.* **17**, 101–111 (1997).
77. S. Priyanto, G. A. Mansoori, and S. Aryadi, *Chem. Eng. Sci.* **56**, 33–39 (2001).
78. D. M. Jewell, E. W. Albaugh, B. E. Davis, and R. G. Ruberto, *Ind. Eng. Chem. Fundam.* **13**(3), 278–282 (1974).
79. ASTM, *Book of ASTM Standards*, ASTM International, West Conshohocken, PA, 2005.
80. R. Lin and Z. A. Wilk, *Fuel* **74**(10), 1512–1521 (1995).
81. APT, *GC and GC-MS Analysis of NIGOGA Reference Samples (NSO-1 and JR-1)*, 2 May **2003**, Applied Petroleum Technology AS, Kjeller, Norway.
82. A. Shimoyama and H. Yabuta, *Geochem. J.* **36**, 173–189 (2002).
83. J. K. Henderson and J. R. Sitzman, U.S. Patent No. 000001185H, May 1993.
84. L. D. Rollmann, L. A. Green, R. A. Bradway, and H. K. C. Timken, *Catalysis Today* **31**, 163–169 (1996).
85. S. I. Zones, Y. Nakagawa, G. S. Lee, C. Y. Chen, and L. T. Yuen, *Microporous & Mesoporous Materials* **21**, 199–211 (1998).
86. M. A. Meador, *Annu. Rev. Mater. Sci.* **28**, 599–630 (1998).

87. A. A. Malik, T. G. Archibald, K. Baum, and M. R. Unroe, *Macromolecules* **24**, 5266–5268 (1991).
88. C. Yaw-Terng and W. Jane-Jen, *Tetrahedron Lett.* **36**, 5805–5806 (1995).
89. L. J. Mathias and G. L. Tullos, *Polymer* **37**, 3771–3774 1996).
90. Y.-T. Chern, *Polymer* **39**, 4123–4127, (1998). See also: Y.-T. Chern and W.-L. Wang, *Polymer* **39**, 5501–5506 (1998)
91. C.-F. Huang, H.-F. Lee, S.-W. Kuo, H. Xu, and F.-C. Chang, *Polymer (Guilford)* **45**, 2261–2269 (2004).
92. G. C. Eastmond, M. Gibas, and J. Paprotny, *Eur. Polymer J.* **35**, 2097–2106 (1999).
93. Y. K. Lee, H. Y. Jeong, K. M. Kim, J. C. Kim, H. Y. Choi, Y. D. Kwon, D. J. Choo, Y. R. Jang, K. H. Yoo, J. Jang, and A. Talaie, *Curr. Appl. Phys.* **2**, 241–244 (2002). See also: H. Y. Jeong, Y. K. Lee, A. Talaie, K. M. Kim, Y. D. Kwon, Y. R. Jang, K. H. Yoo, D. J. Choo, and J. Jang, *Thin Solid Films* **417**, 171–174 (2002).
94. A. K. Kar and D. A. Lightner, *Tetrahedron Asymmetry* **9**, 3863–3880 (1998).
95. J. Zhang, W. B. Lin, Z. F. Chen, R. G. Xiong, B. F. Abrahams, and H. K. Fun, *J. Chem. Soc. Dalton Trans.* **12**, 1806–1808 (2001).
96. R. W. Siegel, E. Hu, and M. C. Roco, *Nanostructure Science & Technology—A Worldwide Study.* Prepared under the guidance of the IWGN, NSTC; WTEC, Loyola Collage in Maryland, 1999.
97. G. A. Mansoori, *Principles of Nanotechnology: Molecular Based Study of Condensed Matter in Small Systems*, World Scientific Publishing Co., New York, 2005.
98. G. A. Mansoori, T. F. George, G. Zhang, and L. Assoufid (eds.), *Molecular Building Blocks for Nanotechnology: From Diamondoids to Nanoscale Materials and Applications* (Topics in Applied Physics Series), Springer-Verlag,New York, 2006.
99. K. E. Drexler, *Engines of Creation.* Anchor Press/Doubleday, Garden City, NY, **1986**.
100. K. E. Drexler, *Nanosystems, Molecular Machinery, Manufacturing, and Computation*, John Wiley & Sons, Hoboken, NJ. 1992.
101. G. C. McIntosh, M. Yoon, S. Berber, and D. Tománek, *Phys. Rev. B* **70**, 045401 (2004).
102. R. C. Merkle, *Trends Biotechnol.* **17**, 271–274 (1999). See also: R. C. Merkle, *Nanotechnology* **11**, 89–99 (2000).
103. G. A. Mansoori, Advances in atomic and molecular nanotechnology, *United Nations Tech. Monitor Special Issue*, 53–59 (2002).
104. R. Rawls, *Material Sci.* **80**, 13 (2002).
105. V. Kovalev, E. Shokova, A. Khomich, and Y. Luzikov, *New J. Chem.* **20**(4), 483–492 (1996).
106. D. N. Chin, D. M. Gordon, and G. M. Whitesides, *J. Am. Chem. Soc.* **116**(26), 12033–12044 (1994).
107. O. Ermer and L. Lindenberg, *Helv. Chim. Acta* **74**(4), 825–877 (1991).
108. O. Ermer and L. Lindenberg, *Helv. Chim. Acta* **71**(5), 1084–1093 (1988)
109. (a) E. Shokova, V. Tafeenko, and V. Kovalev, *Tetrahedron Lett.* **43**, 5153–5156 (2002). See also: (b) K. A. Hirsch, S. R. Wilson, and J. S. Moore, *Chem. Eur. J.* **3**(5), 765–771 (1997).
110. N. D. Drummond, A. J. Williamson, R. J. Needs, and G. Galli, *Phys. Rev. Lett.* **95**(9), 096801 (2005).
111. P. R. Seidl and K. Z. Leal, *J. Mol. Structure: THEOCHEM* **539**, 159–162 (2001).
112. V. V. Martin, I. S. Alferiev, and A. L. Weis, *Tetrahedron Lett.* **40**, 223–226 (1999).
113. Q. Li, A. V. Rukavishnikov, P. A. Petukhov, T. O. Zaikova, and J. F. W. Keana, *Org. Lett* **4**(21), 3631–3634 (2002).

114. O. R. Evans, R.-G. Xiong, Z. Wang, G. K. Wong, and W. Lin, *Angew. Chem. Int. Ed. Engl.* **38**, 536–538 (1999). See also: O. R. Evans and W. B. Lin, *Chem. Materials* **13**(8), 2705–2712 (2001).

115. R. Vaidhyanathan, S. Natarajan, and C. N. R. Rao, *Encyclopedia of Supramolecular Chemistry*, pp. 1–12, 2004.

116. M. J. Zaworotko, *Chem. Soc. Rev.* **23**(4), 283–288 (1994).

117. D. S. Reddy, D. C. Craig, and G. R. Desiraju, *J. Am. Chem. Soc.* **118**(17), 4090–4093 (1996).

118. C. Phoenix, *Design of a primitive nanofactory, J. Evolution & Tech.*, **13**(2), *http://jetpress.org/Volume13/NanoFactory.htm* (2003).

119. R. A. Freitas Jr. and R. C. Merkle, *Kinematic Self-Replicating Machines*, Landes Bioscience, Georgetown, TX, 2004.

120. A. Herman, *Modelling Simul. Mater. Sci. Eng.* **7**(1), 43–58 (1999).

121. J. Peng, R. A. Freitas Jr., and R. C. Merkle, *J. Comp. Theor. Nanosci.*, 62–70 (1 Mar. 2004).

122. K. E. Drexler, C. Peterson, and G. Pergamit, *Unbounding the Future, The Nanotechnology Revolution*, William Morrow & Co., New York, **1991**.

123. K. E. Drexler, *Trends Biotechnol.* **17**, 5–7 (1999).

124. K. Bogunia-Kubik and M. Sugisaka, *Biosystems* **65**, 123–138 (2002).

125. R. A. Freitas, *Nanotechnology* **2**, 8–13 (1996).

126. R. A. Freitas, Jr., *Artificial Cells, Blood Substitutes & Biotechnol.* **26**, 411–430 (1998).

127. A. Herman, *Modelling Simul. Mater. Sci. Eng.* **7**(1), 43–58 (1999).

128. H. Ramezani and G. A. Mansoori, Diamondoids as molecular building blocks for nanotechnology, in *Molecular Building Blocks for Nanotechnology: From Diamondoids to Nanoscale Materials and Applications*, G.A. Mansoori, Th.F. George, L. Assoufid, and G. Zhang (eds.), Springer, New York, 2007.

129. O. Anderzej, *Acta Biochem. Polon.* **47**, 1–7 (2000).

130. J. G. Hardman and L. E. Limbird, *Goodman & Gilman's: The Pharmacological Basis of Therapeutics,* 10th ed., McGraw-Hill, New York, **2001**.

131. N. Tsuzuki, T. Hama, M. Kawada, A. Hasui, R. Konishi, S. Shiwa, Y. Ochi, S. Futaki, and K. Kitagawa, *J. Pharmaceut. Sci.* **83**, 481–484 (1994).

132. Z. Kazimierczuk, A. Gorska, T. Switaj, and W. Lasek, *Bioorg. Med. Chem. Lett.* **11**, 1197–1200 (2001).

133. G. Stamatiou, A. Kolocouris, N. Kolocouris, G. Fytas, G. B. Foscolos, J. Neyts, and E. De Clercq, *Bioorg. Med. Chem. Lett.* **11**, 2137–2142 (2001).

134. M. V. Samoilova, S. L. Buldakova, V. S. Vorobjev, I. N. Sharonova, and L. G. Magazanik, *Neuroscience* **94**, 261–268 (1999).

135. K. V. Bolshakov, D. B. Tikhonov, V. E. Gmiro, and L. G. Magazanik, *Neurosci. Lett.* **291**, 101–104 (2000).

136. S. L. Buldakova, V. S. Vorobjev, I. N. Sharonova, M. V. Samoilova, and L. G. Magazanik, *Brain Res.* **846**, 52–58 (1999).

137. G. Rammes, R. Rupprecht, U. Ferrari, W. Zieglgansberger, and C. G. Parsons, *Neurosci. Lett.* **306**, 81–84 (2001).

138. D. L. Flynn, D. P. Becker, D. P. Spangler, R. Nosal, G. W. Gullikson, C. Moummi, and D.-C. Yang, *Bioorg. Med. Chem. Lett.* **2**, 1613–1618 (1992).

139. A. Baxter, J. Bent, K. Bowers, M. Braddock, S. Brough, M. Fagura, M. Lawson, T. McInally, M. Mortimore, and M. Robertson, *Chem. Lett.* **13**, 4047–4050 (2003).

140. G. Zoidis, I. Papanastasiou, I. Dotsikas, A. Sandoval, R. G. Dos Santos, Z. Papadopoulou-Daifoti, A. Vamvakides, N. Kolocouris, and R. Felix, *Bioorg. Med. Chem.* **13**, 2791–2798 (2005).

141. C. Shen, D. Bullens, A. Kasran, P. Maerten, L. Boon, J. M. F. G. Aerts, G. van Assche, K. Geboes, P. Rutgeerts, and J. L. Ceuppens, *Int. Immunopharmacol.* **4**, 939–951 (2004).

142. S. Reissmann, F. Pineda, G. Vietinghoff, H. Werner, L. Gera, J. M. Stewart, and I. Paegelow, *Peptides* **21**, 527–533 (2000).

143. W. J. Hoekstra, J. B. Press, M. P. Bonner, P. Andrade-Gordon, P. M. Keane, K. A. Durkin, D. C. Liotta, and K. H. Mayo, *Bioorg. Med. Chem. Lett.* **4**, 1361–1364 (1994).

144. N. Tsuzuki, T. Hama, T. Hibi, R. Konishi, S. Futaki, and K. Kitagawa, *Biochem. Pharmacol.* **41**, R5–R8 (1991).

145. K. Kitagawa, N. Mizobuchi, T. Hama, T. Hibi, R. Konishi, and S. Futaki, *Chem. Pharmaceut. Bull. (Tokyo)*, **45**, 1782–1787 (1997).

146. M. Manoharan, K. L. Tivel, and P. D. Cook, *Tetrahedron Lett.* **36**, 3651–3654 (1995).

147. N. Lomadze and H. J. Schneider, *Tetrahedron Lett.* **43**, 4403–4405 (2002).

148. L. Moine, S. Cammas, C. Amiel, P. Guerin, and B. Sebille, *Polymer* **38**, 3121–3127 (1997).

149. I. Habus, Q. Zhao, and S. Agrawal. *Bioconjugate Chem.* **6**, 327–331 (1995). See also: M. Manoharan, K. L. Tivel, and P. D. Cook, *Tetrahedron Lett.* **36**, 3651–3654 (1995).

150. C. L. D. Gibb, X. Li, and B. C. Gibb, ***PNAS*** **99**(8), 4857–4862 (2002). See also: S. Aoki, M. Shiro, and E. Kimura, *Chem. Eur. J.* **8**(4), 929–939 (2002); and S. B. Copp, S. Subramanian, and M. J. Zaworotko, *J. Am. Chem. Soc.* **114**(22), 8719–8720 (1992).

151. D. Ranganathan and S. Kurur, *Tetrahedron Lett.* **38**, 1265–1268 (1997).

152. A. K. Dillow and A. M. Lowman (eds.), *Biomimetic Materials & Design, Biointerfacial Strategies, Tissue Engineering, and Targeted Drug Delivery*, Marcel Dekker, New York, 2002.

153. K. Busch and R. Tampé, *Rev. Mol. Biotechnol.* **82**, 3–24 (2001).

154. J. H. Kim, J.-A. Hong, M. Yoon, M. Y. Yoon, H.-S. Jeong, and H. J. Hwang, *J. Biotechnol.* **96**, 213–221 (2002).

155. C. J. Noren, S. J. Anthony-Cahil, M. C. Griffith, and P. G. Schultz, *Science* **244**, 182–188 (1989).

156. J. A. Piccirilli, T. Krauch, S. E. Moroney, and S. A. Benner, *Nature* **343**, 33–43 (1990).

157. Anon., *Host–Guest Molecular Interactions, from Chemistry to Biology*, CIBA Foundation Symposia Series, No. 158, John Wiley & Sons, Hoboken, NJ, 1991.

158. L. Mandolini and R. Ungaro (eds.), *Calixarenes in Action*, World Scientific Publishing Co., New York, 2000.

159. D. Ranganathan, M. P. Samant, R. Nagaraj, and E. Bikshapathy, *Tetrahedron Lett.* **43**, 5145–5147 (2002).

160. D. Ranganathan, V. Haridas, and I. L. Karle, *Tetrahedron* **55**, 6643–6656 (1999).

161. D. Ranganathan, A. Thomas, V. Haridas, S. Kurur, K. P. Madhusudanan, R. Roy, A. C. Kunwar, A. V. Sarma, M. Vairamani, and K. D. Sarma, *J. Org. Chem.* **64**, 3620–3629 (1999).

162. D. Ranganathan, V. Haridas, S. Kurur, R. Nagaraj, E. Bikshapathy, A. C. Kunwar, A. V. Sarma, and M. Vairamani, *J. Org. Chem.* **65**, 365–374 (2000).

163. D. Ranganathan, V. Haridas, R. Nagaraj, I. L. Karle, and L. Isabella, *J. Org. Chem.* **65**, 4415–4422 (2000).

164. I. L. Karle, *J. Mol. Structure* **474**, 103–112 (1999).

165. I. L. Karle and D. Ranganathan, *J. Mol. Structure* **647**, 85–96 (2003).
166. C. Jaime, J. Redondo, F. Sanchez-Ferrando, and A. Virgili, *J. Mol. Structure* **248**, 317–329 (1991).
167. K. Fujita, W.-H. Chen, D.-Q. Yuan, Y. Nogami, and T. Koga, *Tetrahedron Asymmetry* **10**, 1689–1696 (1999).
168. D. Krois and U. H. Brinker, *J. Am. Chem. Soc.* **120**(45), 11627–11632 (1998).
169. C. Karakasyan, M.-C. Millot, and C. Vidal-Madjar, *J. Chromatogr. B* **808**, 63–67 (2004).
170. F. D. Ayres, S. I. Khan, and O. L. Chapman, *Tetrahedron Lett.* **35**, 8561–8564 (1994).
171. T. Ishizone, H. Tajima, S. Matsuoka, and S. Nakahama, *Tetrahedron Lett.* **42**, 8645–8647 (2001).

TIME-RESOLVED X-RAY DIFFRACTION FROM LIQUIDS

SAVO BRATOS

Laboratoire de Physique Théorique des Liquides, Université Pierre et Marie Curie, 75252 Paris Cedex, France

MICHAEL WULFF

European Synchrotron Radiation Facility, 38043 Grenoble Cedex, France

CONTENTS

Advances in Chemical Physics, Volume 136, edited by Stuart A. Rice

I. INTRODUCTION

Since the discovery of X-rays by Roentgen, X-ray diffraction has always been the major technique permitting the localization of atoms in molecules and crystals. Great scientists such as Bragg, Laue, and Debye made major contributions to its development. Atomic structures of many systems have been determined using this technique. These structures are actually known with great accuracy, and one can hardly imagine a science without this information. The recent progress reached using synchrotron sources of X-ray radiation should be emphasized.

However, considering systems with localized atoms represents only a first challenge. The next challenge consists in monitoring atomic motions in systems that vary in time. Following atomic motions during a chemical process has always been a dream of chemists. Unfortunately, these motions evolve from nano - to femtosecond time scales, and this problem could not have been challenged until ultrafast detection techniques were invented. Spectacular developments in laser technology, and recent progress in construction of ultrafast X-ray sources, have proved to be decisive. Two main techniques are actually available to visualize atomic motions in condensed media.

The first of them is time-resolved optical spectroscopy. The system is excited by an intense optical pulse, and its return to statistical equilibrium is probed by another pulse, which is also optical. Zewail and several other outstanding scientists contributed much to its development. For textbooks describing it, for example, see Refs. [1–4]. The second technique comprises time-resolved X-ray diffraction and absorption. The excitation of the system is optical as before, but the probing is done using an X-ray pulse. Long-range order may be probed by diffraction, whereas short-range order may be monitored by absorption. Unfortunately, the general literature is still scarce in this domain [5–7].

The major advantage of time-resolved X-ray techniques, as compared to optical spectroscopy, is that their wavelength λ as well as the pulse duration τ can be chosen to fit the atomic scales. This is not the case for optical spectroscopy, where the wavelength λ exceeds interatomic distances by three orders of magnitude at least. Unfortunately, X-ray techniques also have their drawbacks. They require large-scale instruments such as the synchrotron. Even much larger instruments based on free electron lasers are actually under construction. The "nonhuman" size of X-ray instrumentation is sometimes an objection against the use of this method, whereas optical spectroscopy is free of this objection.

The purpose of this chapter is to review ultrafast, time-resolved X-ray diffraction from liquids. Both experimental and theoretical problems will be treated. The structure of the chapter is as follows. Section II describes the principles of a time-resolved X-ray experiment and details some of its characteristics. Basic elements of the theory are discussed briefly in Sections III–V. Finally, Section VI presents recent achievements in this domain. The related field of time-resolved X-ray spectroscopy, although very promising, will not be discussed.

II. EXPERIMENT

A. Basic Principles

The system under consideration is a liquid sample, either a pure liquid or a solution. It is pumped by a laser, which promotes a fraction of molecules into one or several excited quantum states. The energy deposited by the laser diffuses into the system in one or several steps, generating several sorts of events. If the excitation is in the optical spectral range, chemical reactions may be triggered and the sample is heated. If it is in the infrared, only the heating is generally present. The return of the excited system to thermal equilibrium is then probed using a series of time-delayed X-ray pulses (Fig. 1). The resulting diffraction patterns consist of circular rings, centered on the forward beam direction. Finally, the collection of diffraction patterns obtained in this way is transformed into a collection of molecular photographs; this step is accomplished by theory. A time-resolved X-ray experiment thus permits one to "film" atomic motions.

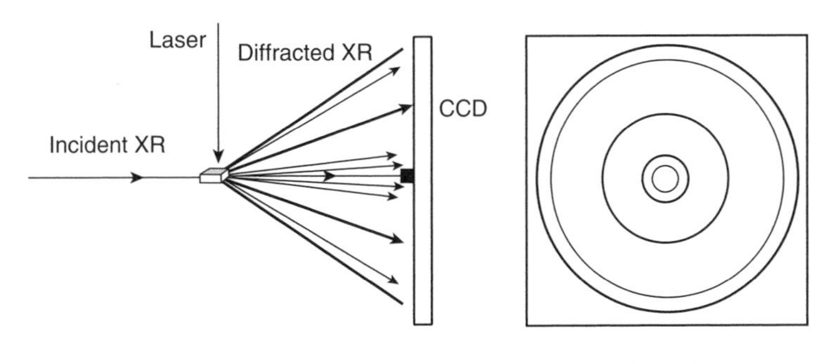

Figure 1. Time-resolved X-ray diffraction experiment (schematic). The liquid sample is excited by a laser pulse, and its temporal evolution is monitored by a time-delayed X-ray pulse. The diffracted radiation is measured by a charge-coupled detector (CCD). In practice, the laser and X-ray beams are not perpendicular to each other, but nearly parallel.

This rough picture can be sharpened by providing some additional information. The experimental setup appropriate for time-resolved diffraction comprises a pulsed synchrotron source, a chopper that selects single X-ray pulses from the synchrotron, a femtosecond laser activating the process to be studied, a capillar jet, and an integrating detector measuring the intensity of the scattered X-ray radiation. What is desired in reality is not the scattered X-ray intensity by itself, but the difference between scattered intensities in the presence or absence of the laser excitation. This difference intensity is very small and is thus particularly difficult to measure. The images must be integrated azimuthally and corrected for polarization and space–angle effects. How is this sort of experiment realized in practice? Some major points are discussed next [5–8]. For new trends in instrumentation, see Refs. [9–11].

B. Practical Realization

1. *X-ray and Optical Sources*

The first—and central—point in this discussion concerns the pulsed X-ray source. The shortest time scales involved when chemical bonds are formed or broken are of the order of a few femtoseconds. An ideal X-ray source should thus be capable of providing pulses of this duration. Unfortunately, generating them represents a heavy technological problem. The best one can do at present is to use a pulsed synchrotron X-ray source (Fig. 2). Electrons are rapidly circulating in its storage ring at speeds close to the speed of light. X-rays are spontaneously

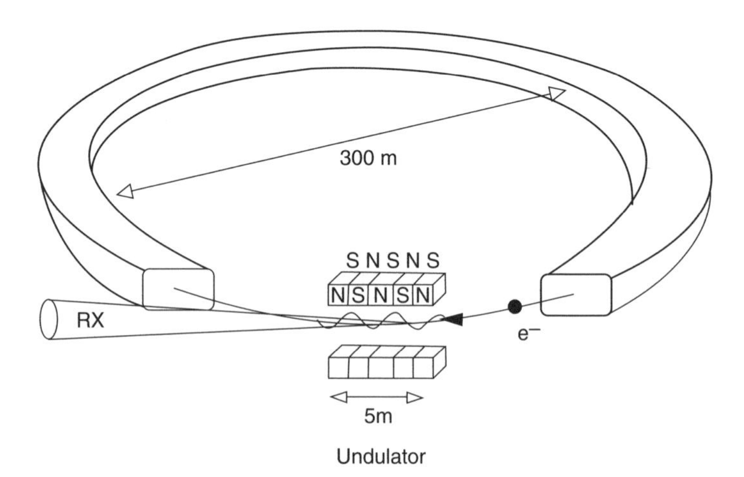

Figure 2. Synchrotron X-ray source (schematic). The electrons execute circular motions in the storage ring and emit intense X-rays along the tangent of the orbit. This radiation is enhanced by undulator magnets that are often placed inside the vacuum vessel for enhanced performance. The storage ring has a number of straight sections for undulators and wigglers (not shown).

emitted longitudinal to the orbit. This emission is amplified in socalled straight sections, where a sinusoidal motion is imposed on the electrons by undulator magnets. The X-rays emitted from successive bends interfere and enhance the radiation. A bunched electron beam then produces an intense X-ray radiation with 100 ps X-ray pulses. Unfortunately, subpicosecond X-ray pulses cannot be generated in this way. There is a gap between what is possible at present and what is needed. Facilities based on the use of free electron lasers are under construction to bridge this gap.

The sources of the optical radiation are much more conventional. One generally employs commercially available lasers, generating 100 fs pulses with pulse energy between 10 and 100 μJ. These lasers run in phase with the chopper. The power density of the optical beam on the sample is typically on the order of 10 GW/mm^2. The time lag between the X-ray and optical pulses is controlled electronically by shifting the phase of the oscillator feedback loop with a digital delay generator. The short time jitter in the delay is on the order of a few picoseconds, but at long times it increases to tens of picoseconds. The angle between the X-ray and laser beam is 10°, making the excitation geometry near colinear.

2. *Detectors*

Several detection techniques have been developed. In the first of them, the scattered X-rays are intercepted by a phosphor screen, which transforms them into optical photons. The latter are then channeled by optical fibers to a charge-coupled CCD device. In this detector the vast majority of X-ray photons are registered. However, it is not straightforward to measure very small relative changes in the CCD signal. The detection should be strictly linear in the photon number, which is not easy to achieve in practice. Finally, the X-ray dose must be kept constant during the data collection. The exposure must thus be prolonged to compensate for the decaying synchrotron current. Another very different detection technique consists in using streak cameras. The incident X-ray pulse is transformed into an electronic pulse, which is "streaked" by an electrostatic field onto a CCD device. Although there are streak cameras with picosecond time resolution, they require very high X-ray intensities and are therefore of limited use. There is no perfect detector; all have strengths and weaknesses, and the optimum choice depends on the exact nature of the measurements.

3. *Data Reduction Procedures*

As emphasized earlier, the weakness of the difference intensity is a specific difficulty of this sort of experiments. The ratio between the difference and the full scattering intensity is on the order of 10^{-2}–10^{-4}. This is particularly problematic in solution work, where the radiation scattered by the solute is buried in that from the solvent. A further complication arises from the interference between X-ray and optical manipulations. In fact, the intensity of the difference radiation depends on the number of excited molecules, which in turn is a function of the solute concentration. If the system is excited with ultrashort optical pulses, this number should not exceed a critical system-specific level. If this restriction is not respected, multiphoton absorption may be activated in the solute and solvent, and this may obscure the interpretation. In practice, the concentration of excited species is thus always very low, typically 1/1000. Once again, the intensity of the X-ray beam is all important.

Another specific difficulty in X-ray experiments is that diffraction images contain contributions not only from the liquid sample but also from the capillary and air. It is often delicate to disentangle these contributions from each other (Fig. 3). Note that the main noise in the difference signal comes from the photon statistics in the X-ray background. This is an intrinsic limitation in solution phase ultrafast X-ray scattering. One has also to take into account the presence of radioactivity and cosmic rays. The parasitic counts generated by these two mechanisms may be eliminated by subtracting two images, the original minus 180° rotated.

All measured intensities can be put on absolute scale by proceeding as follows. At high angles the scattering pattern can be considered as arising from a collection of noninteracting gas molecules rather than from a liquid sample.

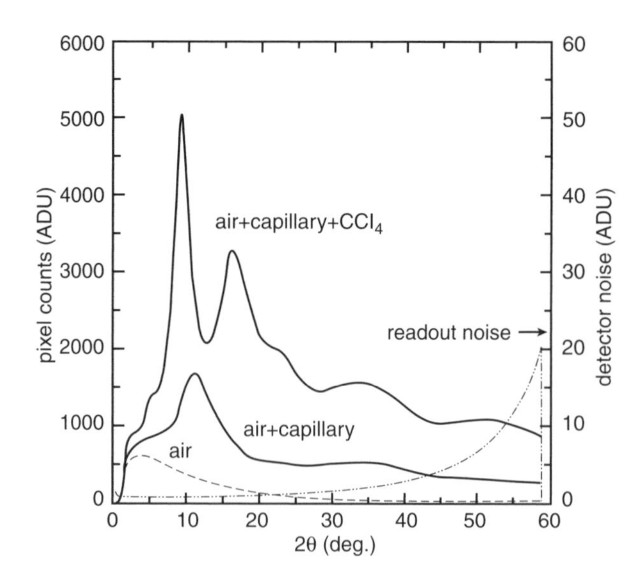

Figure 3. Various contributions to the scattered X-ray intensity. The system under consideration is a dilute I_2/CCl_4 solution.

The Compton scattering cannot be neglected, but it is independent of molecular structure. Then, fitting experimental data to formulas from gas phase theory, the concentration of excited molecules can be determined. Another problem is that the undulator X-ray spectrum is not strictly monochromatic, but has a slightly asymmetric lineshape extending toward lower energies. This problem may be handled in different ways, for example, by approximating its spectral distribution by its first spectral moment [12].

III. THEORY

A. Generalities

The first theoretical attempts in the field of time-resolved X-ray diffraction were entirely empirical. More precise theoretical work appeared only in the late 1990s and is due to Wilson et al. [13–16]. However, this theoretical work still remained preliminary. A really satisfactory approach must be statistical. In fact, macroscopic transport coefficients like diffusion constant or chemical rate constant break down at ultrashort time scales. Even the notion of a molecule becomes ambiguous: at which interatomic distance can the atoms A and B of a molecule A–B be considered to be free? Another element of consideration is that the electric field of the laser pump is strong, and that its interaction with matter is nonlinear. What is needed is thus a statistical theory reminiscent of those from time-resolved optical spectroscopy. A theory of this sort was elaborated by Bratos and co-workers and was published over the last few years [17–19].

An important specific feature of the present experiment is worth noting. The X-ray photons have energies that are several orders of magnitude larger than those of optical photons. The pump and probe processes thus evolve on different time scales and can be treated separately. It is convenient to start with the X-ray probing processes, and treat them by Maxwellian electrodynamics. The pumping processes are studied next using statistical mechanics of nonlinear optical processes. The electron number density $n(\mathrm{r},t)$, supposed to be known in the first step, is actually calculated in this second step.

We shall now focus attention on spatially isotropic liquids. The key quantities of the theory are as follows. An intense optical pulse of frequency Ω_O brings it into an appropriate initial state. τ seconds later, an X-ray pulse of frequency $\Omega_X \gg \Omega_O$ hits the sample and is then diffracted by it. What one measures is the difference signal $\Delta S(\mathbf{q}, \tau)$, defined as the time integrated X-ray energy flux $S(\mathbf{q}, \tau)$ scattered in a given solid angle in the presence of the pump, minus the time integrated X-ray energy flux $S(\mathbf{q})$ in the same angle in the absence of the pump. It depends on two variables: the scattering wavevector $\mathbf{q} = \mathbf{q}_\mathrm{I} - \mathbf{q}_\mathrm{S}$, where $\mathbf{q}_\mathrm{I}$ and $\mathbf{q}_\mathrm{S}$ are wavevectors of the incident and the scattered

X-ray radiation, respectively, and the time delay τ between pump and probe. $\Delta S(\mathrm{q}, \tau)$ is the main quantity to be examined in what follows.

B. Maxwellian Description of X-Ray Probing

The Maxwell theory of X-ray scattering by stable systems, both solids and liquids, is described in many textbooks. A simple and compact presentation is given in Chapter 15 of *Electrodynamics of Continuous Media* [20]. The incident electric and magnetic X-ray fields are plane waves $\mathrm{E}_X(\mathbf{r}, t) = \mathbf{E}_{X0} \exp[i(\mathbf{q}_I\mathbf{r} - \Omega_X t)]$ and $\mathbf{H}(\mathbf{r}, t) = \mathbf{H}_{X0} \exp[i(\mathbf{q}_I\mathbf{r} - \Omega_X t)]$ with a spatially and temporally constant amplitude. The electric field $\mathbf{E}_X(\mathbf{r}, t)$ induces a forced oscillation of the electrons in the body. They then act as elementary antennas emitting the scattered X-ray radiation. For many purposes, the electrons may be considered to be free. One then finds that the intensity $I_X(\mathbf{q})$ of the X-ray radiation scattered along the wavevector q is

$$I_X(\mathbf{q}) = \left(\frac{e^2}{mc^2}\right)^2 \sin^2 \phi I_{0X} f^*(\mathbf{q}) f(\mathbf{q}) \tag{1}$$

where I_{0X} is the intensity of the incident X-ray radiation, ϕ is the angle between $\mathbf{E}_X$ and $\mathbf{q}$, and $f(\mathbf{q}) = \int dx \exp(-\mathbf{iqr}) n(\mathbf{r})$, the Fourier-transformed electron number density $n(\mathbf{r})$; this latter quantity is generally termed a form factor. The success of this theory is immense; see the textbooks by Guinier [21], Warren [22], and Als-Nielsen and Morrow [23]. In ordered systems like crystals, it permits one to determine atomic positions, that is, to "photograph" them. In disordered systems like powders or liquids, the data are less complete but still remain very usable. A large variety of systems have been analyzed successfully using this approach.

How must this theory be modified to describe the effect of the optical excitation? The incident electric and magnetic X-ray fields are now pulses $\mathbf{E}_X(\mathbf{r}, t) = \mathbf{E}_{X0}(t) \exp[i(\mathbf{q}_I\mathrm{r} - \Omega_X t)]$ and $\mathbf{H}_X(\mathbf{r}, \mathbf{t}) = \mathbf{H}_{X0}(t) \exp[i(\mathbf{q}_I\mathbf{r} - \Omega_X t)]$. They still are plane waves with a carrier frequency Ω_X, but their amplitudes $\mathbf{E}_{X0}(t)$ and $\mathbf{H}_{X0}(t)$ vary with time. The same statement applies to the electron density $n(\mathbf{r}, t)$, which also is time dependent. However, these variations are all slow with time scales on the order of $1/\Omega_X$, and one can neglect $\delta\mathbf{E}_{X0}(t)/\delta t$ and $\delta\mathbf{H}_{X0}(t)/\delta t$ as compared to $i\Omega_X\mathbf{E}_{X0}(t)$ and $i\Omega_X\mathbf{H}_{X0}(t)$. Detailed calculations then show that [17]

$$\begin{aligned} S(\mathbf{q}, r) = \left(\frac{e^2}{mc^2}\right)^2 \sin^2 \phi \int_{-\infty}^{\infty} dt\, I_{0X}(t) \\ < f^*(\mathbf{q}, t+\tau) f(\mathbf{q}, t+\tau) > \end{aligned} \tag{2}$$

where $S(\mathbf{q}, \tau)$ is the time-integrated intensity of the X-ray pulse, scattered τ seconds after optical excitation in the direction of the vector $\mathbf{q}$. This expression can be deduced from that for $I_X(\mathrm{q})$ using the following arguments. (1) As the incident X-ray radiation is pulsed, one must replace I_{0X} by $I_{0X}(t)$. (2) Optical excitation brings the system out of thermal equilibrium. It no longer remains stationary, but varies with time; thus $n(\mathbf{r}) \rightarrow n(\mathrm{r}, t)$ and $f(\mathbf{q}) \rightarrow f(\mathbf{q}, t) = \int dx \exp(-\mathbf{iqr}) n(\mathbf{r}, t)$. (3) The quantity $S(\mathbf{q}, \tau)$ is a time-integrated quantity. It is then useful to replace the integration variable t by the variable $t + \tau$, which permits one to introduce the time delay τ between the pump and the probe explicitly. (4) An X-ray diffraction experiment does not permit one to single out a given state of the system. Only an average $< >$ over all these states can be observed.

C. Statistical Description of Optical Pumping

In the previous Maxwellian description of X-ray diffraction, the electron number density $n(\mathbf{r}, t)$ was considered to be a known function of **r**,*t*. In reality, this density is modulated by the laser excitation and is not known a priori. However, it can be determined using methods of statistical mechanics of nonlinear optical processes, similar to those used in time-resolved optical spectroscopy [4]. The laser-generated electric field can be expressed as $\mathbf{E}(\mathbf{r}, t) = \mathbf{E}_{O0}(t)$ $\exp(i(\mathbf{q}_O \mathrm{r} - \Omega_O t))$, where Ω_O is the optical frequency and $\mathbf{q}_O$ the corresponding wavevector. The calculation can be sketched as follows.

The main problem is to calculate $\langle f^*(\mathbf{q}, t + \tau) f(\mathbf{q}, t + \tau) \rangle$ of Eq. (2). To achieve this goal, one first considers **E**(**r**,*t*) as a well-defined, deterministic quantity. Its effect on the system may then be determined by treating the von Neumann equation for the density matrix $\rho(t)$ by perturbation theory; the laser perturbation is supposed to be sufficiently small to permit a perturbation expansion. Once $\rho(t)$ has been calculated, the quantity

$$\langle f^*(\mathbf{q}, t + \tau) f(\mathbf{q}, t + \tau) \rangle = \mathrm{T}_7[\rho(t + \tau) f^*(\mathbf{q}) f(\mathbf{q})] \quad (3)$$

can be determined for a given realization of the electric field **E**(**r**,*t*). In the second step, this restriction to deterministic processes is suppressed and the incident laser field **E**(**r**, *t*) is identified as a stochastic quantity. In reality, **E**(**r**, *t*) is never completely coherent: averaging over this stochastic process is thus necessary. This can be done using theories of transmission of electric signals. The resulting expression is inserted into Eq. (2).

D. Difference Signal ΔS(q, τ)

The theoretical difference signal $\Delta S(\mathbf{q}, \tau)$ is a convolution between the temporal profile of the X-ray pulse $I_{0X}(t)$ and the diffraction signal $\Delta S_{\mathrm{inst}}(\mathbf{q}, t)$ from an infinitely short X-ray pulse. This expression is[17]

$$\Delta S(\mathbf{q},\tau) = \int_{-\infty}^{\infty} dt\, I_{0X}(t-\tau)\Delta S_{\text{inst}}(\mathbf{q},t)$$

$$\Delta S_{\text{inst}}(\mathbf{q},t) = -\left(\frac{e^2}{mc^2\hbar}\right)^2 \sin^2\theta \int_{-\infty}^{\infty}\int_{-\infty}^{\infty} d\tau_2\, d\tau_2$$
$$\times \langle E_i(\mathbf{r},t-\tau_1)E_j(\mathbf{r},t-\tau_1-\tau_2)\rangle_O$$
$$\times \langle[[f(\mathbf{q},\tau_1+\tau_2)f^*(\mathbf{q},\tau_1+\tau_2), M_i(\tau_2], M_j(0)]\rangle_S \qquad (4)$$

where $\mathrm{E} = (E_x, E_y, E_z)$ is the electric field of the optical pulse and $\mathrm{M} = (M_x, M_y, M_z)$ is the dipole moment of the system. Moreover, the indices i, j designate the Cartesian components x, y, z of these vectors; $< >_O$ realizes an averaging over all possible realizations of the optical field E, and $< >_S$ that over the states of the nonperturbed liquid sample. Two three-time correlation functions are present in Eq. (4): the correlation function of $\mathbf{E}(t)$ and the correlation function of the variables $f(\mathbf{q}, t)$ and $\mathbf{M}(t)$. Such objects are typical for statistical mechanics of systems out of equilibrium, and they are well known in time-resolved optical spectroscopy [4]. The above expression for $\Delta S(\mathbf{q},\tau)$ is an exact second-order perturbation theory result.

Its general form can easily be understood. The static intensity $I_X(\mathbf{q})$ contains the factor $f^*(\mathbf{q})f(\mathbf{q})$; its time-dependent analog $\Delta S(\mathbf{q},\tau)$ should then contain the factor $\langle f^*(\mathbf{q},\tau_1+\tau_2)f(\mathbf{q};\tau_1+\tau_2)\rangle$. The remaining quantities present in Eq. (4) describe optical excitation. According to Fermi's golden rule, the rate of the latter is on the order of $\sim 1/\hbar^2(\mathrm{EM}_{IF})^2$. The presence of the quantities $1/\hbar^2$, $E_i(t-\tau_1)$, $E_j(t-\tau_1-\tau_2)$, $M_j(0)$, and $M_i(\tau_2)$ is thus natural. It should finally be noted that the scattering process depends on the properties of the material system (through $f(\mathbf{q}, t)$, $\mathbf{M}(t)$), as well as on those of the laser fields (through $\mathbf{E}(r, t)$). They determine jointly the form of the signal.

IV. LONG- AND SHORT-TIME LIMITS

A. Long-Time Signals

The theory of time-resolved X-ray scattering has a comparatively simple limit if the optical excitation is fast as compared with the process to be investigated. This so-called quasistatic condition is of great practical importance. In fact, optical pumping is generally done on subpicosecond time scales, whereas with the present state-of-the-art, X-ray probing is at least 100 times longer. A new time-scale separation appears. The slow variable is the chemically driven electron density f, and the fast variable is the laser-controlled dipole moment M. The correlation function $\langle[[f^*f, \mathbf{M}], \mathbf{M}]\rangle$ in Eq. (4) thus splits into two factors, a factor involving f, f^* and a factor involving $\mathbf{M}$,$\mathbf{M}$. A quasistatic experiment thus has the intrinsic power to disentangle these two sorts of dynamics. This is the first simplification in this problem.

A second simplification results from introducing the Born–Oppenheimer separation of electronic and nuclear motions; for convenience, the latter is most often considered to be classical. Each excited electronic state of the molecule can then be considered as a distinct molecular species, and the laser-excited system can be viewed as a mixture of them. The local structure of such a system is generally described in terms of atom–atom distribution functions $g_{\mu\nu}(r,t)$ [22, 24, 25]. These functions are proportional to the probability of finding the nuclei μ and ν at the distance r at time t. Building this information into Eq. (4) and considering the isotropy of a liquid system simplifies the theory considerably.

B. "Filming" Atomic Motions

The most spectacular success of the theory in its quasistatic limit is to show how to "film" atomic motions during a physicochemical process. As is widely known, "photographing" atomic positions in a liquid can be achieved in static problems by Fourier sine transforming the X-ray diffraction pattern [22]. The situation is particularly simple in atomic liquids, where the well-known Zernicke–Prins formula provides $g(r)$ directly. Can this procedure be transferred to the quasistatic case? The answer is yes, although some precautions are necessary. The theoretical recipe is as follows. (1) Build the quantity $F(q)q\ \Delta S(q,\tau)$, where $F(q) = [\Sigma\Sigma_{\mu\neq\nu} f_\mu(q) f_\nu(q)]^{-1}$ is the "sharpening factor" and $f_\mu(q), f_\nu(q)$ are atomic form factors. (2) Perform the Fourier sine transform of this quantity for a large set of interatomic distances r. Denote the resulting signal by $\Delta S[r,\tau]$. Then, in the case of an atomic liquid [18, 19],

$$
\begin{aligned}
\Delta S[r,\tau] &= \int_{-\infty}^{\infty} dt\, I_x(t-\tau)\Delta S_{\text{inst}}[r,t] \\
\Delta S_{\text{inst}}[r,t] &= \left(\frac{e^2}{mc^2}\right)^2 \sin^2\theta \left[\left(\frac{1}{V(t)}g(r,t) - \frac{1}{V(0)}g(r)\right)\right. \\
&\quad \left. - \left(\frac{1}{V(t)} - \frac{1}{V(0)}\right)\right]
\end{aligned}
\tag{5}
$$

where $g(r,t)$ is the atom pair distribution function in the presence of the laser excitation, and $g(r)$ is its analog in the laser-free system. The interpretation of the above result is as follows. (1) The first term appearing in the expression for $\Delta S[r,\tau]$ describes the variation in the pair distribution function due to laser excitation. It permits the visualization of molecular dynamics in the laser-excited system. (2) The second term probes the change in the volume $V(t)$ due to the laser heating. It dominates $\Delta S_{\text{inst}}[r,t]$ at small r's, where $g(r,t) \to 0$ and only $1/V(t) - 1/V(0)$ survives. Since $\Delta V/V = -\Delta\rho_M/\rho_M$, the evolution of the mass density ρ_M of the liquid can be monitored in this way. (3) Equation (5) represents a generalization of the Zernicke–Prins formula for time-resolved X-ray experiments. At small r's it permits one to monitor macroscopic variations of the mass density ρ_M, whereas at large r's it offers a visualization of atomic motions in the system.

In molecular liquids the situation is slightly more complicated; the following points merit discussion. (1) The signal $\Delta S[r, \tau]$ is composed of a number of different distribution functions $g_{\mu\nu}(r, t)$, but this is not a real handicap. If the bond μ–ν is broken, only the distribution function $g_{\mu\nu}$ of atoms μ, ν forming this bond stands out in the difference signal. Other terms disappear, partially or completely. (2) The atom–atom distribution functions $g_{\mu\nu}(r, t)$ do not enter into $\Delta S_{\text{inst}}[r, t]$ alone, but are multiplied by the atomic form factors $f_\mu(q)f_\nu(q)$. This sort of coupling blurs the information contained in the signal $\Delta S[r, t]$ to a certain extent. Nevertheless, taking the necessary precautions, monitoring atomic motions is still possible.

C. Short-Time Signals

Contrary to the long time limit of Eq. (4) for the signal $\Delta S(\mathbf{q}, \tau)$, its short-time limit has not yet been explored in detail. The reason is that these times are not yet accessible to the experiment, due to technical difficulties. Nevertheless, some characteristics of short-time signals can be understood without any detailed study. (1) First, the liquid is not isotropic at times $t \ll \tau_R$, where τ_R is molecular rotational relaxation time [13–15]. The reason is that the laser-generated electric field $\mathbf{E}(\mathbf{r}, t)$ induces a partial alignment of molecular transition moments, and of the molecules themselves in its direction. The liquid is closer to being an incompletely ordered crystal than an isotropic liquid. The difference signal $\Delta S(\mathbf{q}, \tau)$ is then expected to depend on the relative orientation of the vectors q and $\mathbf{E}(\mathbf{r}, t)$. The isotropy is recovered again at times $t \gg \tau_R$. (2) The quantity $\langle \exp(-i\mathbf{qr}) \rangle$, omnipresent in the theory of X-ray scattering, does not reduce to the function $\sin(q\tau)/(q\tau)$ in the absence of isotropy. The Zernicke–Prins formula as well as its extension given by Eq. (5) rely heavily on the isotropy of the liquid medium; they thus break down in the short-time limit. (3) If two electronic states of the molecule are close enough to be excited simultaneously by the optical pulse, beating phenomena may occur. However, it is not known how these processes evolve on the very shortest time scales in the presence of molecular rotations. These few statements are by no means exhaustive and a detailed study of the short-time regime remains to be done. One conclusion is certain: time-resolved X-ray diffraction is very different at long and short times.

V. SIMULATIONS AND CALCULATIONS

A. Generalities

The purpose of this section is to show how the previous theory can be applied in practical calculations. For the time being only quasistatic processes have been studied in detail; the subsequent discussion will thus focus on this limit. Two sorts of quantities enter into the theory: the atom–atom distribution functions $g^j_{\mu\nu}(r,t)$ in a given electronic state j and the corresponding populations $n_j(t)$. The total atom–atom distribution function $g_{\mu\nu}(r,t)$is then

$$g_{\mu\nu}(r,t) = \Sigma_j n_j(t) g^j_{\mu\nu}(r,t) \qquad (6)$$

One would prefer to be able to calculate all of them by molecular dynamics simulations, exclusively. This is unfortunately not possible at present. In fact, some indices μ, ν of Eq. (6) refer to electronically excited molecules, which decay through population relaxation on the pico- and nanosecond time scales. The other indices μ, ν denote molecules that remain in their electronic ground state, and hydrodynamic time scales beyond microseconds intervene. The presence of these long times precludes the exclusive use of molecular dynamics, and a recourse to hydrodynamics of continuous media is inevitable. This concession has a high price. Macroscopic hydrodynamics assume a local thermodynamic equilibrium, which does not exist at times prior to 100 ps. These times are thus excluded from these studies.

B. Molecular Dynamics Simulations

The basic principles are described in many textbooks [24, 26]. They are thus only sketchily presented here. In a conventional classical molecular dynamics calculation, a system of N particles is placed within a cell of fixed volume, most frequently cubic in size. A set of velocities is also assigned, usually drawn from a Maxwell–Boltzmann distribution appropriate to the temperature of interest and selected in a way so as to make the net linear momentum zero. The subsequent trajectories of the particles are then calculated using the Newton equations of motion. Employing the finite difference method, this set of differential equations is transformed into a set of algebraic equations, which are solved by computer. The particles are assumed to interact through some prescribed force law. The dispersion, dipole–dipole, and polarization forces are typically included; whenever possible, they are taken from the literature.

Molecular dynamics permits one to determine rapidly varying atom-pair distribution functions $g^j_{\mu\nu}(r,t)$. In order to do that, one counts the number of atoms of the species ν in a spherical shell of radius r and thickness Δr centered on an atom of the species μ. This counting is repeated for a large number of computer-generated configurations. Calculations of this kind are generally well controlled, although they may occasionally generate spurious peaks in the difference signals $\Delta S(\mathbf{q})$. This perturbing effect is due to the finite size of the basic computation cell and to the smallness of difference signals. Particular care is thus necessary in the interpretation of these calculations.

The determination of the laser-generated populations $n_j(t)$ is infinitely more delicate. Computer simulations can certainly be applied to study population relaxation times of different electronic states. However, such simulations are no longer completely classical. Semiclassical simulations have been invented for that purpose, and the methods such as surface hopping were proposed. Unfortunately, they have not yet been employed in the present context. Laser spectroscopic data are used instead: the decay of the excited state populations is written $n_j(t) = n_j \exp(-t/\tau_j)$, where τ_j is the experimentally determined population relaxation time. The laws of chemical kinetics may also be used when necessary. Proceeding in this way, the rapidly varying component of $\Delta S(q, \tau)$ can be determined.

C. Hydrodynamics

The calculation of the slow components in $\Delta S(\mathrm{q}, \tau)$ follows another path. It can be realized as follows. (1) Assuming local equilibrium one can write

$$\Delta S(q,\tau) = (\partial \Delta S(q)/\partial T)_\rho \Delta T(\tau) + (\partial \Delta S(q)/\partial \rho)_T \Delta\rho(\tau) \tag{7}$$

where $\Delta T(\tau)$ and $\Delta\rho(\tau)$ are changes in the temperature and density from their equilibrium values T and ρ [18, 19]. (2) In order to estimate the thermodynamic derivatives contained in $\Delta S(\mathrm{q})$, this signal is calculated by molecular dynamics simulation for two temperatures T_1, T_2, and for two densities ρ_1, ρ_2. Finite differences are used next to calculate the derivatives. (3) The temperature and density increments $\Delta T(\tau)$ and $\Delta\rho(\tau)$ are calculated using equations of hydrodynamics of nonviscous fluids [27]. Under the present conditions, these equations can be linearized. Then, denoting by $T'(t)$ and $\rho'(t)$ the density increments $\Delta T(t)$ and $\Delta\rho(t)$, the following equations of motion can be used to calculate the time-dependent parts of $T'(t)$ and $\rho'(t)$:

$$\begin{aligned} \nabla^2 \rho' - \frac{1}{c^2}\frac{\partial^2 \rho'}{\partial t^2} &= \frac{\alpha_p}{C_p}\nabla^2 Q \\ T' &= \frac{Q}{C_p \rho_0} + \frac{C_p - C_v}{C_v}\frac{1}{\alpha_p \rho_0}\rho' \end{aligned} \tag{8}$$

Here $Q(t)$ denotes the heat input per unit volume accumulated up to time t, C_p is the specific heat per unit mass at constant pressure, C_v the specific heat per unit mass at constant volume, c is the sound velocity, α_p the coefficient of isobaric thermal expansion, and ρ_0 the equilibrium density. (4) The heat input $Q(t)$ is the laser energy released by the absorbing molecule per unit volume. If the excitation is in the visible spectral range, the evolution of $Q(t)$ follows the rhythm of the different chemically driven relaxation processes through which energy is transmitted to the liquid medium. If, on the other hand, vibrations are excited in the near-infrared or infrared, the evolution of $Q(t)$ can be considered impulsive. (5) While the expression for $T'(\tau)$ is obtained immediately from $Q(\tau)$ and $\rho'(\tau)$, the calculation of $\rho'(\tau)$ is not trivial. An approximate solution, valid along the axis of a Gaussian laser beam, was found by Longaker and Litvak [28]. A slightly extended version of it is [19]

$$\rho'(\tau) = \frac{\alpha_p}{C_p} \int_{-\infty}^{\tau} dt \left(\frac{dQ(t)}{dt} \right) [-1 + \exp(-c^2(\tau - t)^2 / R^2))] \tag{9}$$

where R is the radius of the laser beam. The important feature of the Longaker–Litvak [28] solution is that thermal expansion does not set in immediately after a transient heat input, but is delayed because perturbations in a liquid cannot propagate faster than sound waves. There exists an acoustic horizon; the quantity $\tau_a = R/c$ is often called the acoustic transit time. This closes the discussion of the slowly varying component of the difference signal $\Delta S(q, \tau)$.

VI. RECENT ACHIEVEMENTS

A. Diffraction and Absorption

During the last few years, time-resolved X-ray techniques have been employed in several areas. The first problem studied by time-resolved X-ray diffraction was surface melting of crystals. Thermal effects consecutive to the impact of an intense laser pulse on the crystal surface were measured on pico- and subpicosecond time scales. A number of different studies were made: melting of the crystal surface, lattice strain propagation, and onset of lattice expansion. These effects were observed for crystals such as Au [29], Ge [30], GaAs [31], and InSb [32–34]. More complex objects like Languir–Blodgett films were examined too [35]. Another very interesting research area concerns biological systems. Crystals of the myoglobin complex with CO and MbCO were examined with particular care and the positions of CO were determined at different times. Combining Laue X-ray diffraction with time-resolved optical spectroscopy permitted a deeper insight into how to initiate a biochemical reaction in a crystal in a nondestructive way [36, 37]. The studies of the yellow protein in its pR state belong to the same general area [38].

Time-resolved X-ray absorption is a very different class of experiments [5–7]. Chemical reactions are triggered by an ultrafast laser pulse, but the laser-induced change in geometry is observed by absorption rather than diffraction. This technique permits one to monitor local rather than global changes in the system. What one measures in practice is the extended X-ray absorption fine structure (EXAFS), and the X-ray extended nearedge structure (XANES).

Systems like SF_6 [39, 40], H_2O [41], CH_3OH [41], and CBr_4/C_6H_{12} [42] have been examined using this technique. Three recent papers on ruthenium (II) tris-2, 2′-bipyridine, or $[Ru^{2+}(bpy)_3]^{2+}$ [43], on photosynthetic O_2 formation in biological systems [44], and on photoexcitation of NITPP – L_2 [45] in solution also merit attention. Theoretical work advanced at the same time. Early approaches are due to Wilson et al. [46], whereas a statistical theory of time-resolved X-ray absorption was proposed by Mukamel et al. [47, 48]. This latter theory represents the counterpart of the X-ray diffraction theory developed in this chapter.

The purpose of this section is to describe recent achievements in time-resolved X-ray diffraction from liquids. Keeping the scope of the present chapter in mind, neither X-ray diffraction from solids nor X-ray absorption will be discussed. The majority of experiments realized up to now were performed using optical excitation, although some recent attempts using infrared excitation were also reported. The main topics that have been studied are (1) visualization of atomic motions during a chemical reaction, (2) structure of reaction intermediates in a complex reaction sequence, (3) heat propagation in impulsively heated liquids, and (4) chemical hydrodynamics of nanoparticle suspensions. We hope that the actual state-of-the-art will be illustrated in this way.

B. Visualization of Atomic Motions

The first reaction "filmed" by X-rays was the recombination of photodissociated iodine in a CCl_4 solution [18, 19, 49]. As this reaction is considered a prototype chemical reaction, a considerable effort was made to study it. Experimental techniques such as linear [50–52] and nonlinear [53–55] spectroscopy were used, as well as theoretical methods such as quantum chemistry [56] and molecular dynamics simulation [57]. A fair understanding of the dissociation and recombination dynamics resulted. However, a fascinating challenge remained: to "film" atomic motions during the reaction. This was done in the following way.

A dilute I_2/CCl_4 solution was pumped by a 520 nm visible laser pulse, promoting the iodine molecule from its ground electronic state X to the excited states A, A', B, and ${}^1\pi_u$ (Fig. 4). The laser-excited I_2 dissociates rapidly into an unstable intermediate $(I_2)^*$. The latter decomposes, and the two iodine atoms recombine either geminately (a) or nongeminately (b):

$$I_2 + h\nu \rightarrow I_2^* \; I_2^* \rightarrow I_2^*(a), \quad I_2^* \rightarrow 2I \rightarrow I_2(b)$$

The resulting atomic motions were probed by X-ray pulses for a number of time delays. A collection of diffraction patterns were then transformed into a series of real-space snapshots; theory is required to accomplish this last step. When the

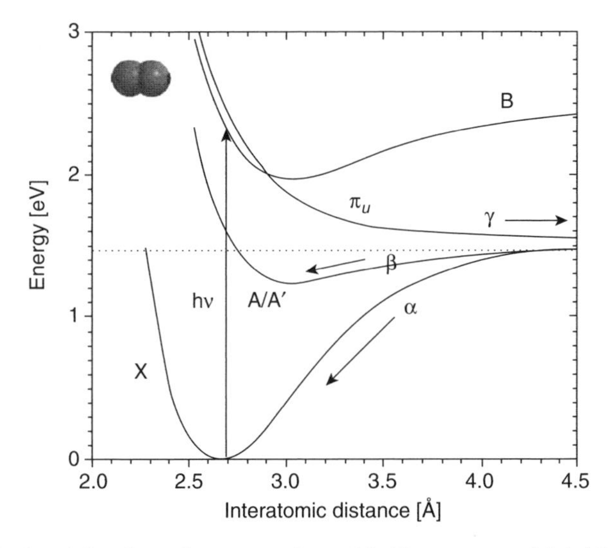

Figure 4. Low-lying electronic energy surfaces of I_2. These states are labeled X, A/A' B, and $^1\pi_u$ states. The processes α, β, and γ denote vibrational cooling along the X potential, geminate recombination through the states A/A', and nongeminate recombination, respectively.

sequence of snapshots were stuck together, it became a film of atomic motions during recombination.

The detailed description of the "film" is as follows. The first minimum in $\Delta S[r, \tau]$ at 2.7 Å at early times is due to the depletion of the X state of molecular iodine from the laser excitation (Fig. 5). The excited molecules then reach the A/A' and higher electronic states, and a maximum appears around 3.2 Å. In addition, the energy transfer from the solute to the solvent induces a rearrangement of the structure of the liquid without any observable thermal expansion. New minima in $\Delta S[r, \tau]$ appear at 4.0 and 6.2 Å. Molecular dynamics simulations assigned them to changes in the intermolecular Cl–Cl distances in liquid CCl_4. At later times, the excited molecules all relax, and the thermal expansion is completed. The strong increase in $\Delta S[r, \tau]$ observed at small r's is due to the decrease in the mass density ρ_M of CCl_4; compare with Eq. (5). In turn, the features observed at large r's reflect the variations in the intermolecular Cl–Cl distances due to the thermal expansion.

"Filming" of atomic motions in liquids was thus accomplished. More specifically, the above experiment provides atom–atom distribution functions $g_{\mu\nu}(r, \tau)$ as they change during a chemical reaction. It also permits one to monitor temporal variations in the mean density of laser-heated solutions. Finally, it shows that motions of reactive and solvent molecules are strongly correlated: the solvent is not an inert medium hosting the reaction [58].

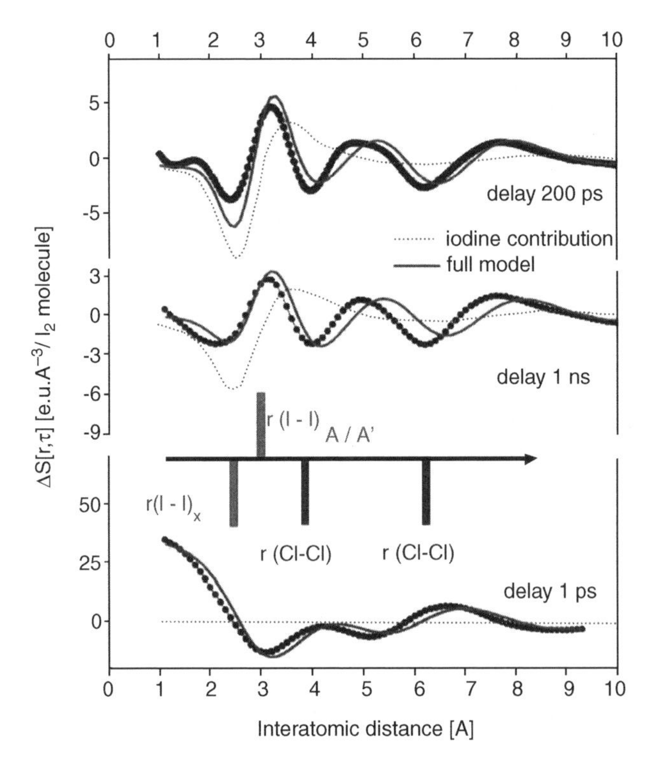

Figure 5. The Fourier transformed signal $\Delta S[r, \tau]$ of I_2/CCl_4. The pump–probe delay times are $\tau = 200\,\text{ps}$, 1 ns, and 1 μs. The green bars indicate the bond lengths of iodine in the X and A/A' states. The blue bars show the positions of the first two intermolecular peaks in the pair distribution function g_{Cl-Cl}. (See color insert.)

C. Structure of Reaction Intermediates

1. *Photodissociation of Diiodomethane*

Determining the geometry of short-lived transients in a complex chemical reaction is a difficult—but very important—problem of chemistry. Time-resolved X-ray diffraction offers new possibilities in this domain, as shown by a recent work on photodissociation of diiodomethane CH_2I_2 in methanol CH_3OH [59]. For many reasons, partially scientific and partially commercial, this reaction was extensively studied in the past. Spectroscopic techniques [60–63] were employed, together with theoretical methods of quantum chemistry [64–66]. In spite of this effort, the reaction mechanism remained poorly understood. For example, Tarnowski et al. [63] suggested the presence of a long-living intermediate (CH_2I—I) in the reaction sequence. In order to prove—or disprove—this statement, the problem was reexamined by time-resolved X-ray diffraction.

The experiment was as follows. A diluted solution of CH_2I_2 was pumped by an optical laser, promoting an electron onto an antibonding orbital of the C—I bond. After excitation this bond is ruptured and the $(I)^\bullet$ and $(CH_2I)^\bullet$ radicals are formed. Several reaction pathways are possible, both geminate or nongeminate. These radicals also react with CH_3OH to form the ions I_2^- and I_3^-. The solution was probed by time-delayed 150 ps long X-ray pulses.

The theory was very similar to that described earlier, but was simplified in view of the complexity of the problem. A number of reaction intermediates were considered explicitly, and the corresponding signals were calculated by molecular dynamics simulation. Kinetic equations governing the reaction sequence were established and were solved numerically. The main simplification of the theory is that, when calculating $\Delta S[r, \tau]$, the lower limit of the Fourier integral was shifted from 0 to a small value q_M. The authors wrote [59]

$$\Delta S[r, \tau] = \int_0^\infty dq\, qF(q)\Delta S(q, \tau) \sin q\tau \rightarrow \int_{q_M}^\infty dq\, qF(q)\Delta S(q, \tau) \sin qr \quad (10)$$

The rationale of this assumption is that low-q contributions to $\Delta S[r, \tau]$ affect this quantity only at large r's. As a consequence, the low-r region, where the reaction intermediates have their signatures, should be solvent free.

The results are presented in Fig. 6, where the measured and calculated difference signals $\Delta S[r, \tau]$ are illustrated. The presence of a strong peak at $r \sim 2.7$–3.0 Å proves the existence of an I_2 bond in different chemical environments; this bond is present in CH_2I—I at times inferior to 10 nanoseconds, and in I_3^- at later times. This study proves, for the first time unambiguously, the presence of the long-living intermediate CH_3I—I.

2. *Photodissociation of Diiodoethane*

Another chemical reaction studied with the same method was the photodissociation of diiodoethane $C_2H_4I_2$ in CH_3OH [67]. Attention was focused on the radical $(CH_2ICH_2)^\bullet$, which appears after release of one iodine atom. This radical was examined carefully by theory [68–70] and experiment [71–74]. In spite of this effort, the geometrical information remained incomplete. A bridged structure was postulated by Skell and co-workers to explain the stereochemistry of free radical addition reactions [75, 76], but direct structural evidence was still lacking. Should the anti configuration $(ICH_2$—$CH_2)^\bullet$ really be excluded? The purpose of this work was to provide the missing information by time-resolved X-ray diffraction.

The experiment was done as follows. A diluted solution of $C_2H_4I_2$ in CH_3OH was pumped by an optical laser, which triggered the elimination of one iodine atom followed by creation of the radicals $(C_2H_4I)^\bullet$ and $(I)^\bullet$. A series of X-ray

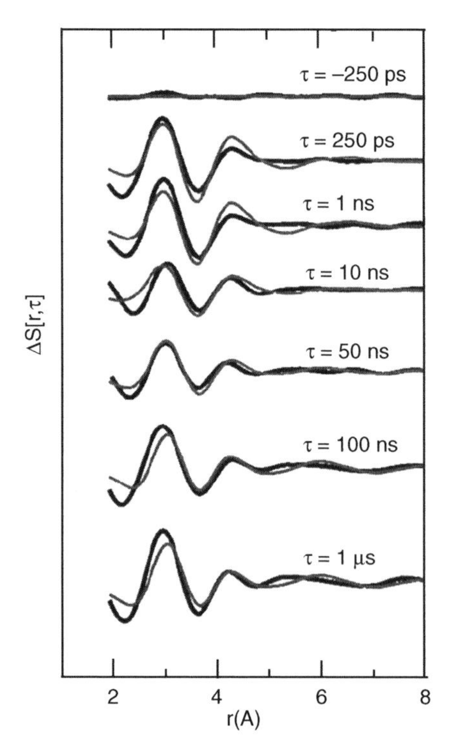

Figure 6. The Fourier transformed signal $\Delta S[r, \tau]$ of CH_2I_2/CH_3OH. The pump–probe time delays vary between $\tau = -250$ ps and 1 μs. The pair distribution function $g_{I\text{–}I}$ peaks in the 3 Å region. If $\tau < 50$ ns, the I—I bond corresponds to the short-lived intermediate ($CH_2I^{\bullet}I$), and if $\tau > 100$ ns it belongs to the $(I_3^-)^{\bullet}$ ion. Red curves indicate the theory, and black curves describe the experiment. (See color insert.)

diffraction pattern were recorded at times between -100 ps and 3 μs. The signals $\Delta S(q, \tau)$ and their Fourier transforms $\Delta S[r, \tau]$ were determined in this experiment. The theory was similar to that described earlier. In particular, the solvent-free signals $\Delta S[r, \tau]$ were calculated using Eq. (10). The main difficulty in this study was that the signature of the bridge and the anti radicals are not very different. Molecular dynamics simulations were thus realized with particular care, and statistical deviation checks (χ^2) were performed.

The results of these investigations are presented in Fig. 7, where the signals $\Delta S[r, \tau]$ are plotted for the bridge and anti form of the radical $(C_2H_2I)^*$. These figures favor neatly the bridge form. The concentrations of different species in the solution at different times were also determined. The crucial role of theory should be emphasized: it would be difficult to extract this information by simple insight of the experimental data.

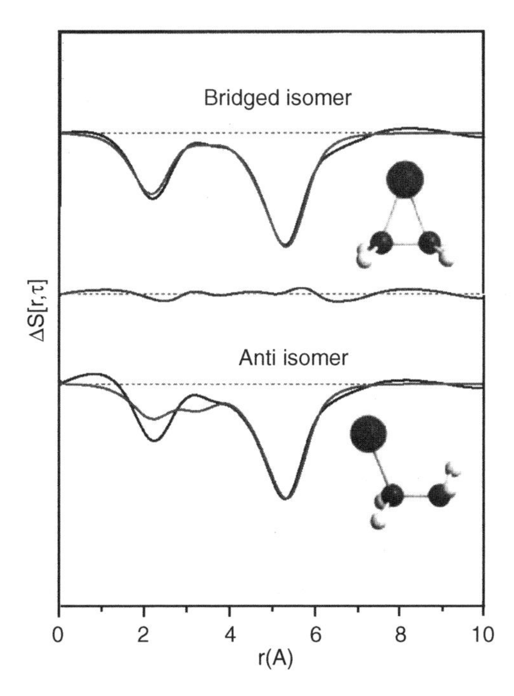

Figure 7. The Fourier transformed signal $\Delta S[r, \tau]$ of the $(C_2H_4I)^{\bullet}$ radical in methanol at $\tau = 100$ ps. The agreement between theory (red curve) and experiment (black curve) is better if the radical is assumed to be bridged (a) rather than to have an anti form (b). (See color insert.)

D. Hydrodynamics of Laser-Heated Liquids

If the system under consideration is chemically inert, the laser excitation only induces heat, accompanied by density and pressure waves. The excitation can be in the visible spectral region, but infrared pumping is also possible. In the latter case, the times governing the delivery of heat to the liquid are those of vibrational population relaxation. They are very short, on the order of 1 ps; this sort of excitation is thus impulsive. Contrary to a first impression, the physical reality is in fact quite subtle. The acoustic horizon, described in Section VC is at the center of the discussion [18, 19]. As laser-induced perturbations cannot propagate faster than sound, thermal expansion is delayed at short times. The physicochemical consequences of this delay are still entirely unknown. The liquids submitted to investigation are water and methanol.

A first study refers to liquid water [77]. The signals $\Delta S(q, \tau)$ and $\Delta S[r, \tau]$ were measured using time-resolved X-ray diffraction techniques with 100 ps resolution. Laser pulses at 266 nm and 400 nm were employed. Only short times τ were considered where thermal expansion was assumed to be negligible, and the density ρ to be independent of τ. To prove this assumption, the authors compared their values of $\Delta S(q, \tau)$ to the values of $\Delta S(q)$ obtained from isochoric (i.e., $\rho = \text{const}$) temperature differential data [78–80]. Their argument is based on the fact that liquid H_2O shows a density maximum at 4 °C. Pairs of temperatures T_1, T_2 thus exist for which the density ρ is the same: constant density conditions can thus be created in this unusual way. The experiment confirmed the existence of the acoustic horizon (Fig. 8).

Another study refers to liquid CH_3OH, where an infrared excitation was employed for the first time [81]. This excitation was realized using the overtone of the νOH mode of methanol and the asymmetric νCH mode. The time was varied over a wide range, between 100 ps and 1 μs. Time-resolved X-ray diffraction techniques employed were similar to those mentioned earlier. The authors then succeeded in extracting the difference signals $\Delta S(q, \tau)$, and in deducing thermodynamic derivatives of the signal $(\partial \Delta S / \partial T)_\rho$ and $(\partial \Delta S / \partial \rho)_T$. These quantities, normally obtainable only by computer simulation, were employed to refine the analysis of the diiodoethane photodissociation, as discussed in Section VIC [67]. The quality of global fits was improved considerably by employing these measured solvent differentials.

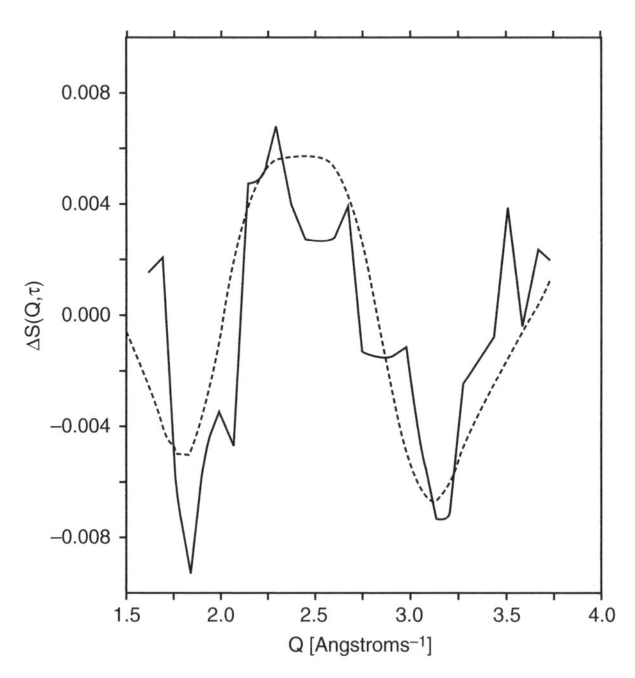

Figure 8. Comparison of the difference signal $\Delta S(q, \tau)$ at $\tau = 700\,\text{ps}$ (solid) with static isochoric data (dashed) for water. The data were collected using a time-resolved avalanche photodiode without area sensitivity.

E. Gold Nanoparticles in Water

The last subject where time-resolved X-ray diffraction techniques proved their exceptional potential concerns chemical physics of gold nanoparticles in water. Belonging jointly to X-ray and nanoparticle physics, this subject is in a certain sense on the borderline of the present Chapter [82–84]. Nevertheless, the possibilities offered by time-resolved X-ray techniques in this domain are fascinating. The heart of the problem is as follows. A suspension of gold nanoparticles is submitted to laser excitation. If the excitation power is low enough, these particles are not damaged by the radiation. If not, the nanoparticles may be transformed or even destroyed. Three sorts of problems were analyzed. The first of them is laser-induced heating and melting after excitation with femtosecond laser pulses [85]. At lower excitation power, the lattice heating is followed by cooling on the nanosecond time scale. The lattice expansion rises linearly with the laser excitation up to a lattice temperature increase ΔT on the order of 500 K. At higher temperatures, the long-range order decreases due to premelting of the particles. At still higher temperatures, complete melting occurs within the first 100 ps after laser excitation.

A second problem in these studies concerns cavitation dynamics on the nanometer length scale [86]. If sufficiently energetic, the ultrafast laser excitation of a gold nanoparticle causes strong nonequilibrium heating of the particle lattice and of the water shell close to the particle surface. Above a threshold in the laser power, which defines the onset of homogeneous nucleation, nanoscale water bubbles develop around the particles, expand, and collapse again within the first nanosecond after excitation (Fig. 9). The size of the bubbles may be examined in this way.

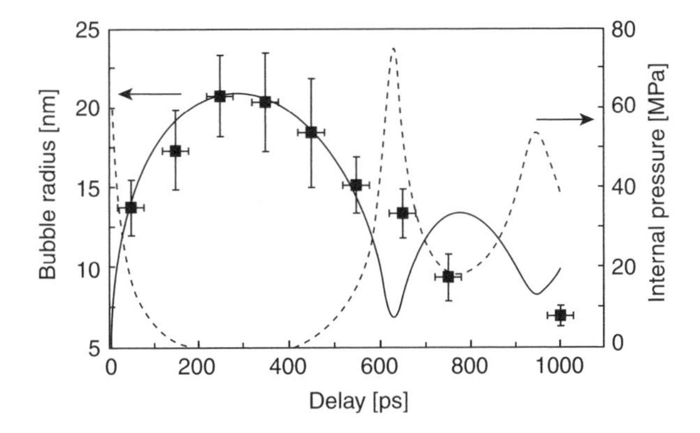

Figure 9. Bubble radius and pressure transients of the water vapor inside the bubbles. The first maximum in pressure at 650 ps marks the collapse of the bubbles. The following modulations are only expected for oscillatory bubble motion.

The last problem of this series concerns femtosecond laser ablation from gold nanoparticles [87]. In this process, solid material transforms into a volatile phase initiated by rapid deposition of energy. This ablation is nonthermal in nature. Material ejection is induced by the enhancement of the electric field close to the curved nanoparticle surface. This ablation is achievable for laser excitation powers far below the onset of general catastrophic material deterioration, such as plasma formation or laser-induced explosive boiling. Anisotropy in the ablation pattern was observed. It coincides with a reduction of the surface barrier from water vaporization and particle melting. This effect limits any high-power manipulation of nanostructured surfaces such as surface-enhanced Raman measurements or plasmonics with femtosecond pulses.

VII. CONCLUSIONS

A major breakthrough has been realized in the last few years in time-resolved X-ray diffraction and absorption. It may be compared to that realized fifteen years ago in time-resolved optical spectroscopy. It offers spatial and temporal resolution of atomic motions directly, contrary to optical spectroscopy where extra information is required to pass from energy to geometry. First molecular "films" were realized and the geometry of a few short-lived reaction intermediates was determined. Surface melting of crystals, laser-induced deterioration or destruction of gold nanoparticles in water and migration of carbon monoxide molecules in myoglobin crystals are further examples of successful applications of this technique. A new avenue is opened in this direction.

However, time-resolved X-ray diffraction remains a young science. It is still impossible, or is at least very difficult, to attain time scales below to a picosecond. General characteristics of subpicosecond X-ray diffraction and absorption are hardly understood. To progress in this direction, free electron laser X-ray sources are actually under construction subject to heavy financial constraints. Nevertheless, this field is exceptionally promising. Working therein is a challenge for everybody!

Acknowledgments

The authors are indebted to Hyotcherl Ihee, Philip Anfinrud, Friedrich Schotte, Anton Plech, Maciej Lorenc, Qingyu Kong, Marco Cammarata, Rodolphe Vuilleumie, and Fabien Mirloup for experimental and theoretical assistance. They would also like to thank Marie-Claire Belissent-Funel and Yann Gauduel for their help at early stages of this research. The EU grant FP6-503641 titled FLASH is also gratefully acknowledged.

References

1. Y. R. Shen, *The Principles of Nonlinear Optics*, John Wiley & Sons, Hoboken, NJ, 2002.
2. A. Yariv, *Quantum Electronics*, 3rd ed., John Wiley & Sons, Hoboken, NJ, 1988.

3. M. D. Levenson and S. S. Kano, *Introduction to Nonlinear Laser Spectroscopy*, Optics and Photonics Series, Academic Press; Revised edition (March 1989), New York, 1988.
4. S. Mukamel, *Principles of Nonlinear Optical Spectroscopy*, Oxford University Press, New York, 1999.
5. J. R Helliwell and P. M. Rentzepis, *Time-Resolved Diffraction*, Oxford Series on Synchrotron Radiation, No. 2, Oxford University Press, Oxford, 1997.
6. C. Bressler and M. Chergui, Ultrafast X-ray absorption spectroscopy. *Chem. Rev.* **104**(4), 1781–1812 (2004).
7. I. V. Tomov, D. A. Oulianov, P. Chen, and P. M. Rentzepis, Ultrafast time-resolved transient structures of solids and liquids studied by means of X-ray diffraction and EXAFS. *J. Phys. Chem. B* **103**(34), 7081–7091 (1999).
8. F. Schotte, S. Techert, P. Anfinrud, V. Srajer, K. Moffat, and M. Wulff, Picosecond structural studies using pulsed synchrotron radiation. In D. M. Mills (ed.), *Third-Generation Hard X-Ray Synchrotron Radiation Sources: Source Properties, Optics, and Experimental Techniques*, Chap.10, p. 345–402. John Wiley & Sons, Hobokon, NJ, 2002.
9. M. Dohlus and T. Limberg, Calculation of coherent synchrotron radiation in the ttf-fel bunch compressor magnet chicanes. *Nucl. Instrum. Methods Phys. Res. Sect. A* **407**(1-3), 278–284 (1998).
10. R. W. Schoenlein, S. Chattopadhyay, H. H. W. Chong, T. E. Glover, P. A. Heimann, C. V. Shank, A. A. Zholents, and M. S. Zolotorev, Generation of femtosecond pulses of synchrotron radiation. *Science* **287**(5461), 2237–2240 (2000).
11. S. M. Gruner and D. H. Bilderback, Energy recovery linacs as synchrotron light sources. *Nucl. Instrum. Methods Phys. Res. Sect. A* **500**(1-3), 25–32 (2003).
12. A. Plech, R. R. A. Geis, and M. Wulff, Diffuse scattering from liquid solutions with white-beam undulator radiation for photoexcitation studies. *J. Synchrotron Radiation* **9**(5), 287–292 (2002).
13. M. Ben-Nun C. P. J. Barty, T. Guo, F. Ràksi, C. Rose-Petruck, J. Squier, K. R. Wilson, V. V. Yakovliev, P. M. Weber, Z. Jiang, A. Ikhlef, and J.-C. Kieffer, Ultrafast X-ray diffraction and absorption. In J. R. Helliwell and P. M. Rentzepis (eds.), *Time-Resolved Diffraction*, Volume 2 of Oxford Series on Synchrotron Radiation, Chap. 2, pp. 44–70. Oxford University Press, Oxford, 1997.
14. M. Ben-Nun, J. Cao, and K. R. Wilson, Ultrafast X-ray and electron diffraction: theoretical considerations. *J. Phys. Chem. A* **101**(47), 8743–8761 (1997).
15. J. Cao and K. R. Wilson, Ultrafast X-ray diffraction theory. *J. Phys. Chem. A.* **102**(47), 9523–9530 (1998).
16. C. H. Chao, S. H. Lin, W. K. Liu, and P. Rentzepis, Theory of ultrafast time-resolved X-ray and electron diffraction. In J. R. Helliwell and P. M. Rentzepis (eds.), *Time-Resolved Diffraction*, Volume 2 of Oxford Series on Synchrotron Radiation, Chap. 11, pp. 260–283. Oxford University Press, Oxford, 1997.
17. S. Bratos, F. Mirloup, R. Vuilleumier, and M. Wulff, Time-resolved X-ray diffraction: statistical theory and its application to the photo-physics of molecular iodine. *J. Chem. Phys.* **116**(24), 10615–10625 (2002).
18. S. Bratos, F. Mirloup, R. Vuilleumier, M. Wulff, and A. Plech, X-ray "filming" of atomic motions in chemical reactions. *Chem. Phys.* **304**(3), 245–251 (2004).
19. M. Wulff, S. Bratos, A. Plech, R. Vuilleumier, F. Mirloup, M. Lorenc, Q. Kong, and H. Ihee,. Recombination of photodissociated iodine: a time-resolved X-ray diffraction study. *J. Chem. Phys.* **124**(3), 034501 (2006).

20. L. D. Landau, E. M. Lifshitz, and L. P. Pitaevskii, *Electrodynamics of Continuous Media*, Volume 8 of Course of Theoretical Physics. Pergamon Press, Oxford, 1984.
21. A. Guinier, *X-Ray Diffraction: In Crystals, Imperfect Crystals, and Amorphous Bodies*, Dover, New York, 1963.
22. B. E. Warren. *X-Ray Diffraction*, Dover, New York, 1990.
23. J. Als-Nielsen and Des McMorrow, *Elements of Modern X-Ray Physics*, John Wiley & Sons, Hoboken, NJ, 2000.
24. J.-P. Hansen and I. R. McDonald, *Theory of Simple Liquids*, Academic Press, London, 1986.
25. A. H. Narten and H. A. Levy, Observed diffraction pattern and proposed models of liquid water. *Science* **165**(3892), 447–454 (1969).
26. M. P. Allen and D.-J. Tildesley, *Computer Simulation of Liquids*, Oxford University Press, Oxford, 1989.
27. L. D. Landau, E. M. Lifshitz, and L. P. Pitaevskii, *Fluid Mechanics*, Volume 8 of Course of Theoretical Physics. Pergamon Press, Oxford, 1987.
28. P. R. Longaker and M. M. Litvak, Perturbation of the refractive index of absorbing media by a pulsed laser beam. *J. Appl. Phys.* **40**(10), 4033–4041 (1969).
29. P. Chen, I. V. Tomov, and P. M. Rentzepis, Time resolved heat propagation in a gold crystal by means of picosecond X-ray diffraction. *J. Chem. Phys.*, **104**(24), 10001–10007 (1996).
30. C. W. Siders, A. Cavalleri, K. Sokolowski-Tinten, C. Tóth, T. Guo, M. Kammler, M. Horn von Hoegen, K. R. Wilson, D. von der Linde, and C. P. J. Barty, Detection of nonthermal melting by ultrafast X-ray diffraction. *Science* **286**(5443), 1340–1342 (1999).
31. C. Rose-Petruck, R. Jimenez, T. Guo, A. Cavalleri, C. W. Siders, F. Rksi, J. A. Squier, B. C. Walker, K. R. Wilson, and C. P. J. Barty, Picosecond-milliangström lattice dynamics measured by ultrafast X-ray diffraction. *Nature* **398**, 310–312 (1999).
32. A. H. Chin, R. W. Schoenlein, T. E. Glover, P. Balling, W. P. Leemans, and C. V. Shank, Ultrafast structural dynamics in InSb probed by time-resolved X-ray diffraction. *Phys. Rev. Lett.* **83**, 336–339 (1999).
33. D. A. Reis, M. F. DeCamp, P. H. Bucksbaum, R. Clarke, E. Dufresne, M. Hertlein, R. Merlin, R. Falcone, H. Kapteyn, M. M. Murnane, J. Larsson, T. Missalla, and J. S. Wark. Probing impulsive strain propagation with X-ray pulses. *Phys. Rev. Lett.* **86**, 3072–3075 (2001).
34. A. Rousse, C. Rischel, S. Fourmaux, I. Uschmann, S. Sebban, G. Grillon, P. Balcou, E. Forster, J. P. Geindre, P. Audebert, J. C. Gauthier, and D. Hulin, Nonthermal melting in semiconductors measured at femtosecond resolution. *Nature* **410**, 65–68 (2001).
35. C. Rischel, A. Rousse, I. Uschmann, P. A. Albouy, J.-P. Geindre, P. Audebert, J.-C. Gauthier, E. Froster, J.-L. Martin, and A. Antonetti, Femtosecond time-resolved X-ray diffraction from laser-heated organic films. *Nature* **390**, 490–492 (1997).
36. V. Srajer, T. Teng, T. Ursby, C. Pradervand, Z. Ren, S. Adachi, W. Schildkamp, D. Bourgeois, M. Wulff, and K. Moffat, Photolysis of the carbon monoxide complex of myoglobin: nanosecond time-resolved crystallography. *Science* **274**, 1726–1729 (1996).
37. F. Schotte, M. Lim, T. A. Jackson, A. V. Smirnov, J. Soman, J. S. Olson, Jr., G. N. Phillips, M. Wulff, and P. A. Anfinrud, Watching a protein as it functions with 150-ps time-resolved X-ray crystallography. *Science* **300**(5627), 1944–1947 (2003).
38. B. Perman, V. Srajer, Z. Ren, T. Teng, C. Pradervand, T. Ursby, D. Bourgeois, F. Schotte, M. Wulff, R. Kort, K. Hellingwerf, and K. Moffat, Energy transduction on the nanosecond time scale: early structural events in a xanthopsin photocycle. *Science* **279**, 1946–1950 (1998).

39. F. Raksi, K. R. Wilson, Z. Jiang, A. Ikhlef, C. Y. Cote, and J.-C. Kieffer, Ultrafast X-ray absorption probing of a chemical reaction. *J. Chem. Phys.* **104**(15), 6066–6069 (1996).

40. H. Nakamatsu, T. Mukoyama, and H. Adachi, Theoretical X-ray absorption spectra of SF_6 and H_2S. *J. Chem. Phys.* **95**(5), 3167–3174 (1991).

41. K. R. Wilson, R. D. Schaller, D. T. Co, R. J. Saykally, B. S. Rude, T. Catalano, and J. D. Bozek, Surface relaxation in liquid water and methanol studied by X-ray absorption spectroscopy. *J. Chem. Phys.* **117**(16), 7738–7744 (2002).

42. I. V. Tomov and P. M. Rentzepis, Ultrafast X-ray determination of transient structures in solids and liquids. *Chem. Phys.* **299**(2-3), 203–213 (2004).

43. M. Saes, C. Bressler, R. Abela, D. Grolimund, S. L. Johnson, P. A. Heimann, and M. Chergui, Observing photochemical transients by ultrafast X-ray absorption spectroscopy. *Phys. Rev. Lett.* **90**(4), 047403 (2003).

44. M. Haumann, P. Liebisch, C. Muller, M. Barra, M. Grabolle, and H. Dau, Photosynthetic O_2 formation tracked by time-resolved X-ray experiments. *Science* **310**(5750), 1019–1021 (2005).

45. L. X. Chen, W. J. H. Jager, G. Jennings, D. J. Gosztola, A. Munkholm, and J. P. Hessler, Capturing a photoexcited molecular structure through time-domain X-ray absorption fine structure. *Science* **292**(5515), 262–264 (2001).

46. F. L. H. Brown, K. R. Wilson, and J. Cao, Ultrafast extended X-ray absorption fine structure (EXAFS)—theoretical considerations. *J. Chem. Phys.* **111**(14), 6238–6246 (1999).

47. S. Tanaka, V. Chernyak, and S. Mukamel, Time-resolved X-ray spectroscopies: nonlinear response functions and Liouville space pathways. *Phys. Rev. A* **63**(6), 063405 (2001).

48. S. Tanaka and S. Mukamel, X-ray fourwave mixing in molecules. *J. Chem. Phy.* **116**(5), 1877–1891 (2002).

49. A. Plech, M. Wulff, S. Bratos, F. Mirloup, R. Vuilleumier, F. Schotte, and P. A. Anfinrud, Visualizing chemical reactions in solution by picosecond X-ray diffraction. *Phys. Rev. Lett.* **92**(12), 125505 (2004).

50. J. Franck and E. Rabinowitsch, Some remarks about free radicals and the photochemistry of solutions. *Trans. Faraday Soc.* **30**, 130 (1934).

51. E. Rabinowitch and W. C. Wood, Properties of illuminated iodine solutions. I. Photochemical dissociation of iodine molecules in solution. *Trans. Faraday Soc.* **32**, 547–555 (1936).

52. J. Zimmerman and R. M. Noyes, The primary quantum yield of dissociation of iodine in hexane solution. *J. Chem. Phys.* **18**(5), 658–666 (1950).

53. G. W. Hoffman T. J. Chuang, and K. B. Eisenthal, Picosecond studies of the cage effect and collision induced predissociation of iodine in liquids. *Chem. Phys. Lett.* **25**(2), 201–205 (1974).

54. N. A. Abul-Haj and D. F. Kelley, Geminate recombination and relaxation of molecular iodine. *J. Chem. Phys.* **84**(3), 1335–1344 (1986).

55. A. L. Harris, J. K. Brown, and C. B. Harris, The nature of simple photodissociation reactions in liquids on ultrafast time scales. *Annu. Rev. Phys. Chem.* **39**, 341–366 (1988).

56. R. S. Mulliken, Iodine revisited. *J. Chem. Phys.* **55**(1), 288–309 (1971).

57. J. P. Bergsma, M. H. Coladonato, P. M. Edelsten, J. D. Kahn, K. R. Wilson, and D. R. Fredkin, Transient X-ray scattering calculated from molecular dynamics. *J. Chem. Phys.* **84**(11), 6151–6160 (1986).

58. S. A. Rice, Atom tracking. *Nature* **429**, 255–256 (2004).

59. J. Davidsson, J. Poulsen, M. Cammarata, P. Georgiou, R. Wouts, G. Katona, F. Jacobson, A. Plech, M. Wulff, G. Nyman, and R. Neutze, Structural determination of a transient isomer of CH_2I_2 by picosecond X-ray diffraction. *Phys. Rev. Lett.* **94**(24), 245503 (2005).

60. J. Zhang and D. G. Imre, CH_2I_2 photodissociation: emission spectrum at 355 nm. *J. Chem. Phys.* **89**(1), 309–313 (1988).

61. B. J. Schwartz, J. C. King, J. Z. Zhang, and C. B. Harris, Direct femtosecond measurements of single collision dominated geminate recombination times of small molecules in liquids. *Chem. Phys. Lett.* **203**(5-6), 503–508 (1993).

62. W. M. Kwok and D. L. Phillips, Solvation effects and short-time photodissociation dynamics of CH_2I_2 in solution from resonance Raman spectroscopy. *Chem. Phys. Lett.* **235**(3-4), 260–267 (1995).

63. A. N. Tarnovsky, V. Sundstrom, E. Akesson, and T. Pascher, Photochemistry of diiodomethane in solution studied by femtosecond and nanosecond laser photolysis. Formation and dark reactions of the CH_2I-I isomer photoproduct and its role in cyclopropanation of olefins. *J. Phys. Chem. A* **108**(2), 237–249 (2004).

64. M. Odelius, M. Kadi, J. Davidsson, and A. N. Tarnovsky, Photodissociation of diiodomethane in acetonitrile solution and fragment recombination into iso-diiodomethane studied with ab initio molecular dynamics simulations. *J. Chem. Phys.* **121**(5), 2208–2214 (2004).

65. A. E. Orel and O. Kühn, Cartesian reaction surface analysis of the CH_2I_2 ground state isomerization. *Chem. Phys. Lett.* **304**(3-4), 285–292 (1999).

66. D. L. Phillips, W.H. Fang, and X. Zheng, Isodiiodomethane is the methylene transfer agent in cyclopropanation reactions with olefins using ultraviolet photolysis of diiodomethane in solutions: a density functional theory investigation of the reactions of isodiiodomethane, iodomethyl radical, and iodomethyl cation with ethylene. *J. Am. Chem. Soc.* **123**(18), 4197–4203 (2001).

67. H. Ihee, M. Lorenc, T. K. Kim, Q. Y. Kong, M. Cammarata, J. H. Lee, S. Bratos, and M. Wulff. Ultrafast X-ray diffraction of transient molecular structures in solution. *Science* **309**(5738), 1223–1227 (2005).

68. B. Engels and S. D. Peyerimhoff, Theoretical study of the bridging in β-halo ethyl. *J. Mol. Structure THEOCHEM* **138**(1-2), 59–68 (1986).

69. F. Bernardi and J. Fossey, An ab initio study of the structural properties of β-substituted ethyl-free radicals. *J. Mol. Structure THEOCHEM* **180**, 79–93 (1988).

70. H. Ihee, A. H. Zewail, and W. A. Goddard, Conformations and barriers of haloethyl radicals (CH_2XCH_2, X = F, Cl, Br, I): ab initio studies. *J. Phys. Chem. A* **103**(33), 6638–6649 (1999).

71. M. Rasmusson, A. N. Tarnovsky, T. Pascher, V. Sundstrom, and E. Akesson, Photodissociation of CH_2ICH_2I, CF_2ICF_2I, and CF_2BrCF_2I in solution. *J. Phys. Chem. A* **106**(31), 7090–7098 (2002).

72. A. J. Bowles, A. Hudson, and R. A. Jackson, Hyperfine coupling constants of the 2-chloroethyl and related radicals. *Chem. Phys. Lett.* **5**(9), 552–554 (1970).

73. D. J. Edge and J. K. Kochi, Effects of halogen substitution on alkyl radicals: conformational studies by electron spin resonance. *J. Am. Chem. Soc.* **94**(18), 6485–6495 (1972).

74. S. P. Maj, M. C. R. Symons, and P. M. R. Trousson, Bridged bromine radicals. an electron spin resonance study. *J. Chem. Soc. Chem. Commun.*, 561–562 (1984).

75. J. Fossey, D. Lefort, and J. Sorba, *Free Radicals in Organic Chemistry*, John Wiley & Sons, Hoboken, NJ, 1995.

76. P. S. Skell, D. L. Tuleen, and P. D. Readio, Stereochemical evidence of bridged radicals. *J. Am. Chem. Soc.* **85**(18), 2849–2850 (1963).

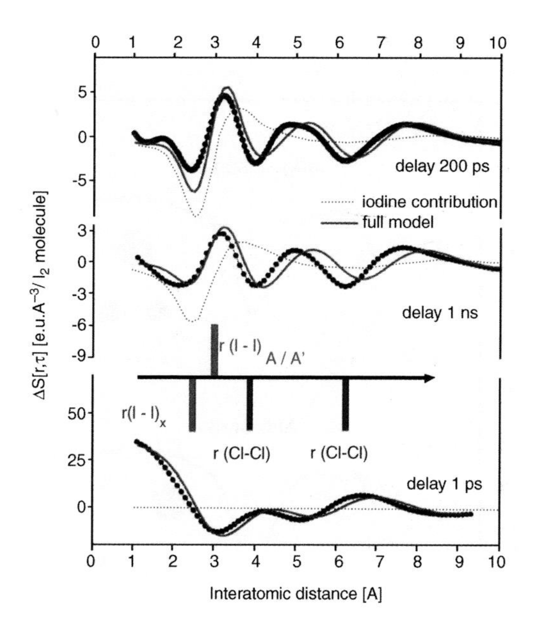

Figure 5. The Fourier transformed signal $\Delta S[r, \tau]$ of I_2/CCl_4. The pump–probe delay times are $\tau = 200\,ps$, 1 ns, and 1 μs. The green bars indicate the bond lengths of iodine in the X and A/A' states. The blue bars show the positions of the first two intermolecular peaks in the pair distribution function g_{Cl-Cl}.

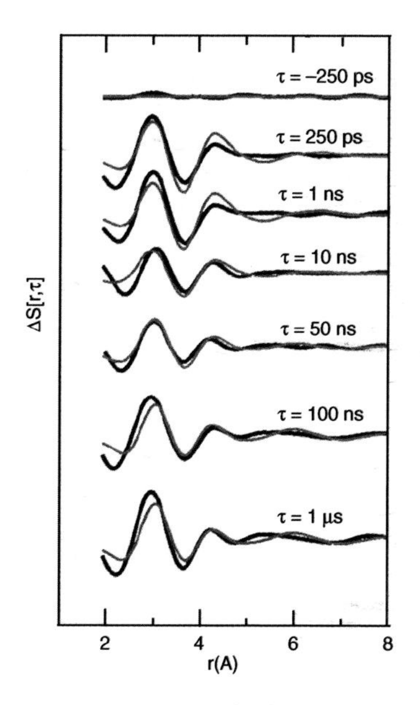

Figure 6. The Fourier transformed signal $\Delta S[r, \tau]$ of CH_2I_2/CH_3OH. The pump–probe time delays vary between $\tau = -250\,ps$ and 1 μs. The pair distribution function gI–I peaks in the 3 Å region. If $\tau < 50\,ns$, the I—I bond corresponds to the short-lived intermediate ($CH_2I^{\bullet}I$), and if $\tau > 100\,ns$ it belongs to the $(I_3^-)^{\bullet}$ ion. Red curves indicate the theory, and black curves describe the experiment.

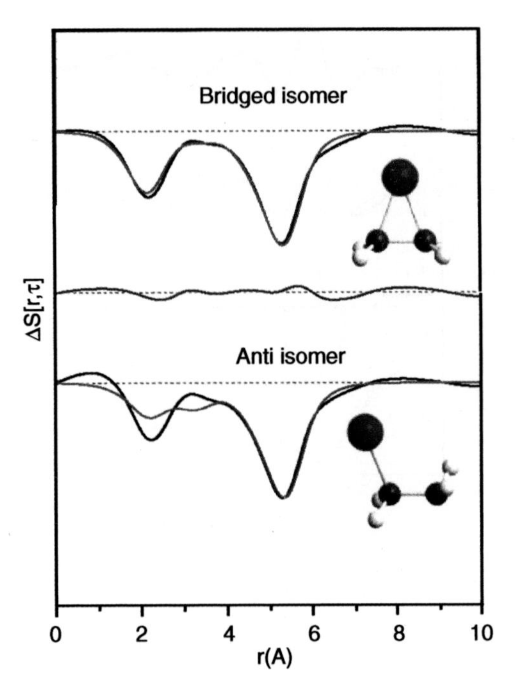

Figure 7. The Fourier transformed signal $\Delta S[r, \tau]$ of the $(C_2H_4I)^\bullet$ radical in methanol at $\tau = 100\,\mathrm{ps}$. The agreement between theory (red curve) and experiment (black curve) is better if the radical is assumed to be bridged (a) rather than to have an anti form (b).

77. A. M. Lindenberg, Y. Acremann, D. P. Lowney, P. A. Heimann, T. K. Allison, T. Matthews, and R. W. Falcone, Time-resolved measurements of the structure of water at constant density. *J. Chem. Phys.* **122**(20), 204507 (2005).

78. L. Bosio, S.-H. Chen, and J. Teixeira, Isochoric temperature differential of the X-ray structure factor and structural rearrangements in low-temperature heavy water. *Phy. Rev. A* **27**(3), 1468–1475 (1983).

79. J. A Polo and P. A. Egelstaff, Neutron-diffraction study of low-temperature water. *Phys. Rev. A* **27**(3), 1508–1514 (1983).

80. J. C. Dore, M. A. M. Sufi, and M. Bellissent-Funel, Structural change in D_2O water as a function of temperature: the isochoric temperature derivative function for neutron diffraction. *Phys. Chem. Chem. Phys.* **2**, 1599–1602 (2000).

81. M. Cammarata, M. Lorenc, T. K. Kim, J. H. Lee, Q. Y. Kong, E. Pontecorvo, M. Lo Russo, G. Schiro, A. Cupane, M. Wulff, and H. Ihee, Impulsive solvent heating probed by picosecond X-ray diffraction. *J. Chem. Phys.* **124**(12), 124504 (2006).

82. U. Kreibig and M. Vollmer, *Optical Properties of Metal Clusters*, Springer Series Materials Science. Springer, New York, 1995.

83. J. R. Krenn, A. Leitner, and F. R. Aussenegg, Metal nano-optics. In H. Singh Nalwa (ed.), *Encyclopedia of Nanoscience and Nanotechnology*, Volume 5, pp. 411–419. American Scientific Publishers, 2004.

84. S. Link and M. A. El-Sayed, Shape and size dependence of radiative, non-radiative and photothermal properties of gold nanocrystals. *Int. Rev. Phy. Chem.* **19**(3), 409–453 (2000).

85. A. Plech, V. Kotaidis, S. Gresillon, C. Dahmen, and G. von Plessen, Laser-induced heating and melting of gold nanoparticles studied by time-resolved X-ray scattering. *Phy. Rev. B* **70**(19), 195423 (2004).

86. V. Kotaidis and A. Plech, Cavitation dynamics on the nanoscale. *Appl. Phys. Lett.* **87**(21), 213102 (2005).

87. A. Plech, V. Kotaidis, M. Lorenc, and J. Boneberg, Femtosecond laser near-field ablation from gold nanoparticles. *Nature Phys.* **2**, 44–47 (2006).

AUTHOR INDEX

Numbers in parentheses are reference numbers and indicate that the author's work is referred to although his name is not mentioned in the text. Numbers in *italic* show the page on which the complete references are listed.

Advances in Chemical Physics, Volume 136, edited by Stuart A. Rice

SUBJECT INDEX

Advances in Chemical Physics, Volume 136, edited by Stuart A. Rice